The Projects

of

the *AA Honours nominees*
with a foreword by
Christopher CM Lee

Edited by
Minseok Kim

2008

Spacetime Publisher

The Projects

A selection of *four* projects from the Architectural Association Diploma Honours' nominees in 2008

Max Kahlen | Deena Fakhro | Keita Tajima | Erlend Bakke-Eidsaa

Edited by Minseok Kim

5F, 200 Hyeonjeo-dong, Seodaemun-gu, Seoul, Korea
T. 82 0 3147-1212 F. 82 2 3147-2626
Cover & Document Design : good position

ISBN 978-89-5592-123-6

Print in Korea
www.spacetime.co.kr | spacetime@korea.com

1판 발행일 2009년 8월 25일

저 자_Max Kahlen | Deena Fakhro | Keita Tajima | Erlend Bakke-Eidsaa
편집자_김민석

발행인_김기현
등 록_1993. 3. 12일
주 소_(120-796) 서울시 서대문구 현저동 200 극동 5층
전 화_02 3147 1212 · 2323
전 송_02 3147 2626

정 가 55,000원

The Projects

A collection of four projects from the Architectural Association Diploma Honours' nominees in 2008

Edited by

Minseok Kim

Contents

Foreword

What is immediately obvious from the four projects presented here is the absence of the 'final image', invariably being the glossy digital renderings that pervade the majority of architectural publications today and increasingly and sadly, even reputable architectural schools. What you will find here instead is a complete portfolio. The four portfolios presented here were nominated for the AA Diploma Honours - an award given out to only a few outstanding project.

The AA Diploma portfolio is the unique outcome of a pedagogical approach of the AA Diploma School. The portfolio, worked through a whole academic year - not a term or a semester - is first and foremost a thought process, a document of the discovery of architecture undertaken by an individual student. Invariably it contains a body of work that is rigorous, coherent and methodical. It is however not only a document of the process, where process can often become a refuge from making any concrete propositions. If at all, the portfolios here are courageously propositional. And this second part of what makes an outstanding AA Diploma portfolio is for me the most crucial: the portfolio is a conjecture. In that, the portfolio contains not only a project that is thoroughly investigated and coherently argued but is used to project an idea for the future. And this idea, with its own promise and validity, will form the next question that can be worked through in years to come. To put it most simply, an outstanding AA Diploma portfolio starts with a question and ends with a meta-question.

Christopher CM Lee
Unit master of diploma unit 6 at AA school
Principal of Serie Architects

About the AA

The Architectural Association was founded in 1847, the oldest architectural school in Britain. It has been located on the Georgian block in Central London, Bedford Square since 1917 and home to numerous students, lecturers and architects all over the world.

For more information,
http://www.aaschool.ac.uk

Preface

This book aims to discover and introduce the quality of work from an architectural school, owing to the success and great efforts of its participants. The first issue features in full length four projects that have been carefully selected nominees from *the AA school* Diploma Honours 2008. Such outstanding yet diverse projects demonstrate unique methodology to architectural ideas and visions for the city. Herewith the book leaves potential debate and inspiration that is open to the readers across the projects.

I would like to thank to all the participants – *Max Kahlen, Deena Fakhro, Keita Tajima* and *Erlend Bakke-Eidsaa*, and the publisher, *Spacetime* for their great efforts on the realization of this publication, to *Mike Weinstock*, Academic Head of the AA school and *Christopher CM Lee*, Unit master of diploma unit6 for their practical advices and the foreword, and finally to my parent for their endless support all the time.

Minseok Kim

London, Spring 2009

Rising Masses

Max Kahlen
Diploma unit 5

AA Diploma Honour, 2008
High Pass in Technical Design Thesis, 2008

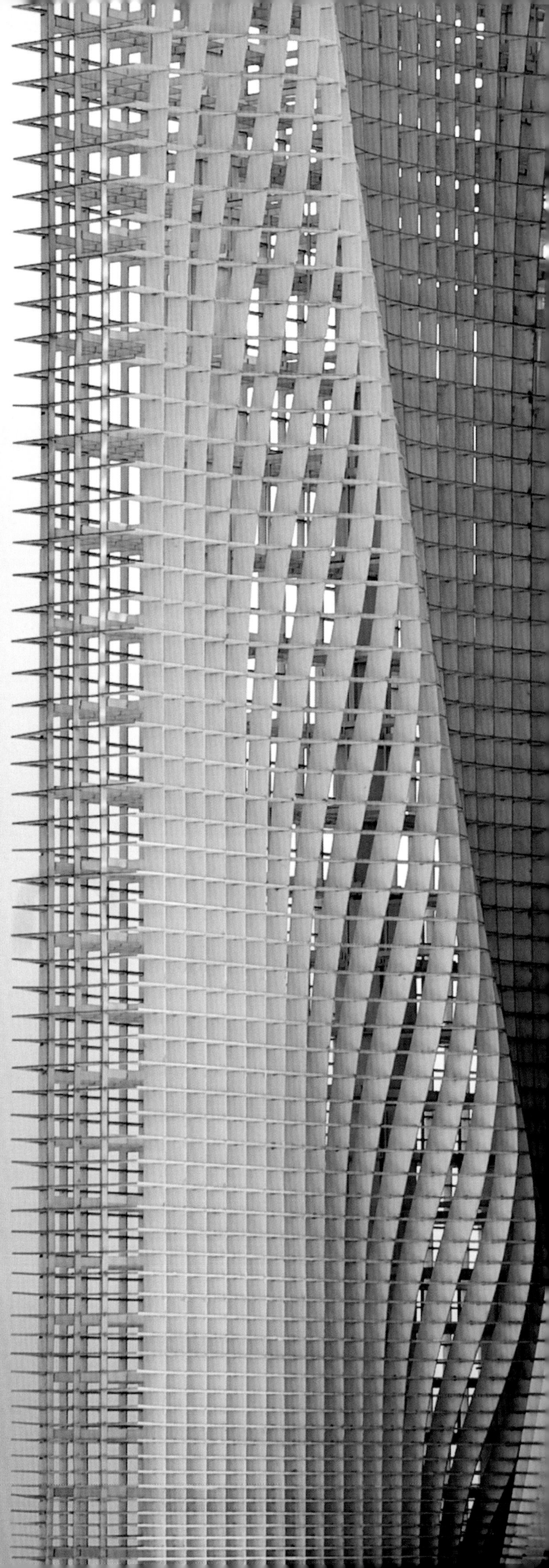

Max Kahlen

AA Dipl. (Hons), ARB/RIBA Part II

Max Kahlen studied at the Stuttgart Academy of Art and Design and the Architectural Association where he received the AA Diploma with Honours in 2008.

During the past years he worked in offices in London, Amsterdam and Germany.

Currently he is Unit Master of Diploma Unit 3 in the Architectural Association.

More about his work on www.gardeneden.net

About Unit 5 & Rising Masses

The project "Rising Masses" ran during the final year of the Diploma School unit set up by George L. Legendre and Lluis Viu-Rebes at the Architectural Association in London from 2001 onwards. Max Kahlen joined Diploma unit 5 in his 4th and 5th undergraduate years and quickly established himself as its most visionary member.

The ethos and the design techniques pioneered by Diploma unit 5 were refined over a number of years with key contributions from generations of architecture students at the Harvard Design School and the Architectural Association. Other key contributions include the development of the original math template and design software Surf_TM by George L. Legendre from 1997 to 2008, which eventually enabled the practical application of parametric equations to the design of tall buildings. The theoretical framework presiding over the unit's thread-based approach to surface design was devised and first introduced in IJP: The Book of Surfaces (AA Publications London, 2003).

Table of Contents

1. Introduction

1.1 Singapore, Housing Districts
1.2 Concept of Surface Porosity
1.3 Case Study, Pearlbank Tower

Fragments of the city of Singapore set the atmosphere and ground for this project.

They represent the mindset of a population where height and density is not questioned and where continuity as well as repetition is believed in as an expression of functionality and productivity.

The particular expression of monotonous continuity defines the seed for the development of this project *Rising Masses*.

This seed should be understood as a critical attempt to grow a building that on the one side camouflages with the cityscape through an appearance that is brutally shaped by the industries' performance. At the same time it sprawls like a weed - trying to escape its' performative bounding box.

Residential High-Rise, Singapore

Shophouse

Pearlbank Tower, Singapore

1.1 Singapore, Housing Districts

The project Rising Masses thematises verticality and the madness of repetition.

Views such as those shown below represent only a fragment of Singapores' endless cityscapes. They uncover the inhabitants' comity of density and build height - resulting in a surreal landscape of efficient monotony.

Images show housing districts in singapore, Ten Mile Junction

1.2 Concept of Surface Porosity

This project thematises the surface and celebrates continuity.

The idea of *surface porosity* guides the core-concept of this development, while the city of Singapore is the ground of the investigation.

Zooming into details of the cities' landscape uncovers various grades of porosities that define and differentiate the building envelopes. The images below (from left to right) show different examples ranging from the traditional Singaporean shophouse that blurrs and interlocks the boundry of the private and public space; to parking houses or external staircases that unveil parts of their inner spaces and thereby engrave a depth into their building surface; all the way to the paper thin curtainwall facades that do not provide any differentiation between inside and outside - those buildings eliminate any hint of scale from the outer appearance. Instead they hide behind the iconic shine of their glass surface.

Shophouse

Underground Train Facility

Overground Parking Garage

Overground Parking Garage

Max Kahlen

Office Builsing, External Staircase

Residential Building, External Walkways

Office District, Singapore

Office District, Singapore

1.3 Case Study, Pearlbank Tower

The Pearlbank Tower is situated in the centre of Singapore and used to be the highest apartment building in the 80's.
In an interesting way it articulates different grades of porosity through the building surface.

From the outside it appears almost closed and rather brutal, while the inner facade opens up towards the interstitial courtyard space. Each apartment is split over two floors with an external staircase that connects both levels. Thereby the inhabitants are continuously forced to travel from the very private to the very public space while an external walkway on every second floor creates an alternative half-private zone, interlocking the apartment entrances with the public spaces.

In other words, the surface appears flat on the outside while, on the inside, it unfolds various layers of privacy. Seen from the entrance of one of the apartments and facing the inner courtyard, this articulation creates a feeling of intimacy and neighbourhood even on the 35th floor.

In contrast to the flat outer facade, the high porosity of the inner facade and the spatial depth that it provides, gradually connects the inner spaces with the outside space. The mad repetition of floors and the extreme height is compensated by a detail-articulation that adapts to the human scale. The porosity of the inner facade induces the resident to step outside the building surface and engages him with the exterior in a more manifold was than a simple terrace.

Outside Facade Inside Facade

Inner Courtyard Inner Courtyard

External Staircases Entrance Space

External Corridor & Staricase

2. Indexical Model

The Indexical Model is dedicated to the problem of form only and is developed through the use of mathematical writing. This writing is based on the purest notation of functions and equations to create a continuous parametric surface as the seed for a vertical typology.

The aim is to create an abstract syntax of form, free from any constrain, which is developed through a set of analytical drawings into one physical model. This model is used as index of ideas for the final proposal in Singapore.

Translating the diagram, represented through drawings, into a physical object is key to this chapter. The idea is not to achieve a 1:1 relation of line and material but the model should be understood as a very personal physical interpretation of a drawn concept developed in a rigorous analytic manner.

The mathematical numerical precision together with the precision of analytical line drawings define the essential ground for this purely formal discourse of the Indexical Model. The detail is as important as the overall form and part of an overall novel.

This precision is used as the common language of the whole project and allows the transition and coexistence of abstract thinking and numerical reality.

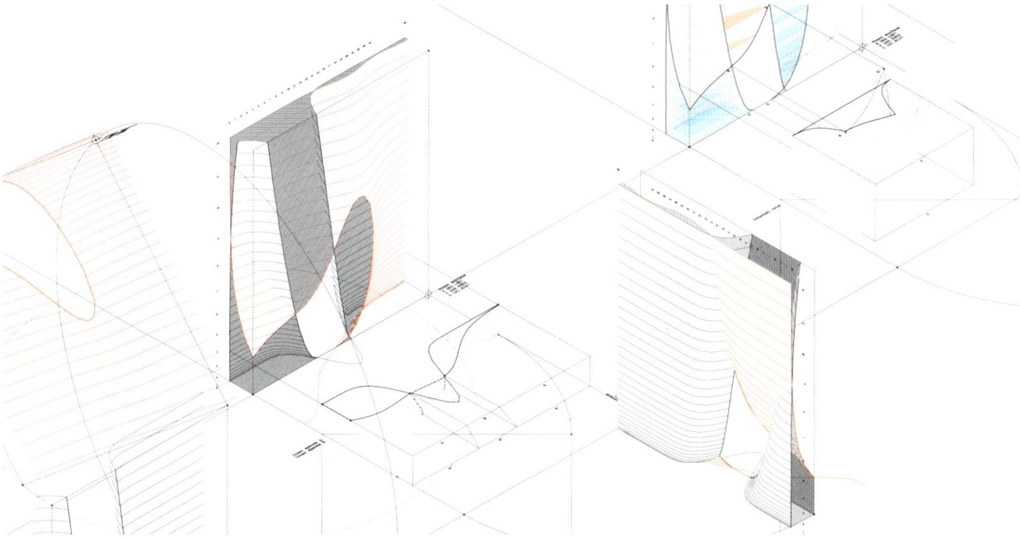

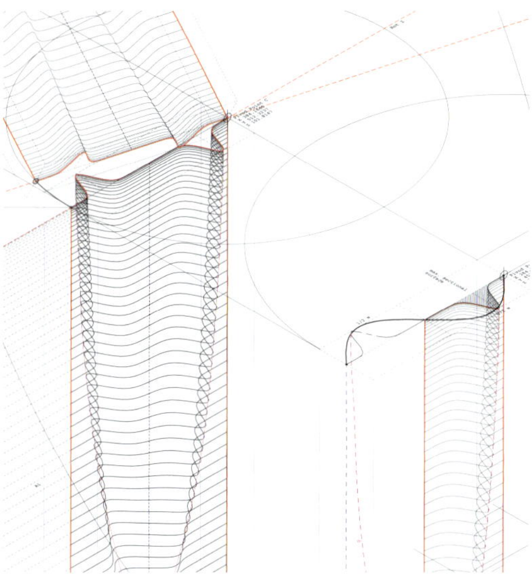

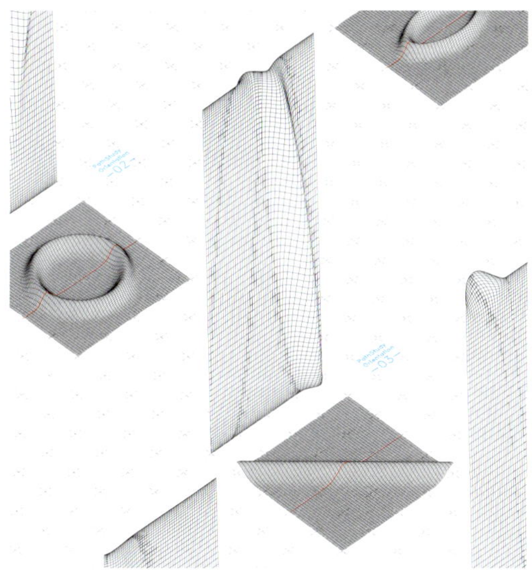

Analysis of Form

Mathematical Concept Drawing

Form Palette

Physical Model

2.1 Writing Form

This is the first of two mathematical concepts. The aim is to develop form by using the purest notation of mathematical functions and equations.

There is no scale, no program, no constrain.

The form is written by the use of a parametric equations that describe a continuous surface which folds up and down again, while further differentiation is achieved through additional functions. One function defines the spacing of the horizontal members, the other function applies curves that deform the shape of the surface.

The square located in front of the vertical plane functions as a control panel that references the original form. The drawing shows how the transition to the final vertical form is achieved mathematically. The two exploded fragments exemplify various stages of deformation.

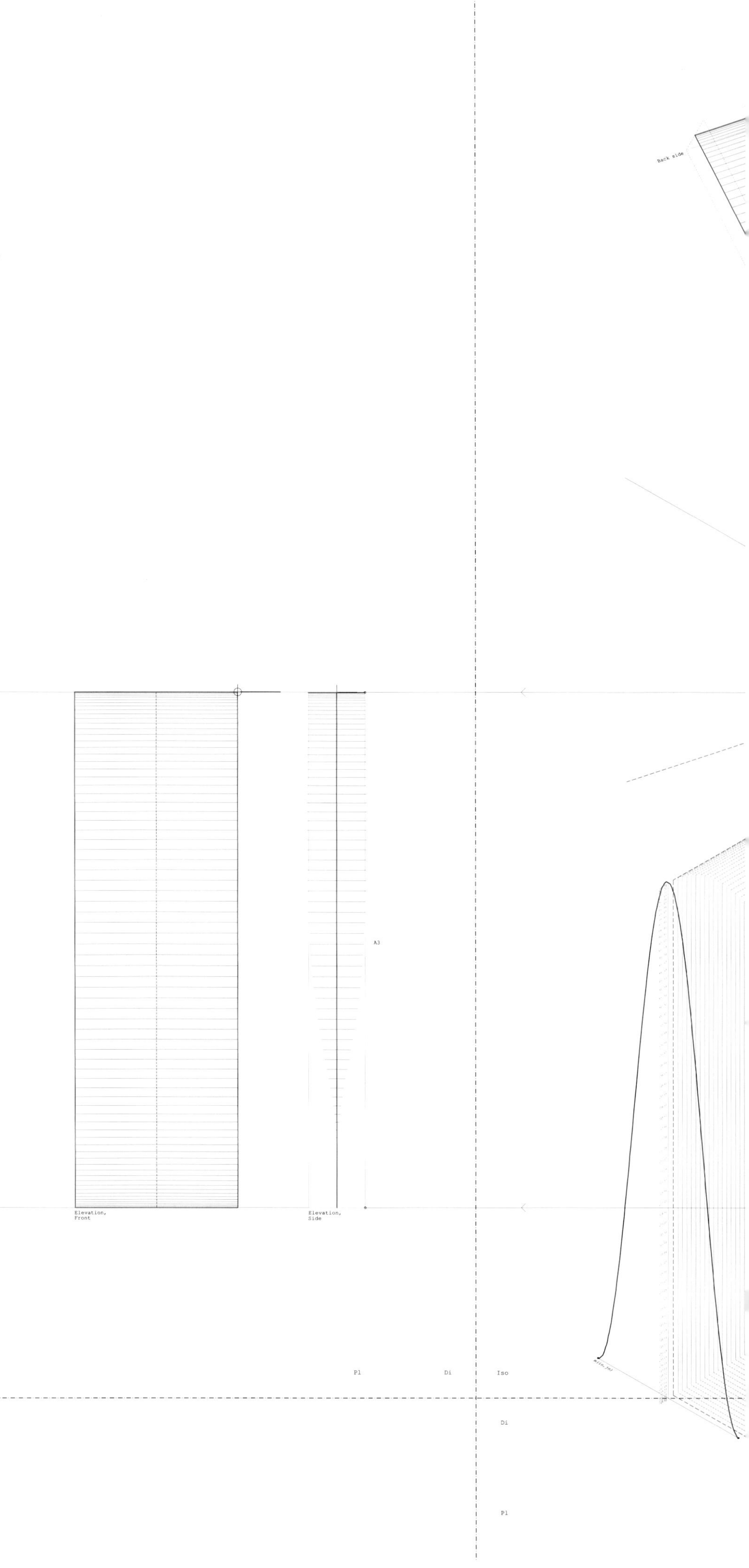

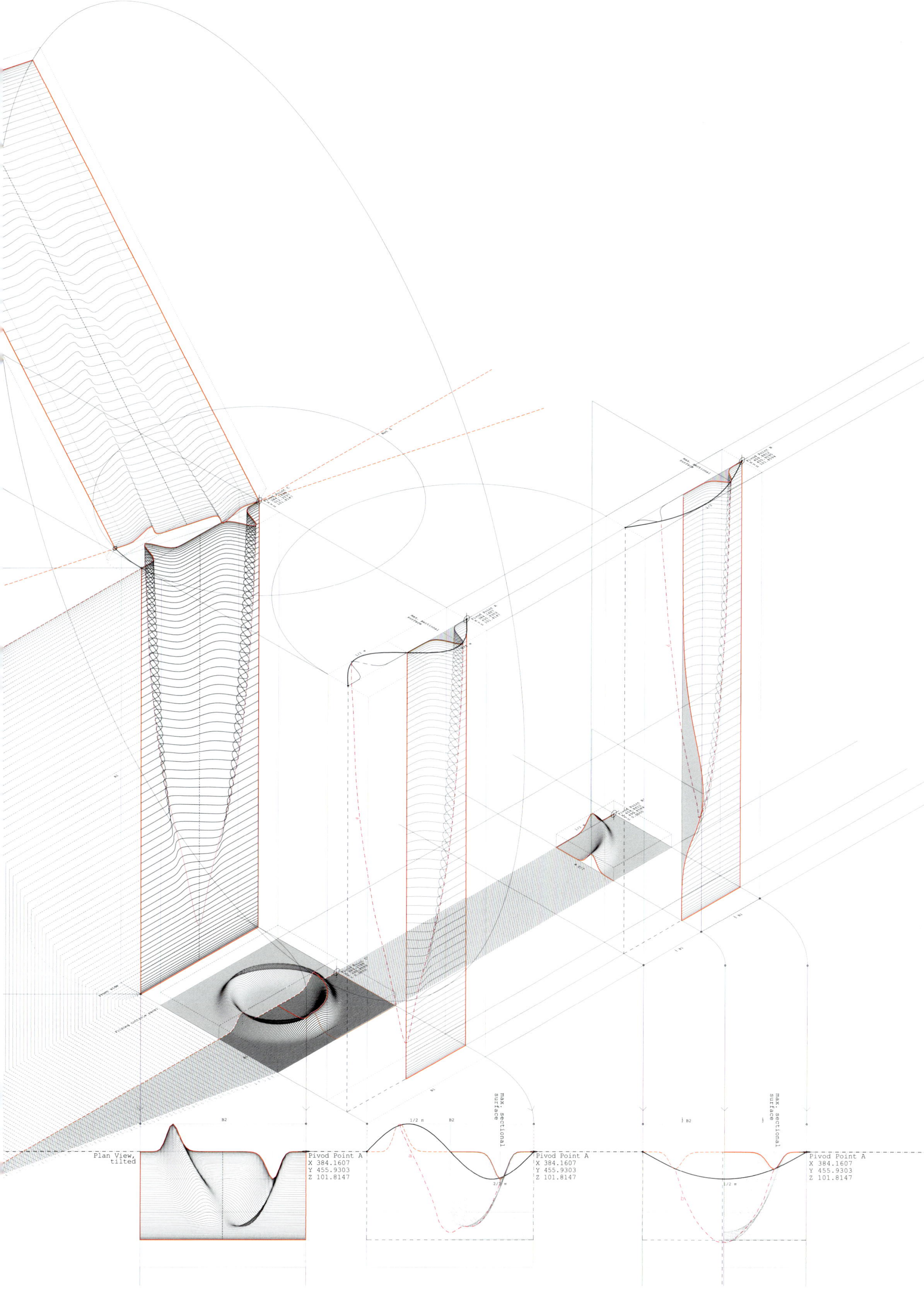
Plan View,
tilted
Pivod Point A
X 384.1607
Y 455.9303
Z 101.8147
max. sectional
surface
1/2 m
Pivod Point A
X 384.1607
Y 455.9303
Z 101.8147
max. sectional
surface
1/2 m
Pivod Point A
X 384.1607
Y 455.9303
Z 101.8147

2.2 Form Palette

Left page: The surface equation allows a palette of different forms that can be achieved by the change of only few parameters. The folds that are applied onto the surface can be articulated in their amount, angle, smoothness and curvature. They can be symmetrical or asymmetrical, they can overlap each other or run parallel.

Right page: If several folds with different sets of parameters are applied onto the same surface the patterns of deformations gain a more random appearance.

The concept of the final diagram (version 02_B) is to underline the thinness and flatness of the straight plane but to use folds with a rather random character to establish a depth within the surface. This depth should help to stabilise the plane and to make it stand on the ground.

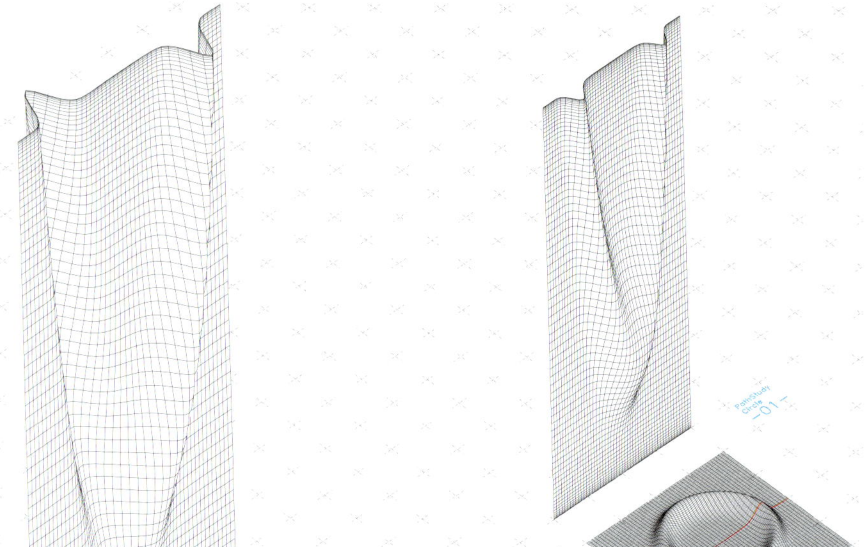

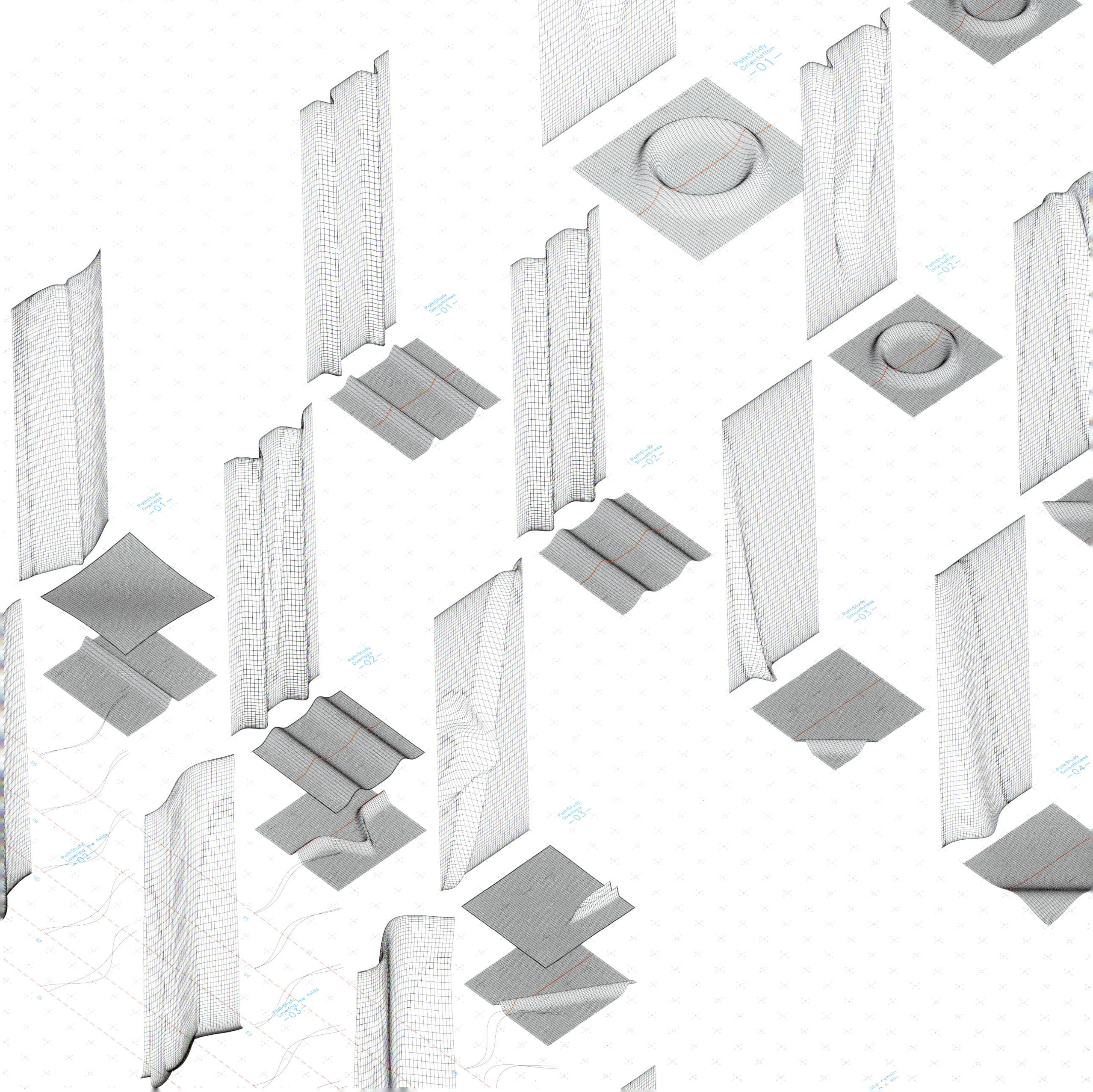

-02_B-
-03_B-
-04_B-
-09_B-
-08_B-
-14_B-
-01_B-
-07_B-
-13_B-
-06_B-
-12_B-
-02_A-
-05_B-
-11_B-
-01_A-
-10_B-
-16_B-
-15_B-
-05_A-
-04_A-

2.3 Developing the Final Diagram

Once the basic setup of position and orientation of folds within the surface is established the functions that define the contour of the footprint are adjusted in order to achieve maximum stability towards the ground. The drawing shows three different surface plots with varying footprint shapes.

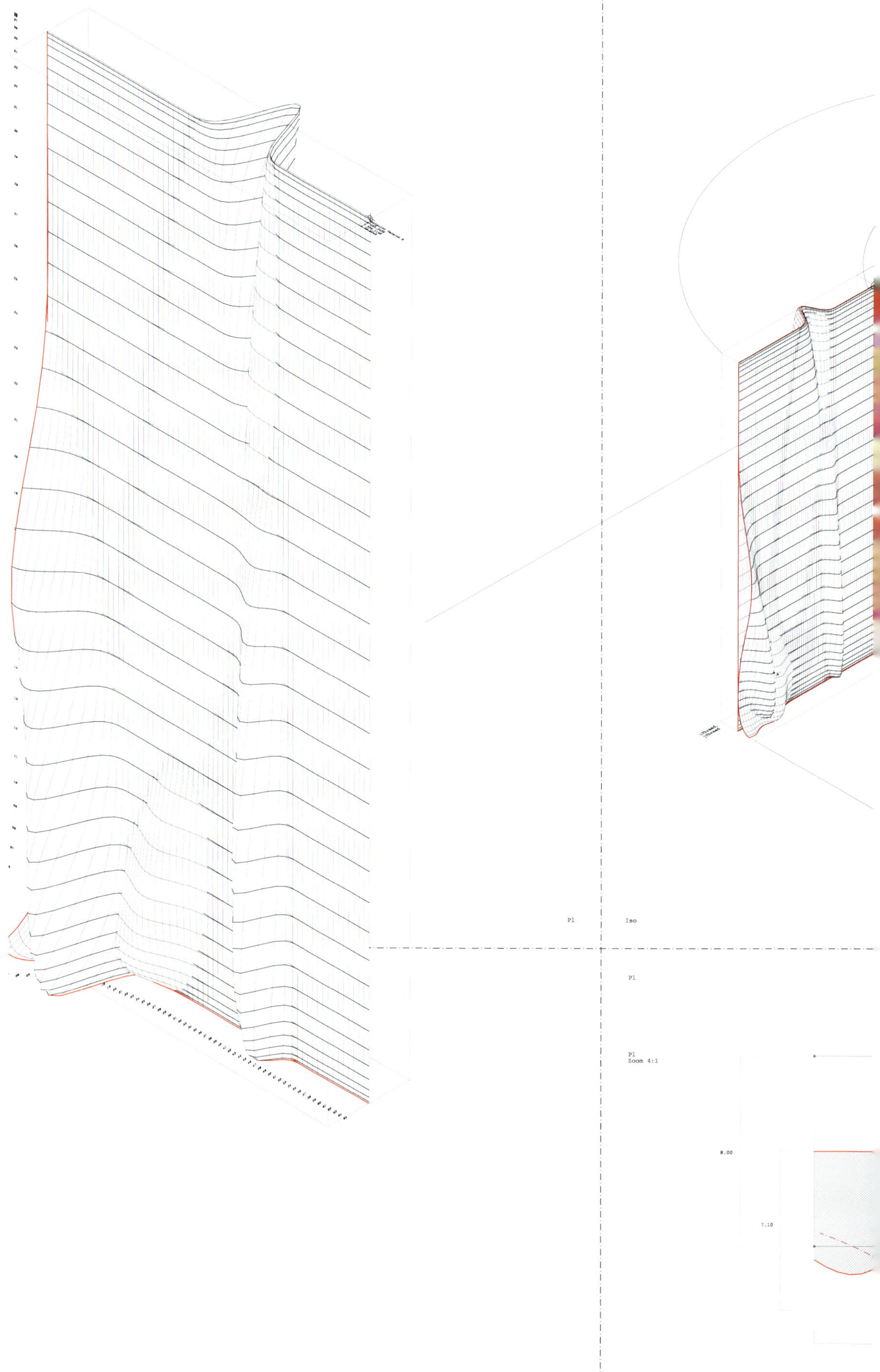

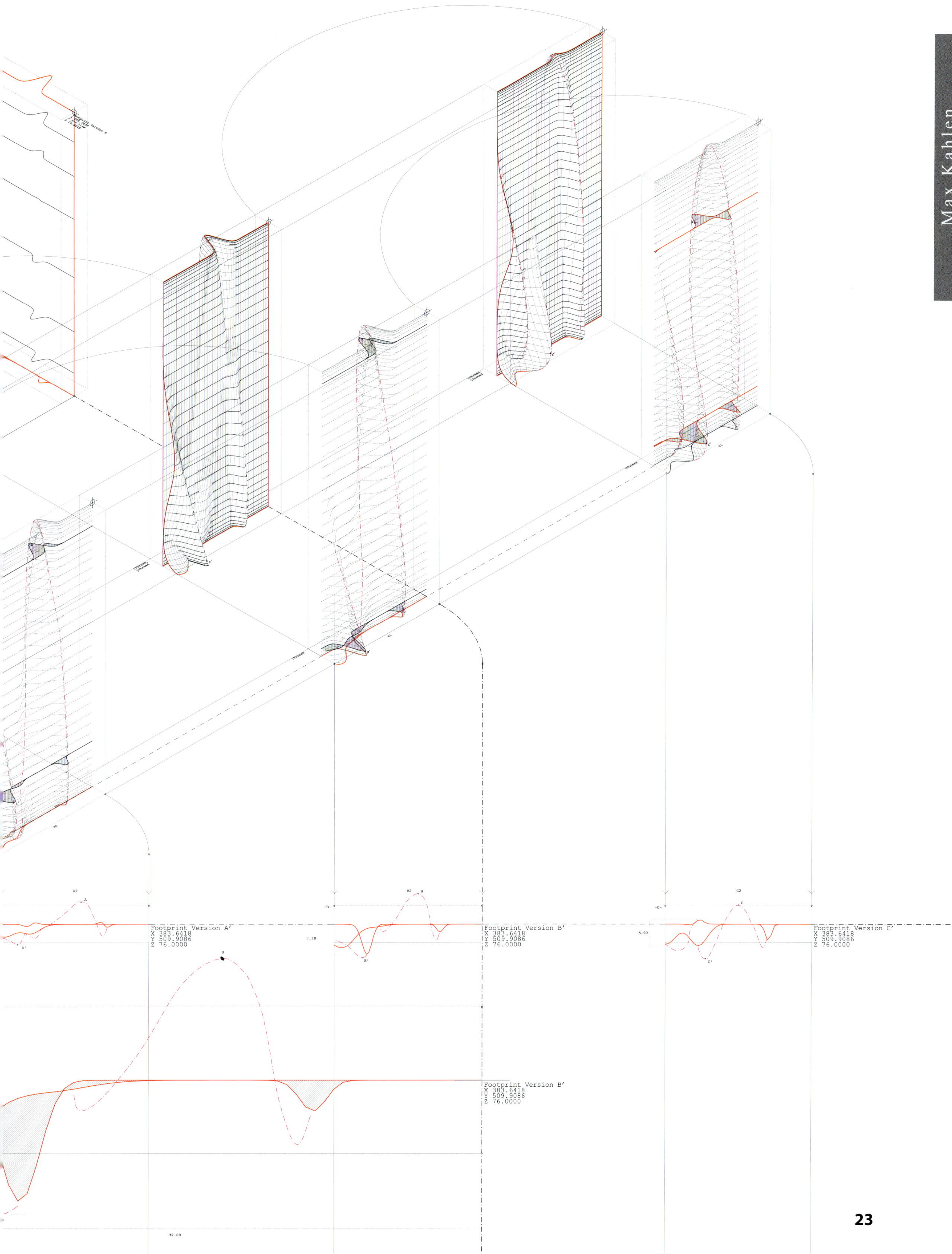
A2
A
A'
Footprint Version A'
X 383.6418
Y 509.9086
Z 76.0000
7.10
B2
B
B'
Footprint Version B'
X 383.6418
Y 509.9086
Z 76.0000
5.90
C2
C
C'
Footprint Version C'
X 383.6418
Y 509.9086
Z 76.0000
B
Footprint Version B'
X 383.6418
Y 509.9086
Z 76.0000
32.00

Physical model study, eroding the surface

2.4 Analysis of Form

To develop a material concept for the physical model of the diagram several sectional studies are made. The aim is to discover the essential character that represents the diagram through a serious of analytic drawings.

The drawing on the left shows an array of sections taken horizontally, parallel to the orientation of the surface grid. While the drawing on the right shows sections taken parallel to the surface plane, against the rhythm of its' grid.

Both sectional studies capture different readings of the spaces enclosed by the folds. Those pocket spaces can be organized in two groups: the *positive* pockets in areas where the two surfaces diverge from each other and the *negative* pockets in areas where the two surfaces intersect each other.

2.4 Analysis of Form

The drawing on the right shows a detailed fragment of the surface. Red indicates a negative pocket, blue indicates a positive pocket.
But the fact that the two surfaces simply enclose spaces and establish depth within the thin plane seams less interesting than the lines of intersection that are the result of the back-and-forth movement of the surfaces - in other words, the change from the negative to the positive. Those lines which indicate where the surfaces intersect reflect the structural integrity of the different pockets that stabilise the plane surface and represent in a different way the superior logic that interacts with the surface.

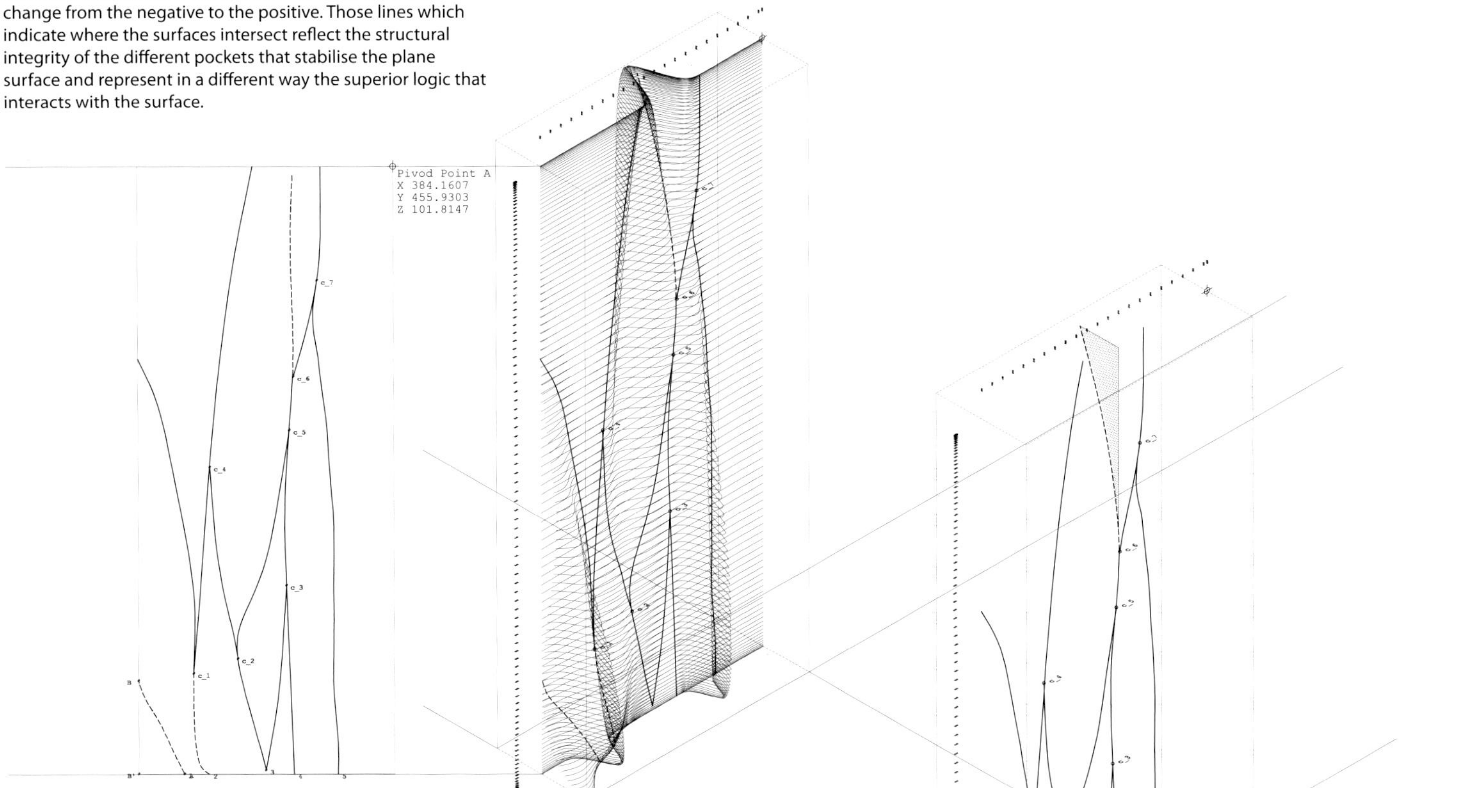

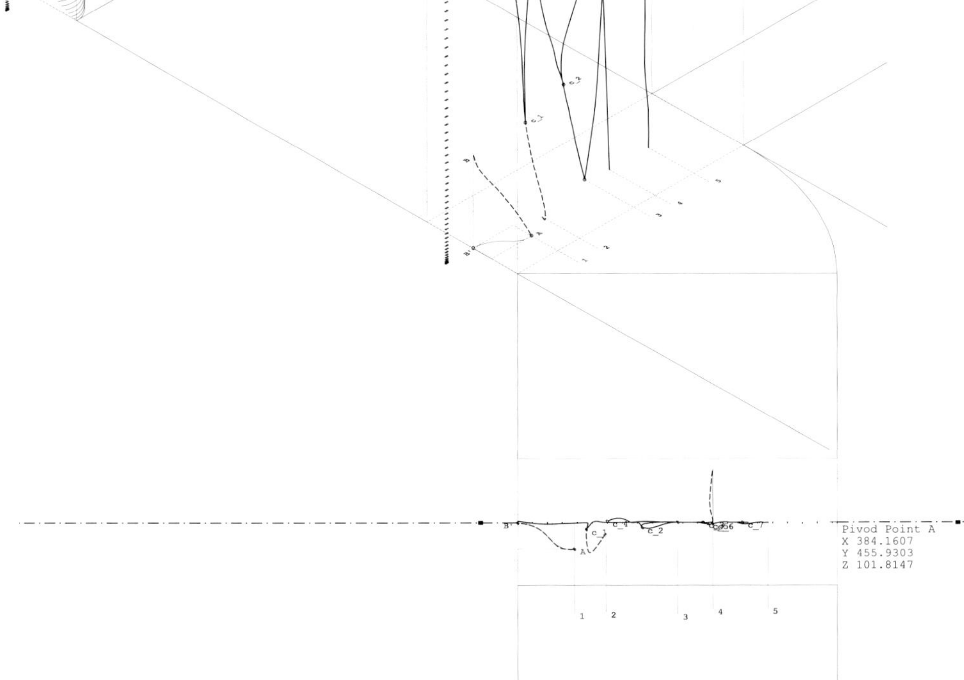

Physical model study, eroding the surface

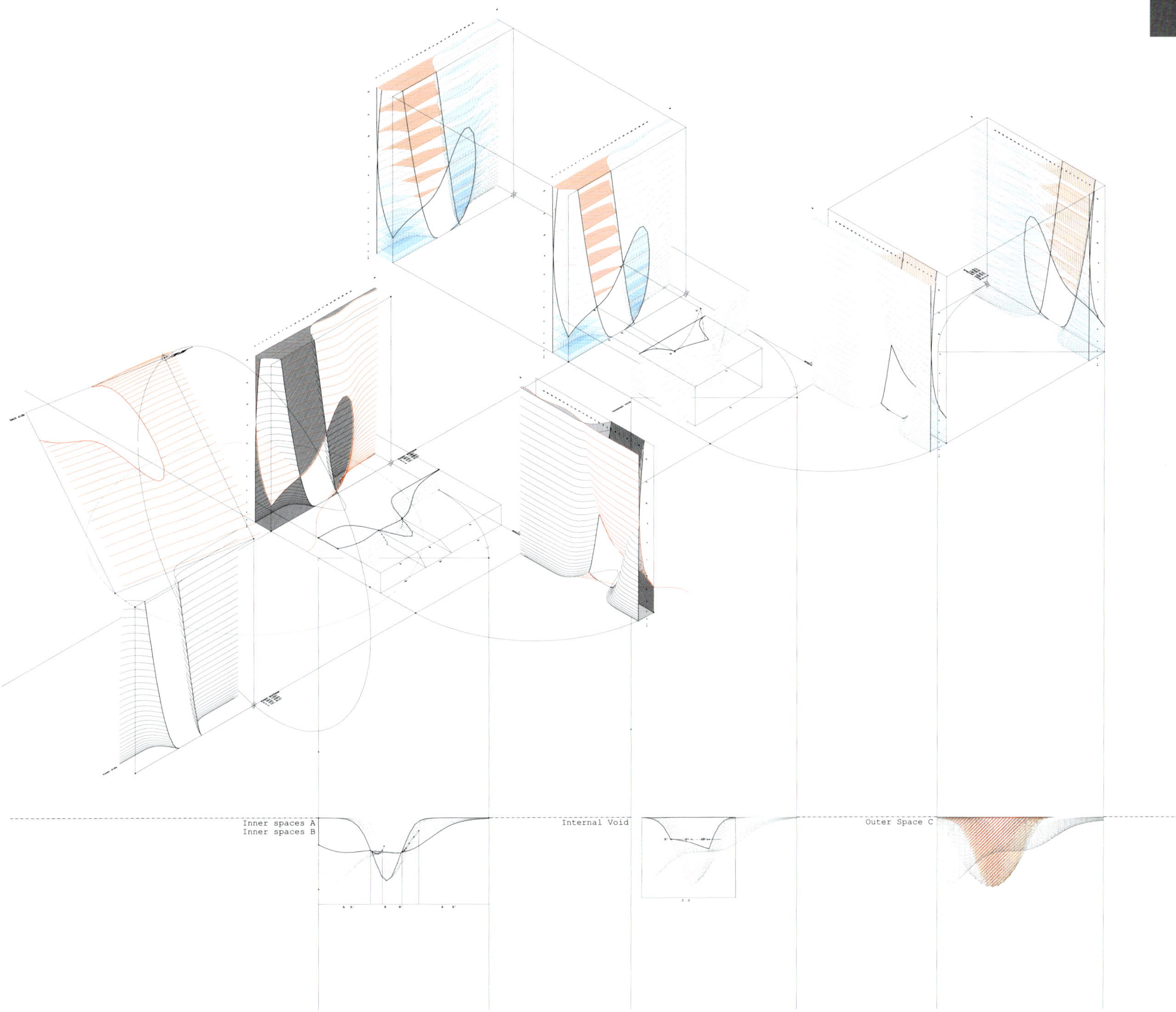
Inner spaces A
Inner spaces B
Internal Void
Outer Space C

2.5 Developing the Indexical Model

In order to translate the drawn representation of the diagram into a physical object, material details need to be developed to satisfy the expression and character of the original drawing as well as they need to be part of an assembly process that should be functional and feasible in its production.
For this translation the articulation of the lines of intersection becomes an integral part.

Mock-up models are used to test different materials and assembly methods.

The image on the far right shows a model where the lines of intersection are expressed explicitly through a thicker element carefully embedded within the grid structure.

The image on the bottom of this page shows the final detail where this line is articulated implicitly. The lines of intersection are annotated through engraved trenches that appear as erosion within the material. Thereby the surface gains lightness. The expression of the grid remains pure while the subtle trenches within the thin members seam like traces of the forces that shaped the folds.

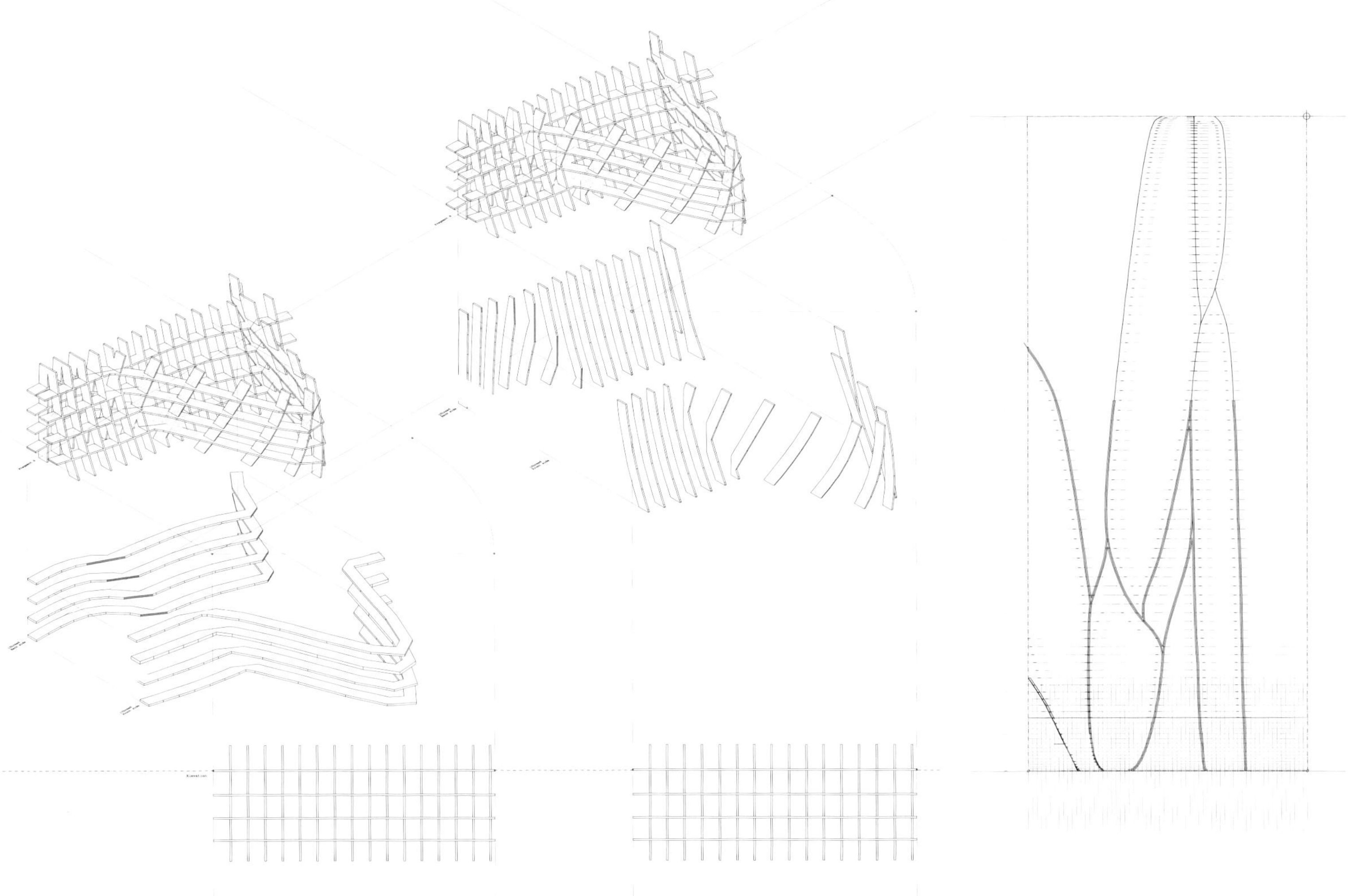

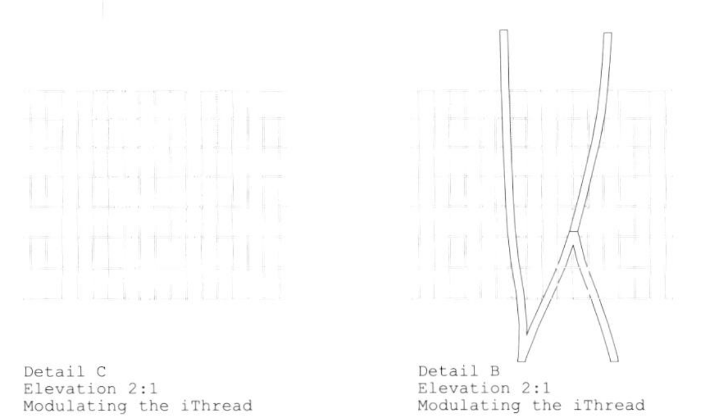

Physical model study, eroding the surface

Physical model study,
embedding additional threads within the surface

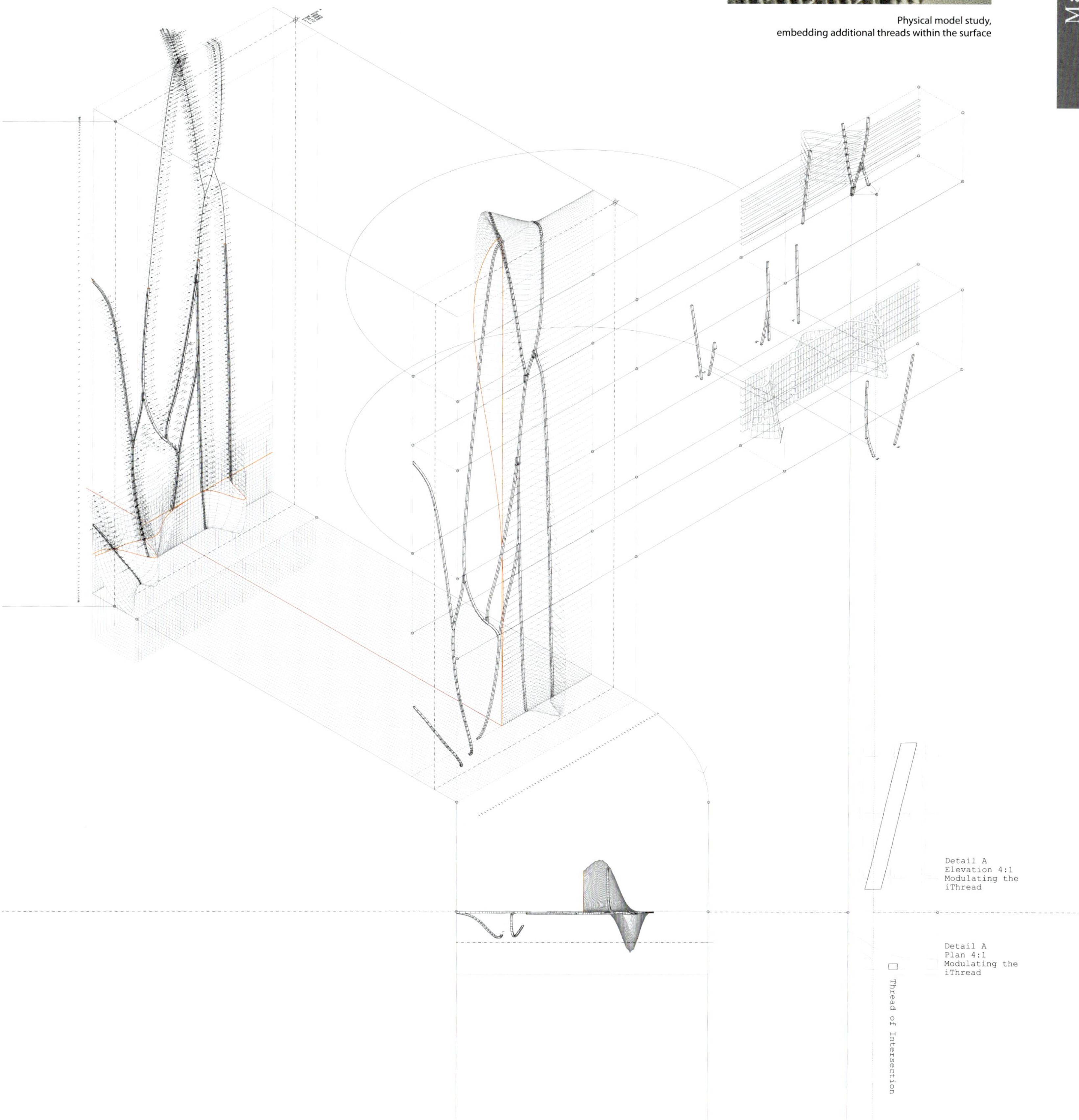

2.6 Evaluating the Indexical Model

Elevation Front & Side
Plan

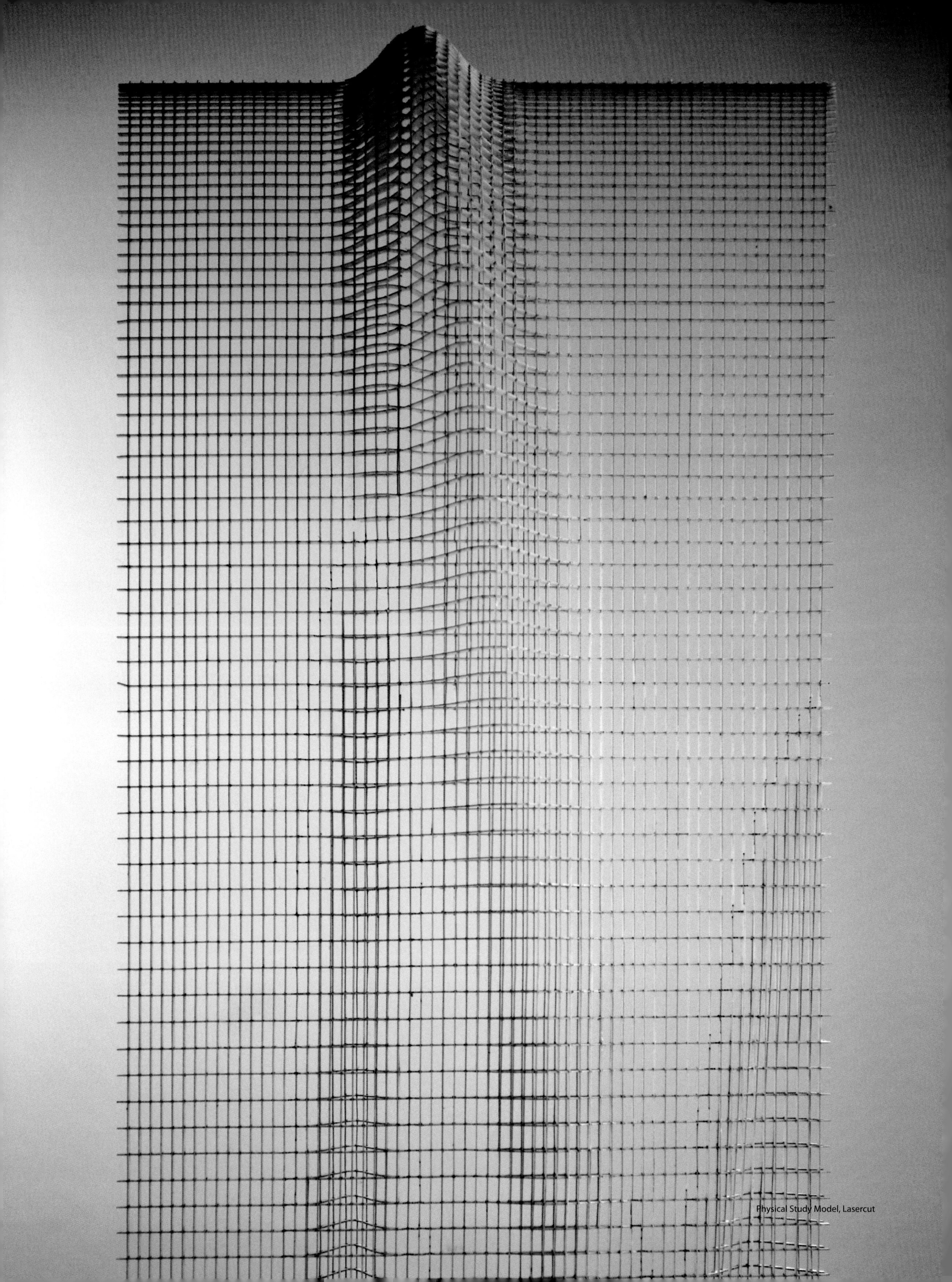

Physical Study Model, Lasercut

2.6 Evaluating the Indexical Model

For the development of the forthcoming final proposal there are two main ideas that characterize this indexical model that should be identified and carried forward as potentials: The contrast of the thin surface versus the depth of the massive base and the depth of the folds (top images, this page & opposite page). As well as the fact that this contrast can be read from one perspective (top image, this page), while from the other perspective this differentiation remains completely unreadable and the surface appears mute (bottom image, this page).
Another potential is the gradual change of intervals of the horizontal members that define the reading of the surface texture.

Model Detail 2 - Solid Footprint

Model Detail 1 - Thinness

Model Detail 3 - Varying Depth

Model Detail 4 - Densities

3. Massing Studies

3.1 The Site, Singapore
3.2 Automating Form
3.3 Introducing Industrial Performance Ratios
3.4 Massing Studies
3.5 Final Massing

For the massing study of the final building proposal in Singapore -again- mathematical writing defines the basis. But in contrast to the first rather abstract mathematical concept of the indexical model, in this case, the equations are derivated purely from real constrains of industry performance ratios and the site.

The concept is to represent restrictions and constrains obtruded by the building authority and the industry as parameters and to join as many as possible of these parameters into one complex equation to create an interface that automates and speeds up the process of massing. This interface does not provide one final result but it renders different alternatives with varying qualities that help to negotiate the final massing.

Furthermore this massing tool is applied to the specific site in Singapore. Diagrams and models document the evolution of the massing concept which is based on a building-slab typology. The equations are used to drive the proportions of those building slabs to its' limits while maintaining the maximum floor area that is achievable on that specific site.

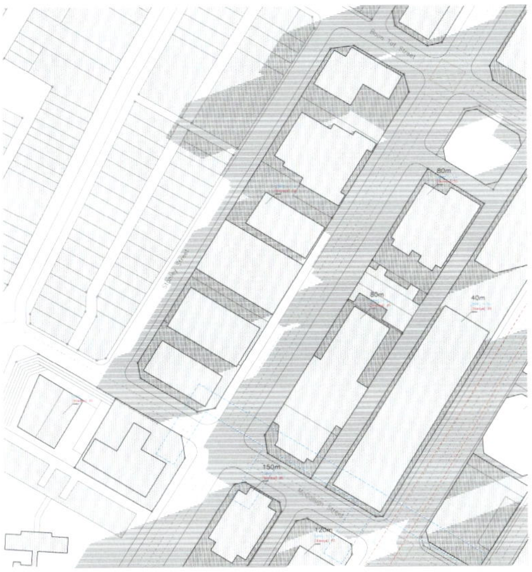

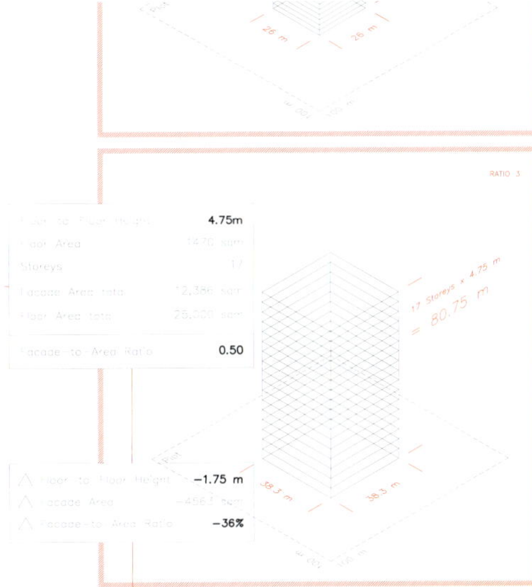

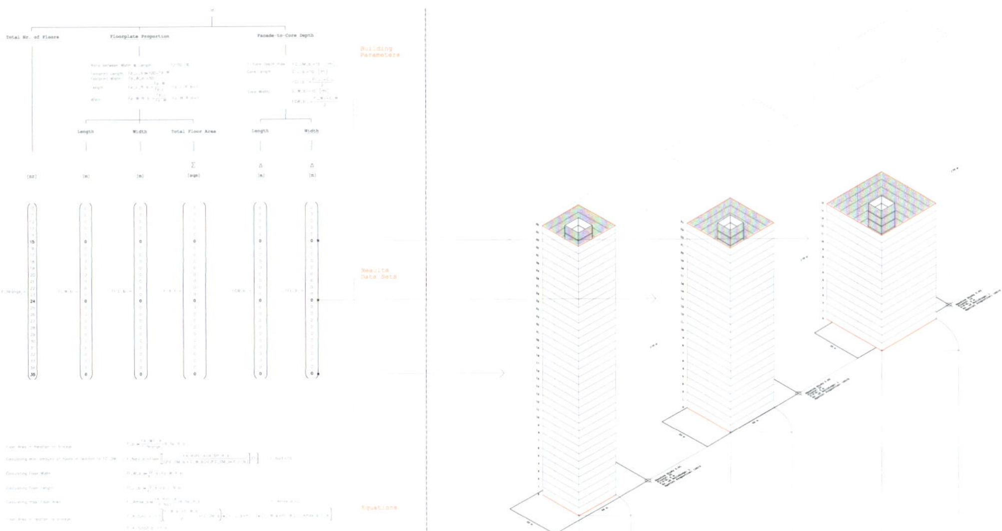

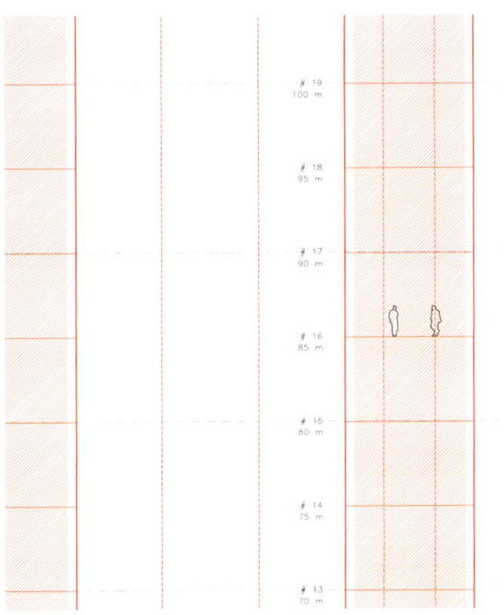

Site map Massing interface

Massing calculation

Final Massing proportion Final massing on site

3.1 The Site, Singapore

The site for the high-rise building proposal is located in the centre Singapores' office district (outlined in red). The planing regulations ask for a development with mainly residential use, commercial spaces in the podium and office spaces in the storeys above.

Site size: 30 x 70 m
Max. building height: 180 m
FAR: 12.5

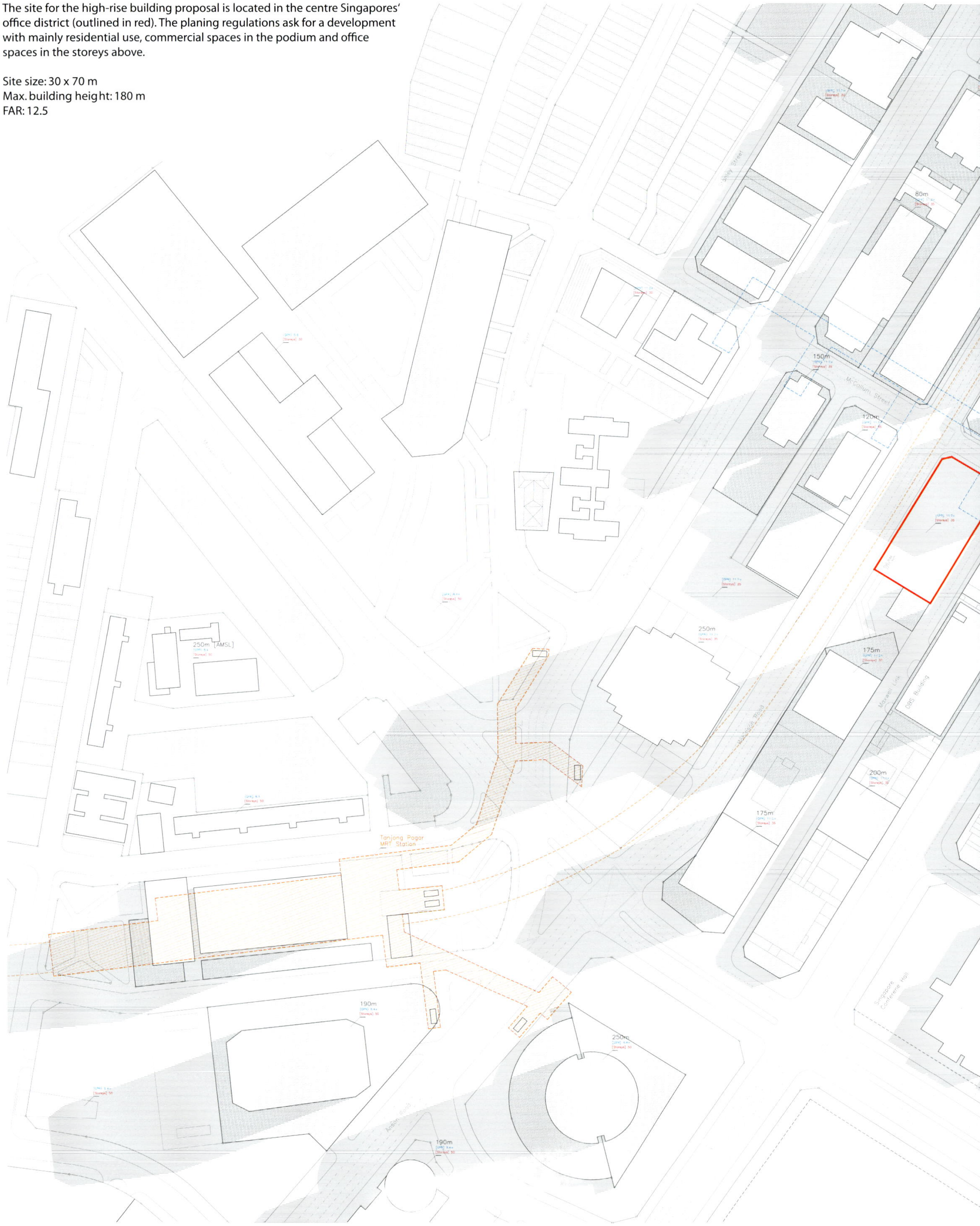

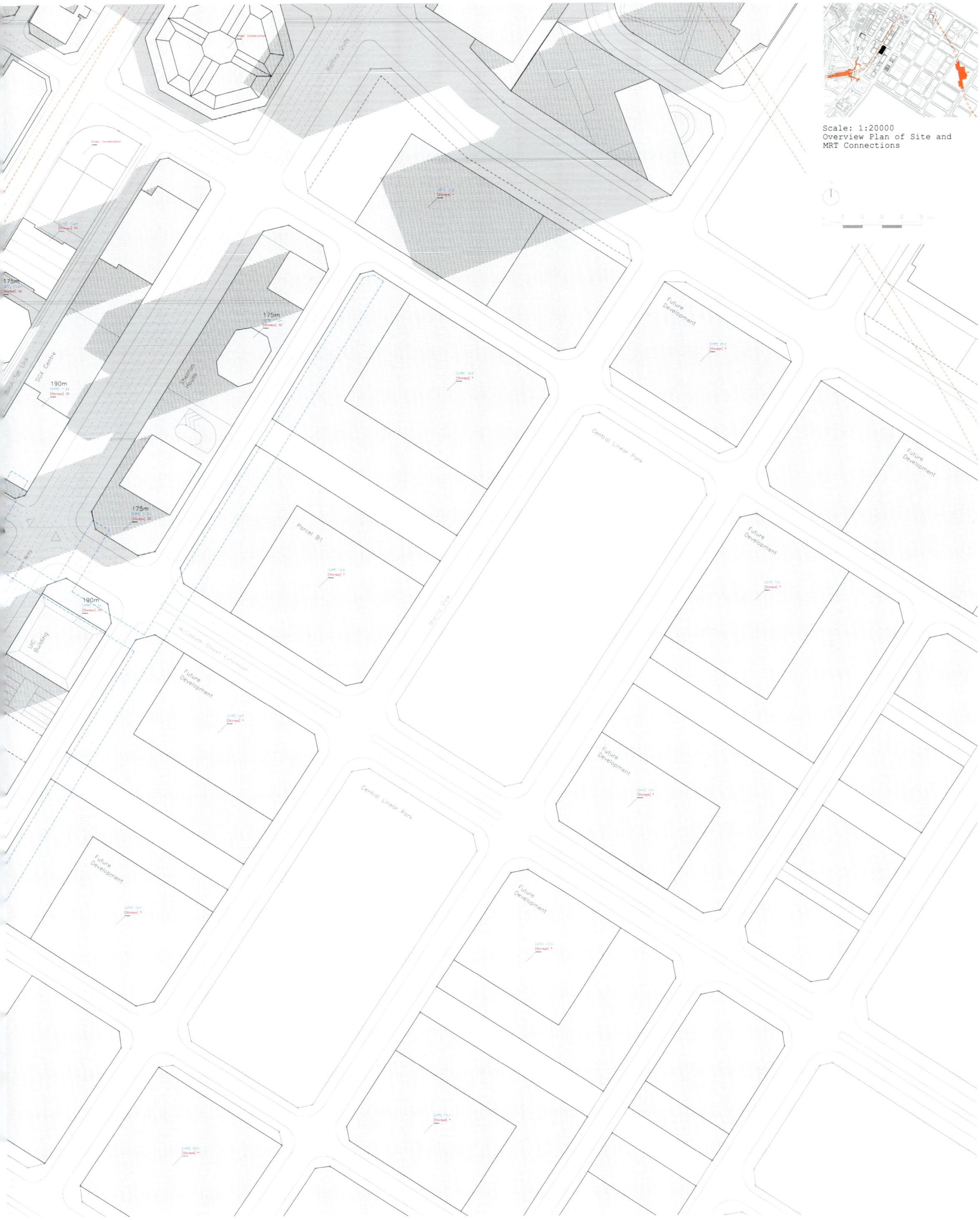

Scale: 1:20000
Overview Plan of Site and MRT Connections

Plot Size
Total Nr. of Floors
GPR
Distribution of Space in two Buildings (A + B)
100 %
0 %
Positioning
Building A
Building B
Total Nr. of Floors
Floorplate Proportion
Facade-to-Core Depth
Length
Width
Total Floor Area
Length
Width
[nr]
[m]
[m]
[sqm]
[m]
[m]
Floor Area in Relation to Nr. of Floors
[Storeys]
5.01
5.02
5.03
[sqm per Floor]

3.2 Automating Form

This second mathematical concept is, in contrast to the first concept (chapter 2.1), dedicated to one fixed scale and very specific constrains.

The idea is to generate form automatically and purely derived from numerical constrains related to building regulations and industry performance ratios such as plot size, maximum building height, floor-area-ratio (FAR), maximum and minimum amount of floors, amount of buildings... etc.

Equations and functions set the basis for this concept to tackle the idea of „automatic writing" and to test how far the authorship of the designer can be reduced during the process of developing form.
Although this process of massing cannot be automated completely, this equation helps negotiate alternatives and speeds up the process of massing.

The equation is set up with the aim to achieve the maximum floor space that can be achieved on the specific plot. Once the main parameters for building and site are set it automatically renders different versions depending on building proportion, number of floors and floor-to-ceiling height. Those arrays of numbers are further processed to evaluate facade-to-core distances in relation to the use as well as the facade-to-area ratio represented through charts.

The drawing below describes how constrains are combined into one equation and how the numerical output is represented through charts and three dimensional building diagrams.

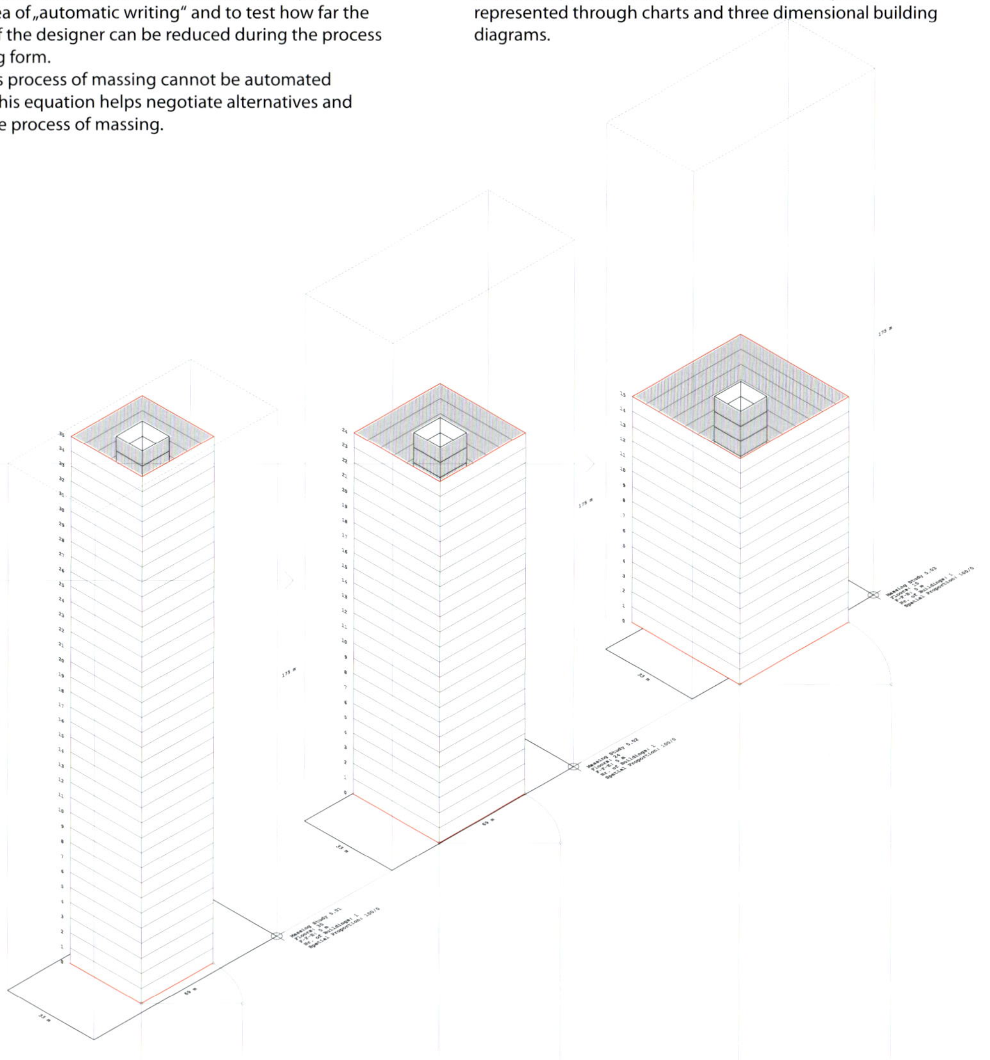

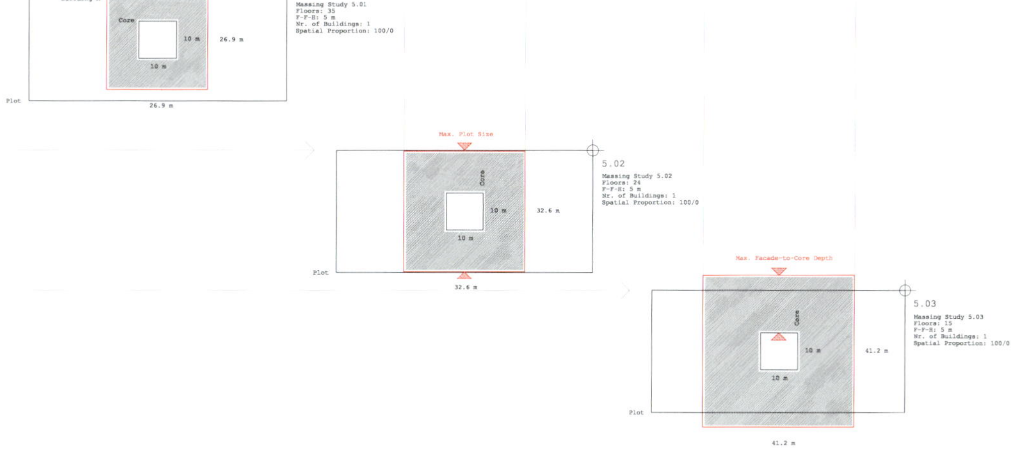

3.2 Automating Form

This page shows the numerical output of the massing equation. The first row indicates the number of floors, the second & third row the width & length of the floor plates, the fourth & fifth row indicate the facade-to-core depth related to the number of floors.

The graph on the right represents those numbers and helps to visualize and conclude.
The diagrams in the right page render the massing proportions and floors in relation to the site contours.

Nr. of Buildings:
Distance Length 0 [m]
Distance Width 0 [m]
Distribution of Space:
Amount of Space -B- 0 [%]

Floorplate Proportion:
Footprint Width 29 [%]
Service Core:
Fassade-to-Core Depth max. 15 [m]
Core Width 8 [m]
Core Length 15 [m]

Floorplate Proportion:
Footprint Width -B- 50 [%]
Service Core:
Fassade-to-Core Depth max. -B- 10 [m]
Core Width -B- 0 [m]
Core Length -B- 0 [m]

10	20.6	123.6	12.6	108.6	0	0	0	0
11	19.7	117.9	11.7	102.9	0	0	0	0
12	18.8	112.9	10.8	97.9	0	0	0	0
13	18.1	108.4	10.1	93.4	0	0	0	0
14	17.4	104.5	9.4	89.5	0	0	0	0
15	16.8	100.9	8.8	85.9	0	0	0	0
16	16.3	97.7	8.3	82.7	0	0	0	0
17	15.8	94.8	7.8	79.8	0	0	0	0
18	15.4	92.2	7.4	77.2	0	0	0	0
19	15	89.7	7	74.7	0	0	0	0
20	14.6	87.4	6.6	72.4	0	0	0	0
21	14.2	85.3	6.2	70.3	0	0	0	0
22	13.9	83.4	5.9	68.4	0	0	0	0
23	13.6	81.5	5.6	66.5	0	0	0	0
24	13.3	79.8	5.3	64.8	0	0	0	0
25	13	78.2	5	63.2	0	0	0	0
26	12.8	76.7	4.8	61.7	0	0	0	0
27	12.6	75.2	4.6	60.2	0	0	0	0
28	12.3	73.9	4.3	58.9	0	0	0	0
29	12.1	72.6	4.1	57.6	0	0	0	0
30	11.9	71.4	3.9	56.4	0	0	0	0
31	11.7	70.2	3.7	55.2	0	0	0	0
32	11.5	69.1	3.5	54.1	0	0	0	0
33	11.4	68.1	3.4	53.1	0	0	0	0
34	11.2	67.1	3.2	52.1	0	0	0	0
35	**11**	**66.1**	**3**	**51.1**	**0**	**0**	**0**	**0**

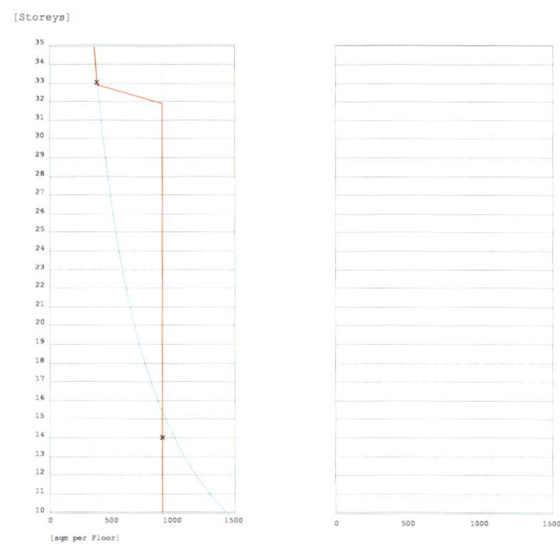

Nr. of Buildings:
Distance Length 0 [m]
Distance Width 12 [m]
Distribution of Space:
Amount of Space -B- 50 [%]

Floorplate Proportion:
Footprint Width 22 [%]
Service Core:
Fassade-to-Core Depth max. 12 [m]
Core Width 3 [m]
Core Length 20 [m]

Floorplate Proportion:
Footprint Width -B- 22 [%]
Service Core:
Fassade-to-Core Depth max. -B- 12 [m]
Core Width -B- 3 [m]
Core Length -B- 20 [m]

Nr. of Floors:	Floor Width:	Floor Length:	F-C Width:	F-C Length:	Floor Width -B-:	Floor Length -B-:	F-C Width:	F-C Length:
10	10.1	126.6	7.1	106.6	10.1	126.6	7.1	106.6
11	9.6	120.7	6.6	100.7	9.6	120.7	6.6	100.7
12	9.2	115.6	6.2	95.6	9.2	115.6	6.2	95.6
13	8.8	111	5.8	91	8.8	111	5.8	91
14	8.5	107	5.5	87	8.5	107	5.5	87
15	8.2	103.4	5.2	83.4	8.2	103.4	5.2	83.4
16	8	100.1	5	80.1	8	100.1	5	80.1
17	7.7	97.1	4.7	77.1	7.7	97.1	4.7	77.1
18	7.5	94.4	4.5	74.4	7.5	94.4	4.5	74.4
19	7.3	91.8	4.3	71.8	7.3	91.8	4.3	71.8
20	7.1	89.5	4.1	69.5	7.1	89.5	4.1	69.5
21	7	87.4	4	67.4	7	87.4	4	67.4
22	6.8	85.4	3.8	65.4	6.8	85.4	3.8	65.4
23	6.6	83.5	3.6	63.5	6.6	83.5	3.6	63.5
24	6.5	81.7	3.5	61.7	6.5	81.7	3.5	61.7
25	6.4	80.1	3.4	60.1	6.4	80.1	3.4	60.1
26	6.2	78.5	3.2	58.5	6.2	78.5	3.2	58.5
27	6.1	77	3.1	57	6.1	77	3.1	57
28	6	75.7	3	55.7	6	75.7	3	55.7
29	5.9	74.3	2.9	54.3	5.9	74.3	2.9	54.3
30	5.8	73.1	2.8	53.1	5.8	73.1	2.8	53.1
31	5.7	71.9	2.7	51.9	5.7	71.9	2.7	51.9
32	5.6	70.8	2.6	50.8	5.6	70.8	2.6	50.8
33	5.5	69.7	2.5	49.7	5.5	69.7	2.5	49.7
34	5.5	68.7	2.5	48.7	5.5	68.7	2.5	48.7
35	**5.4**	**67.7**	**2.4**	**47.7**	**5.4**	**67.7**	**2.4**	**47.7**

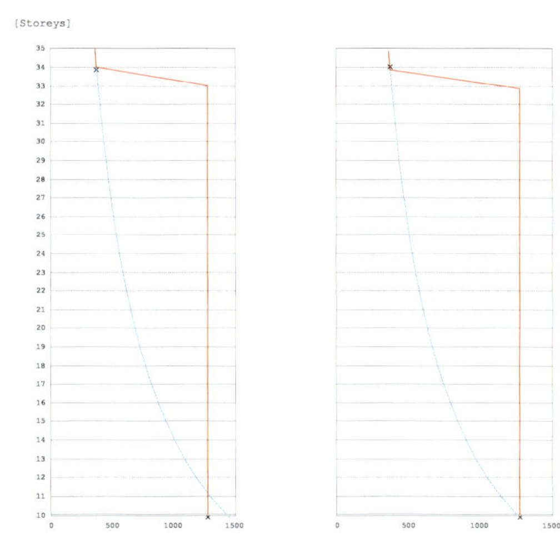

FAR:

Floor-to-Area Ratio	11.2	[GPR]

Plot:

Plot Width	33	[m]
Plot Length	69	[m]

Floors:

Floor Nrumber max.	35	[nr]
Floor Number min.	10	[nr]

Floor-to-Floor Height:

Floor-to-Floor Height max.	5	[m]
Floor-to-Floor Height min.	2.5	[m]
Height intervals	0.25	[m]

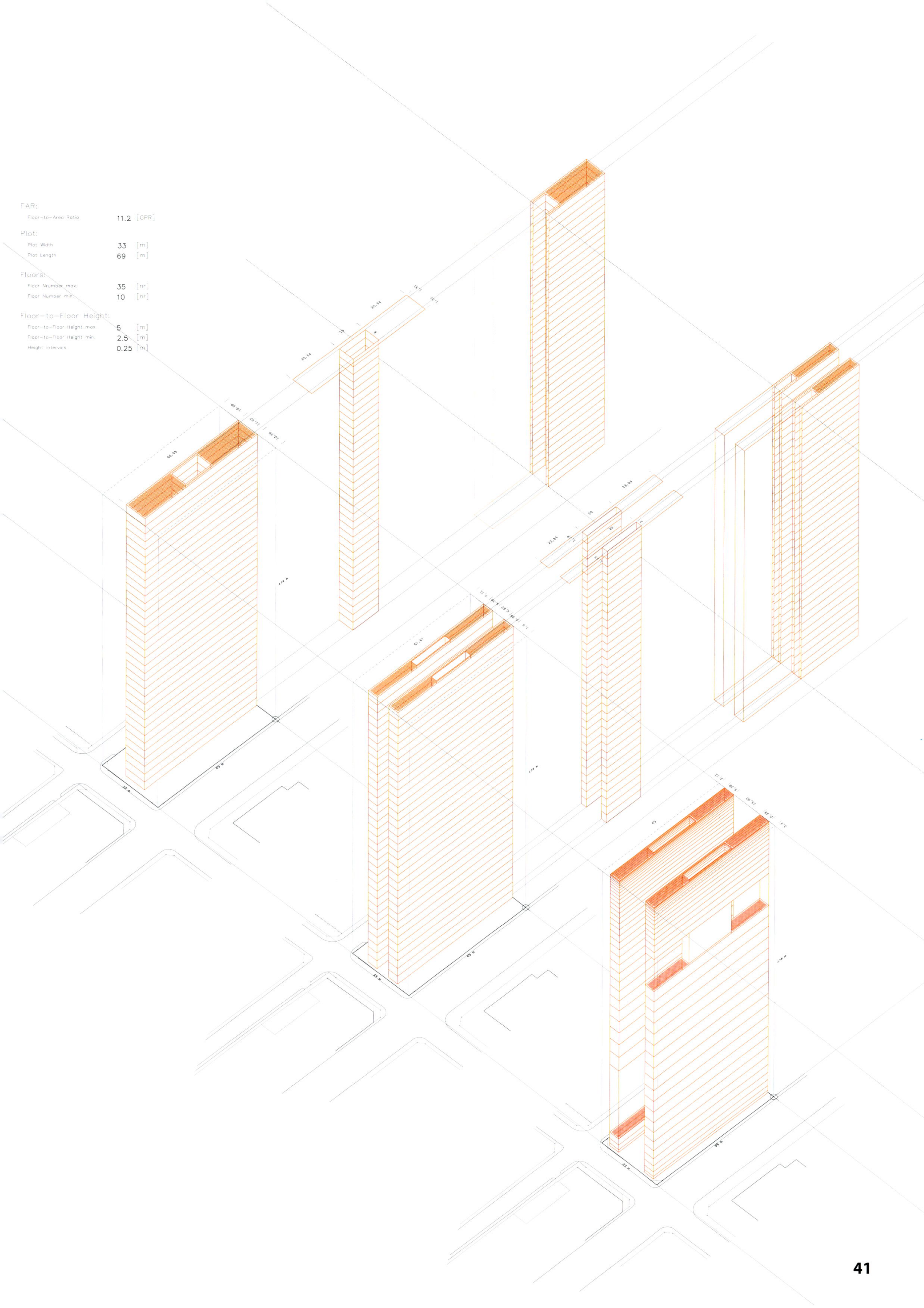

3.3 Introducing Industrial Performance Ratios

These two pages show the interface that guides the user step-by-step through the process of massing. The pages describe the essential constrains and ratios while those numbers can be filled in forms which are placed below and indicated through red frames.

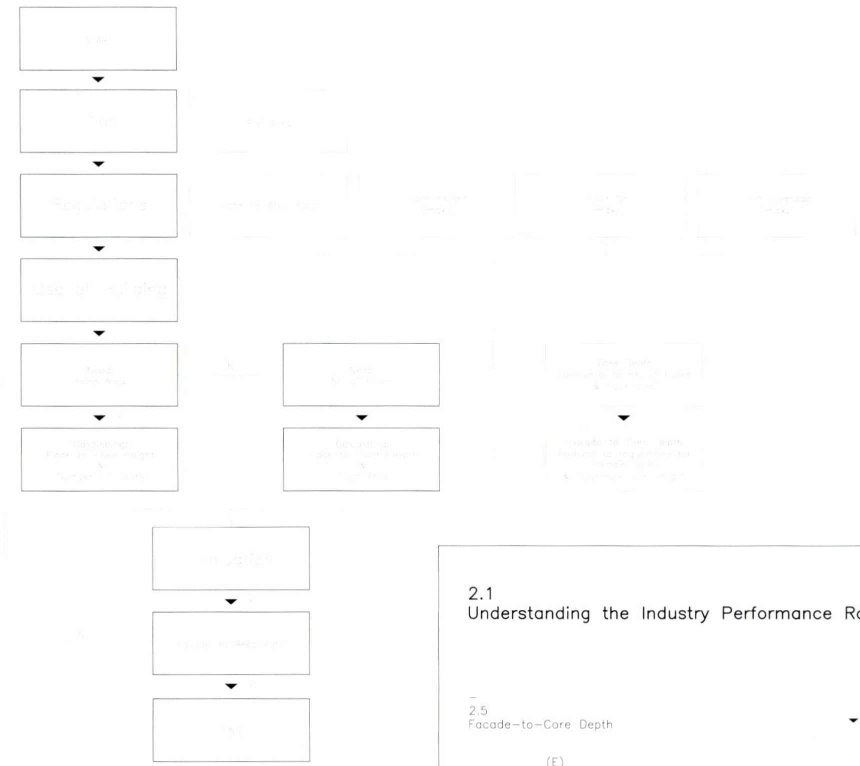

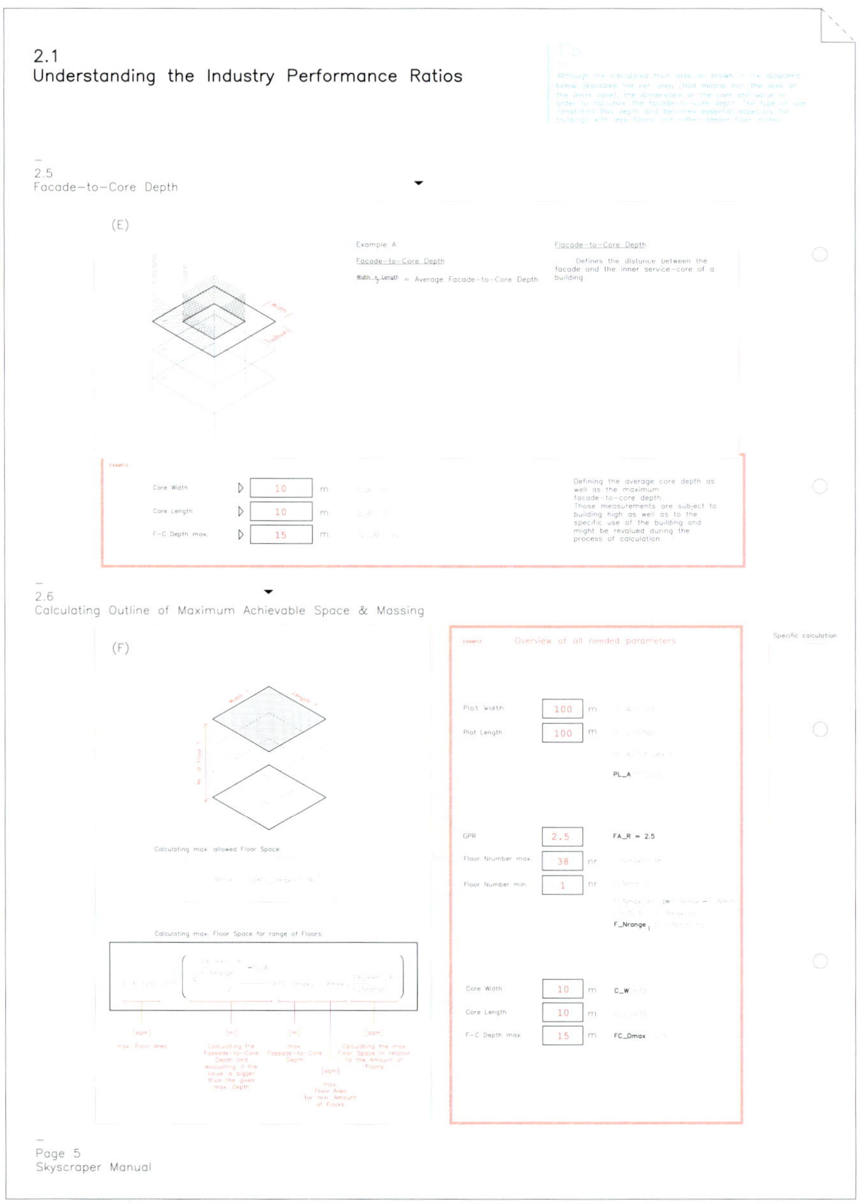

Comment

—

The diagram that outlines the maximum achievable space and massing should be understood as a simple tool to speed up the decision-process. It does not provide one specific solution but a range of alternatives. Furthermore those are subject to parameters such as floor-to-floor height.

—

2.7
Outline of Maximum Achievable Space & Massing

(G)

Diagram:

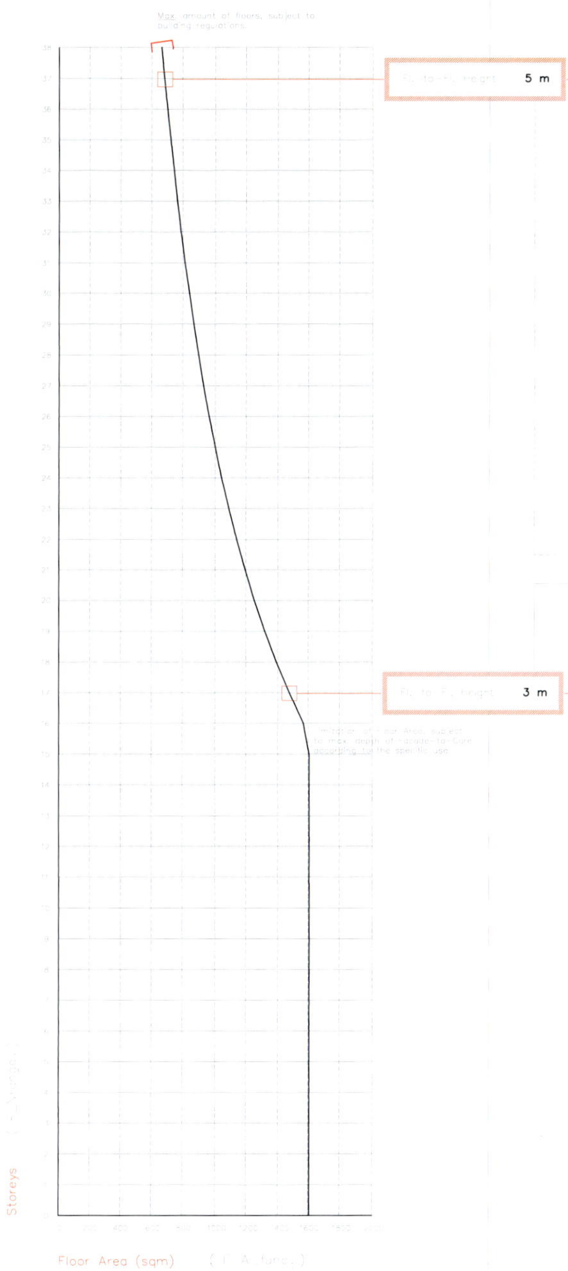

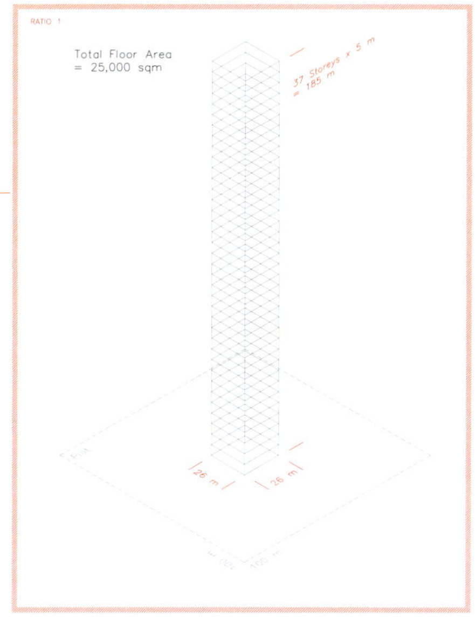

The floor-to-floor height of each storey is constrained to max. 5 meters. Otherwise there is no further specification. The particular floor-to-floor height is subject to economical issues and will be evaluated by the facade-to-area ratio.

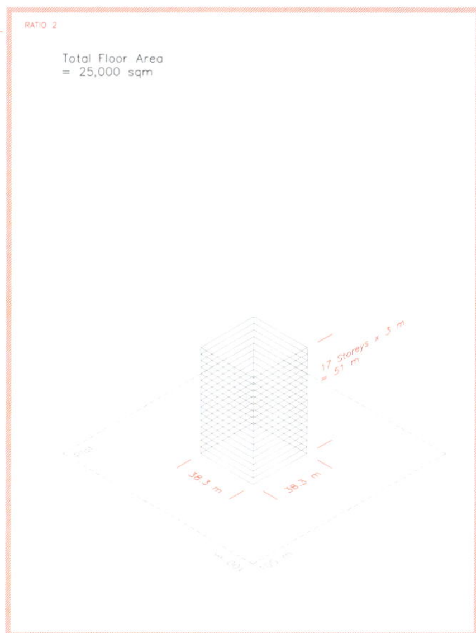

2.1 Understanding the Industry Performance Ratios

—

2.0
Applying the Ratios

The following four pages explain varying parameters and ratios that are essential for the evaluation of the economic performance of a high rise. Those parameters are partly subject to building regulations as well as to the specific use and program of the building. Furthermore those definitions come along together with a form that you can use as a step-by-step guide. This guide allows you to quickly calculate varying building proportions and evaluate the chosen parameters and ratios.

—

2.1
Plot

(A)

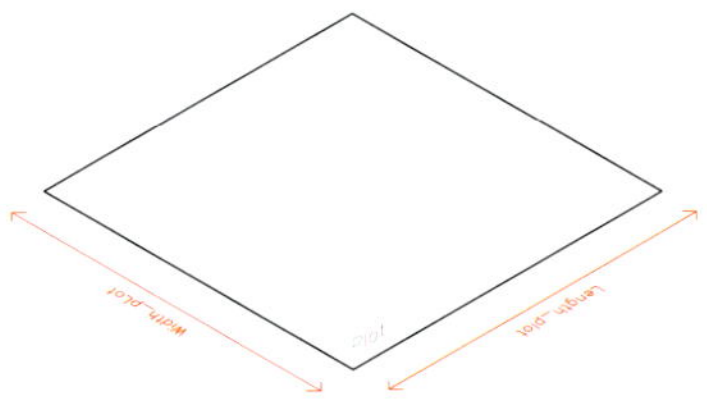

Plot Area

= Plot_Length x Plot_Width = Plot_Area

EXAMPLE

Plot Width	▷	100	m
Plot Length	▷	100	m

Defining the width and length and thereby the total area of the plot, measured from border to border.

—

2.2
Gross Plot Ratio

(B)

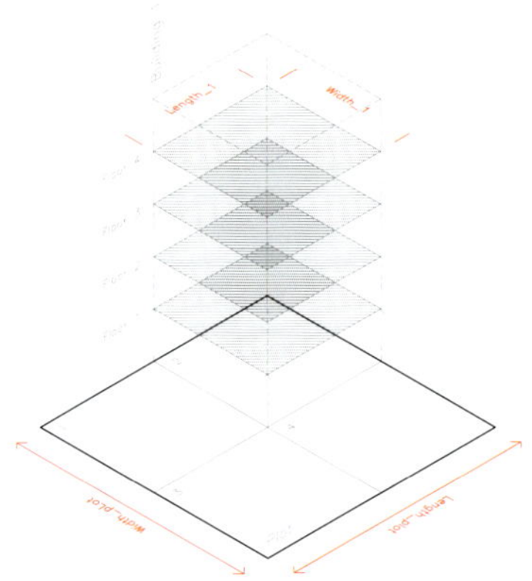

Example A

Plot Area

= Width_plot x Length_plot = Plot_Area

Floor Area

Building 1 = Width_1 x Length_1 x 4 = Building_Area_1

GPR

$= \frac{\text{Floor Area}}{\text{Plot Area}} = \frac{\text{Building\_1\_Area}}{\text{Plot\_Area}} = \text{GPR}$

Gross Plot Ratio (GPR)

GPR is the total building square footage, divided by the site square footage.

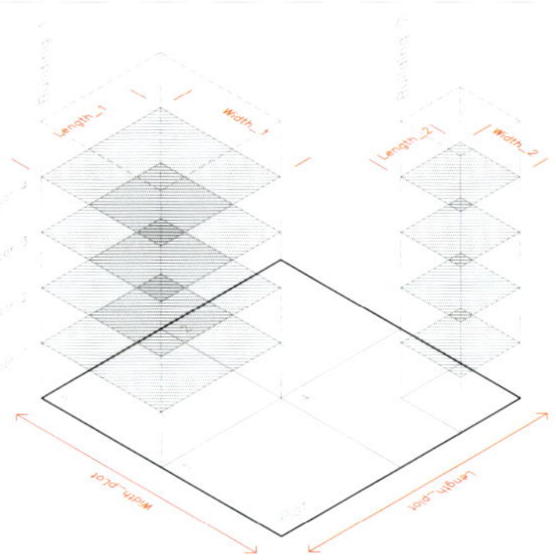

Example B

Plot Area

= Width_plot x Length_plot = Plot_Area

Floor Area

Building 1 = Width_1 x Length_1 x 4 = Building_Area_1
Building 2 = Width_2 x Length_2 x 4 = Building_Area_2

GPR

$= \frac{\text{Floor Area}}{\text{Plot Area}} = \frac{(\text{Building\_Area\_1} + \text{Building\_Area\_2})}{\text{Plot\_Area}} = \text{GPR}$

EXAMPLE

GPR	▷	2.5

Defining the Gross Plot Ratio that is assigned to the specific plot.

—

Page 3
Skyscraper Manual

Comment

The properties for the minimum and the maximum amount of floors as well as the floor-to-floor height are subject to building regulations but also to the specific use and program within the building. During the process of evaluation you might change those numbers.

2.3 Floors

(C)

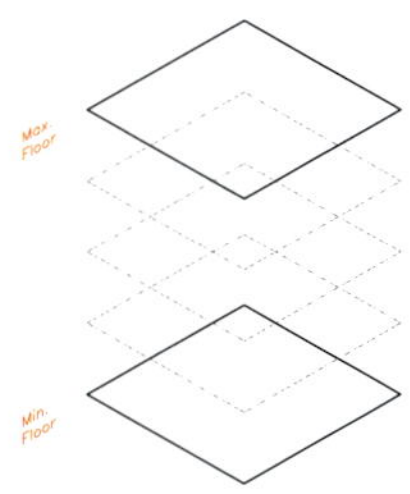

Number of Floors

EXAMPLE

Floor Nrumber max.	38	nr	F_Nmax := 38
Floor Number min.	1	nr	F_Nmin := 3

Defining the maximum and minimum amount of floors, subject either to building regulations or specific use.

F_Nmax_inc := F_Nmax − F_Nmin

i := 0, 1 .. F_Nmax_inc

$F\_Nrange_i := F\_Nmin + i$

F_Nrange_i = (1 2 3 4 5 6 7 8 9 10 11 12 13 14 15 16 17 18 19 20 21 22 23 24 25 26 27 28 29 30 31 32 33 34 35 36 37 38)

2.4 Floor-to-Floor Height

(D)

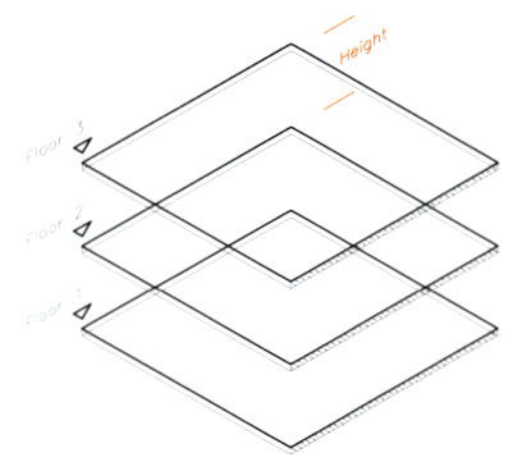

Example A

Floor-to-Floor Height

= Floor_2 − Floor_1

Floor-to-Floor Height

Distance between the upper surfaces of a floor the next floor located above or below.

EXAMPLE

F-F Height max.	5	m	FF_Hmax := 5
F-F Height min.	2.5	m	FF_Hmin := 2.5
Height intervals	0.25	m	Inc := 0.25

Defining the maximum and minimum floor height.
The intervals define the range by which the equation will be optimised.

j := 0, 1 .. FF_Hmax_Inc

$$FF\_Hmax\_Inc := \frac{FF\_Hmax - FF\_Hmin}{Inc}$$

$FF\_Hrange_j := FF\_Hmin + Inc \cdot j$

FF_Hrange_j = (2.5 2.75 3.0 3.25 3.5 3.75 4.0 4.25 4.5 4.75 5.0 5.25 5.5)

2.1
Understanding the Industry Performance Ratios

—

3.0
Evaluating the Ratios

Once all fixed parameters – defined by varying building regulations as well as by the specific type of use – are given and first decisions about the number of floors or a specific floor area are made; ratios help to evaluate those sets of information.
Those ratios distill the different building dimensions and properties to one number. If changes are made within the previously set parameters, for example the floor area is reduced, the number of floors or the dimensions of the core are changed, those ratios help to easily compare the differences.

One of the essential ratios is the facade-to-area ratio. Because the facade is one of the most expensive building-parts of a development the aim is to reduce its area. This ratio describes the area of facade that is used in order to envelop one square meter. The total facade area depends on the number of floors, the floor-to-floor height and the area of each floor.

The function "FA_R" and the resulting diagrams as shown below bring all those parameters together. Because this function always considers the maximum achievable total floor area it simply depends on the number of floors.
The example below shows two versions that proceed from different numbers of floors. The resulting diagrams describe the facade-to-area ratio in relation to the floor-to-floor height.
The following isometric diagrams render the proportional massing of the resulting building volume together with the amount of floors. Those examples show that even with very different settings of building parameters similar facade-to-area ratios can be achieved while maintaining the same total floor space.

—

3.1
Facade-to-Area Ratio

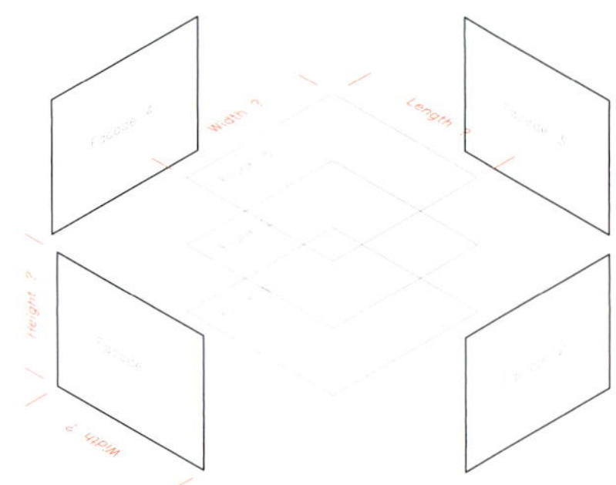

Facade-to-Area Ratio

$= \frac{\text{surface of the facade}}{\text{gross floor area}}$

Evaluating how much sqm of facade is needed to enclose a sqm of floor area.
High value > less efficient
Low value > more efficient

Facade-to-Area Ratio

$= \frac{(\text{Fassade\_1} + \text{Fassade\_2} + \text{Fassade\_3} + \text{Fassade\_4})}{(\text{Floor\_1} + \text{Floor\_2} + \text{Floor\_3})}$

Calculating Facade-to-Area Ratio in relation to fixed Number of Floors:

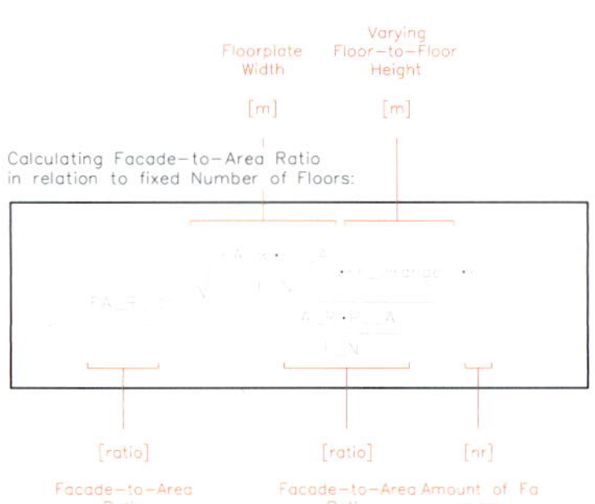

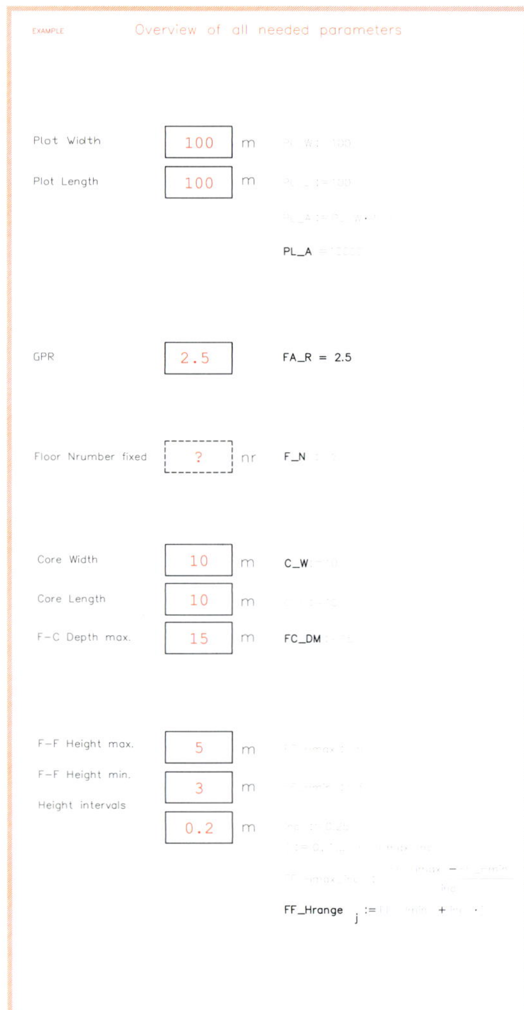

—
3.2 Facade–to–Area Ratio in Relation to Floor–to–Floor Height

Example A

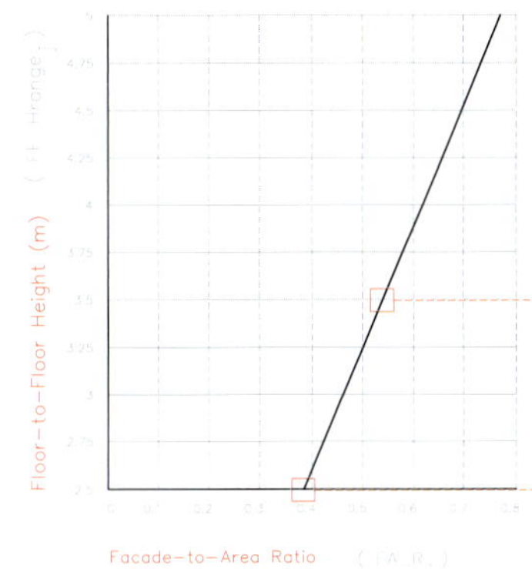

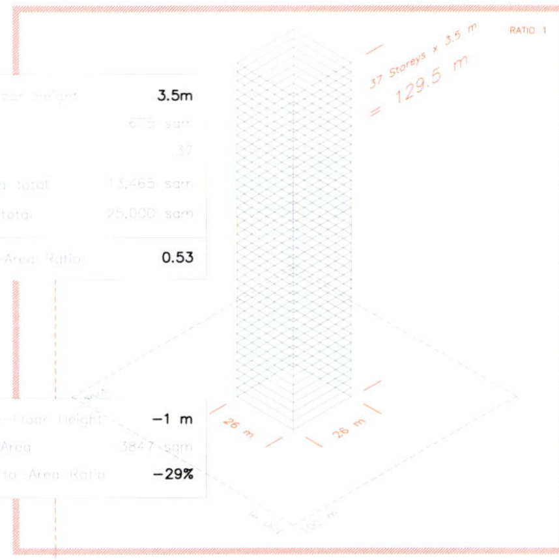

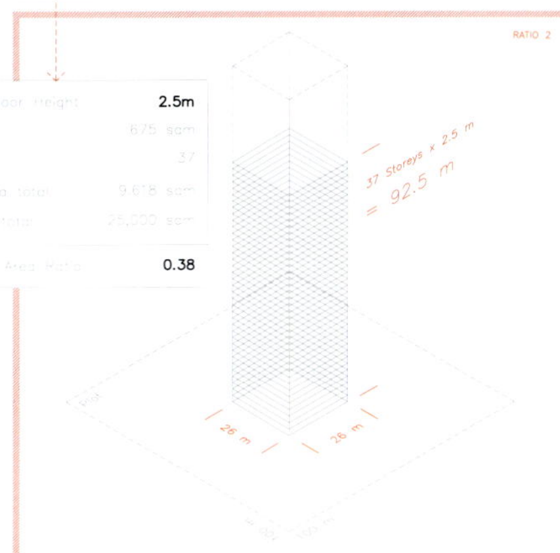

The facade–to–area ratio is subject to multiple parameters. There is no golden rule to for a certain proportion.
Because, the facade is usually the most expensive part within a building, the ratio of the facade–area needed to enclose one sqm of usable space becomes an expression of economic efficiency. The facade–to–area ratio helps to evaluate and compare sets of different parameters (like floor–to–floor height, nr. of storeys, facade–to–core depth, etc...) that are subject to the specific use and overall concept of the building.

Example B

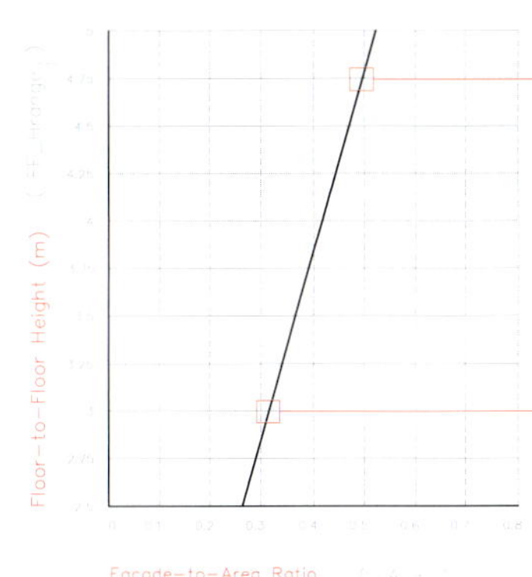

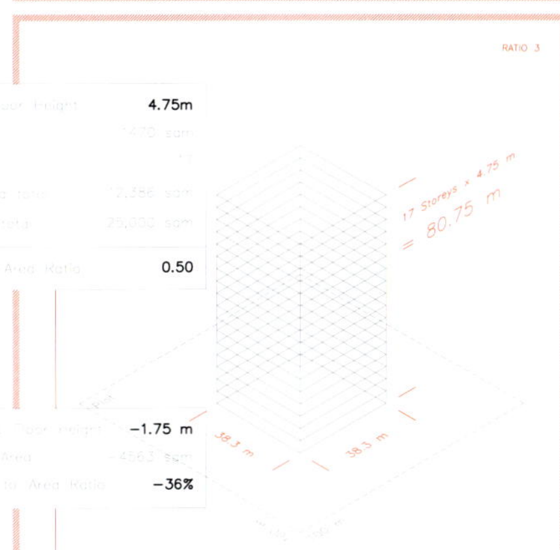

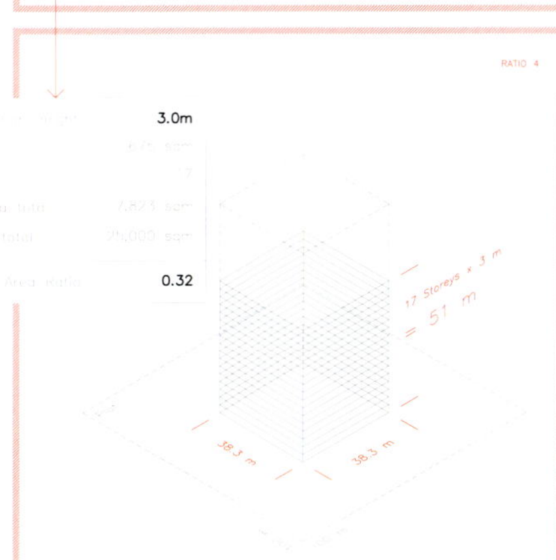

3.4 Massing Studies

The following diagrams represent only a selection of the massing studies that have been generated by the massing equation as described on the previous pages. Each version is optimized to provide exactly the maximum total floor area that can be achieved within the constrains of the specific site.

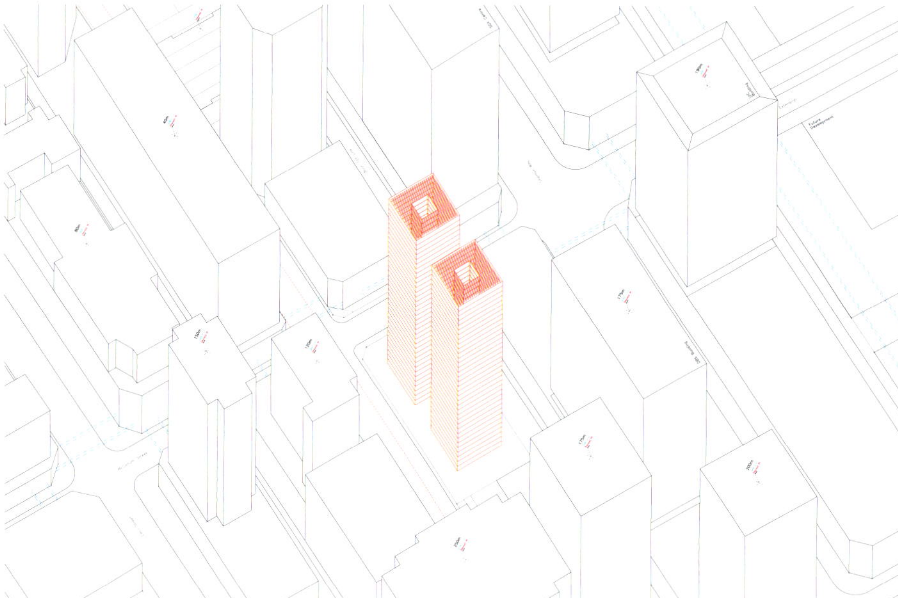

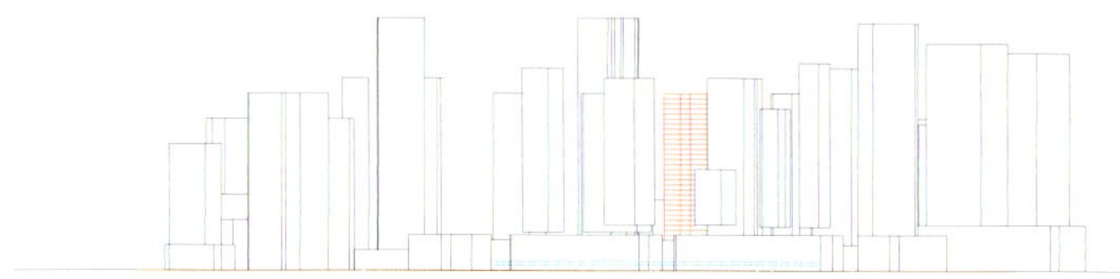

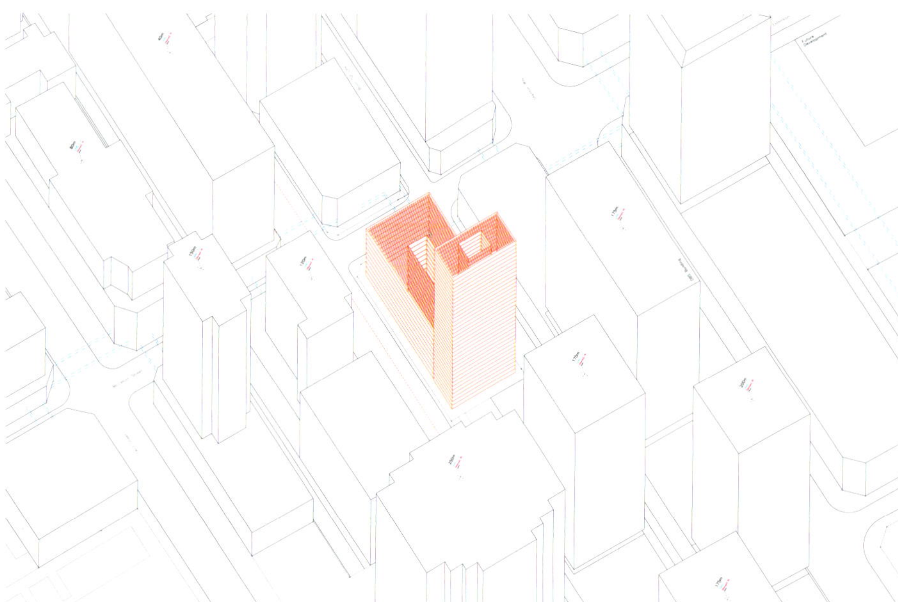

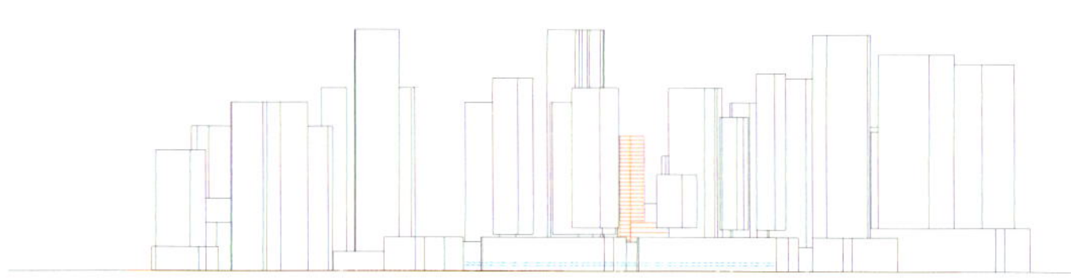

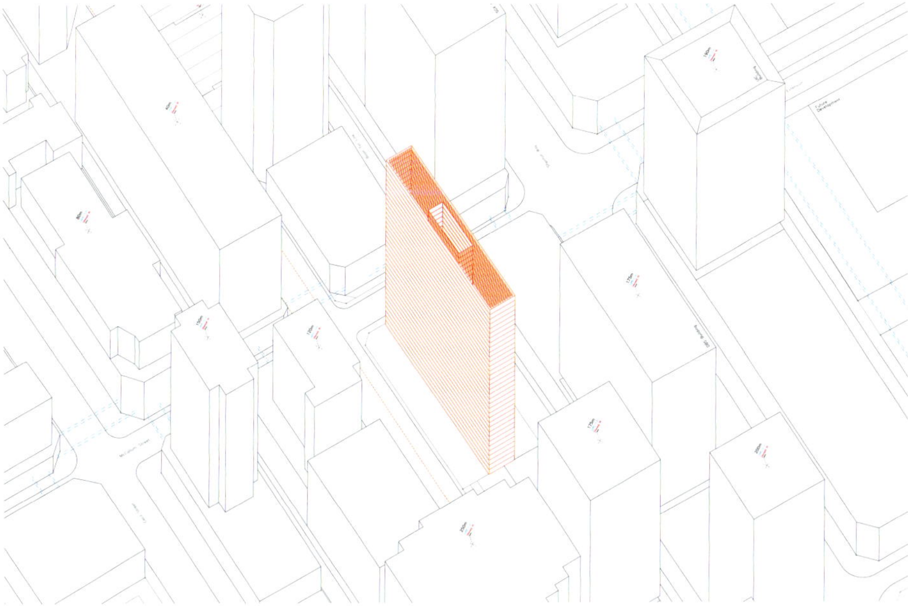

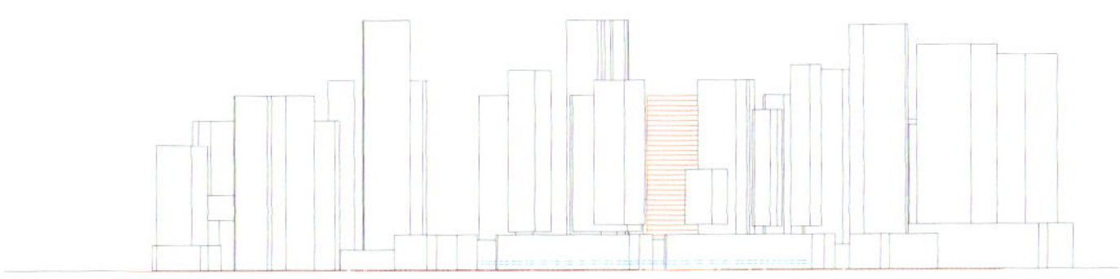

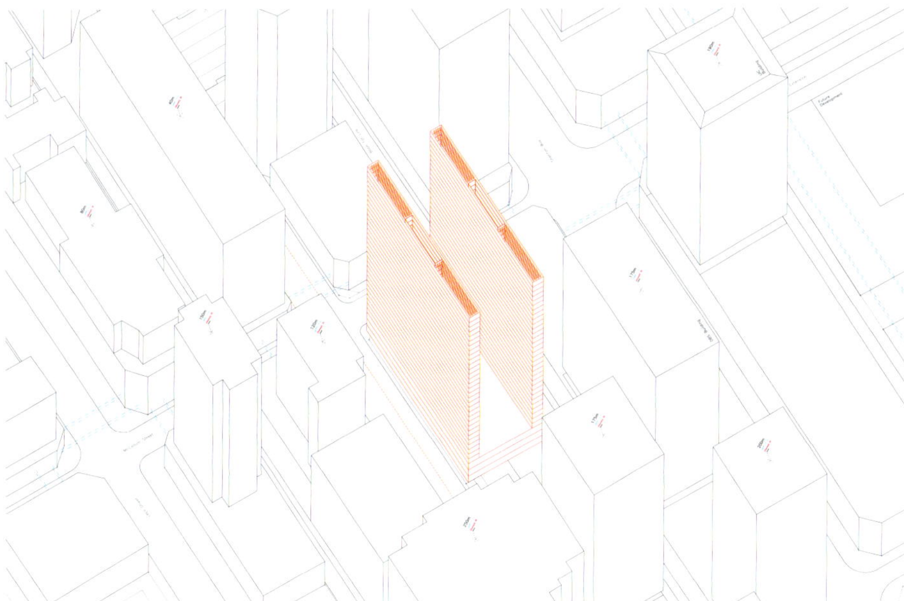

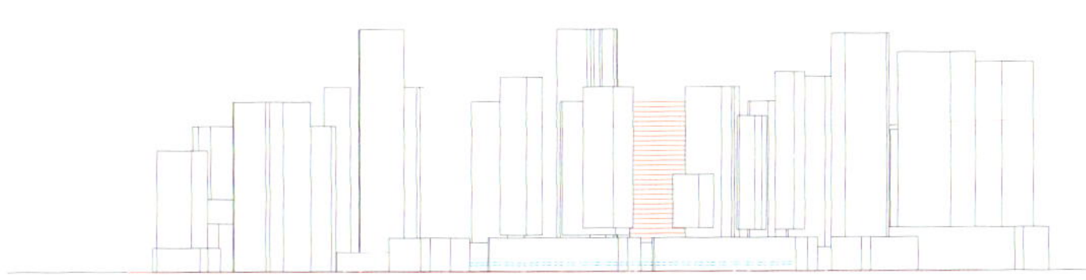

3.4 Massing Studies

The overview below shows the evolution of the massing process.
The concept is to focus on a building-slab typology, based on the idea of thinness as it has been developed in the indexical model. But in this case the idea of thinness should not be understood as a simple material property. Instead, thinness should be understood as the phenomena of a thin space - a configuration of spaces within the building that reach from one façade to the other to develop an interiority that highlights, literally, where the building starts and where it ends and that initiates a living that is constantly confronted with the extreme and dense surrounding urban fabric.

To develop the maximum floor area and at the same time satisfy the right building proportions the mass is split in two slabs and one podium. Both slabs are slightly shifted against each other to interlock the interstitial space, in-between the two slabs, with the neighbouring buildings.

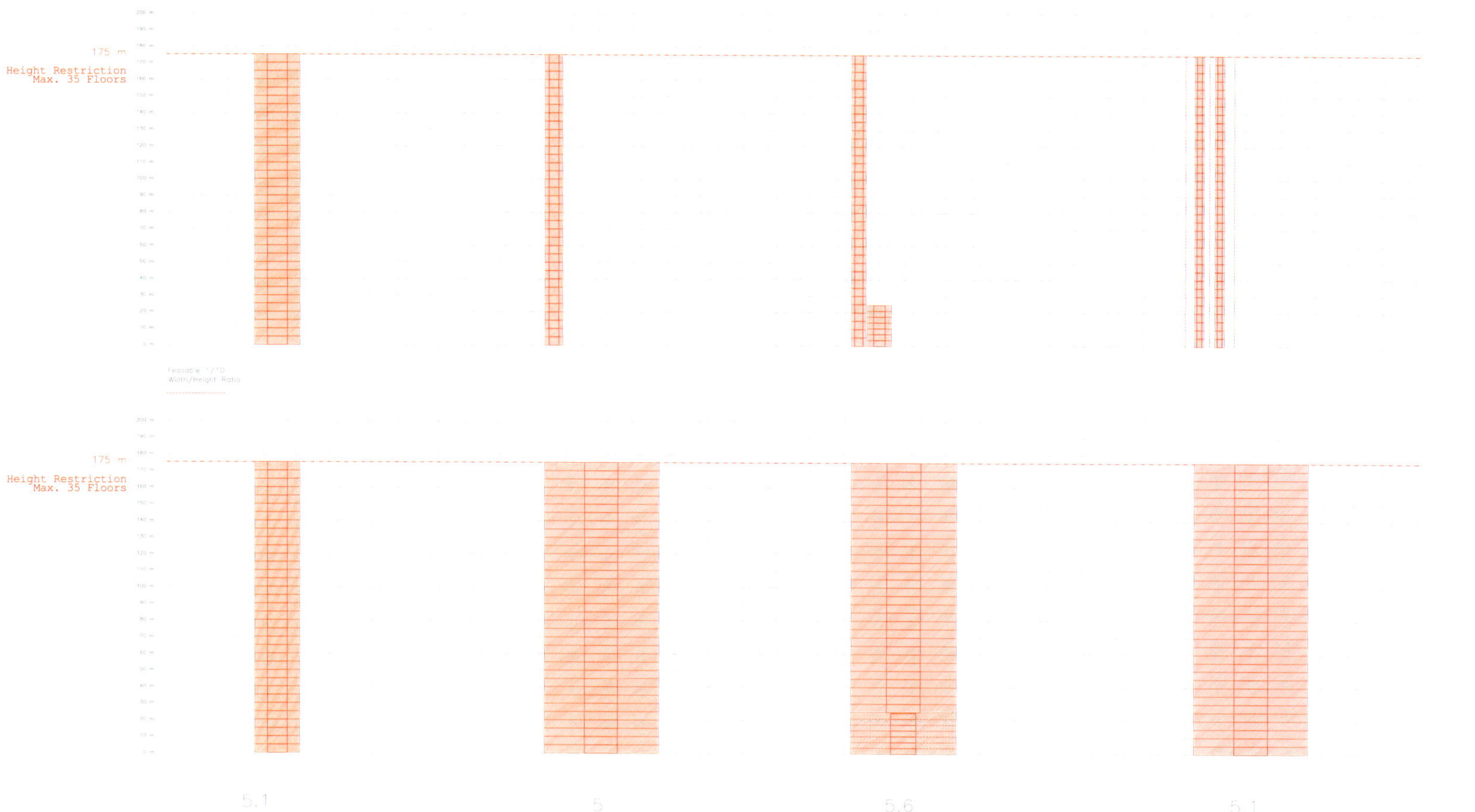

	5.1		5		5.6 (A)		5.6 (B)		5.1 (A)		5.1 (B)	
Storeys	35		35		35		7		35		35	
F-F-Height	5.0 m		5.0 m		5.0 m		3.5 m		5.0 m		5.0 m	
Width	27 m	28.2 %	10.60 m	28.2 %	8.64 m	27.0 %	14.33 m	32.0 %	5.38 m	22.0 %	5.38 m	22.0 %
Length	27 m	71.8 %	68.72 m	71.8 %	63.20 m	73.0 %	63.20 m	68.0 %	67.67 m	78.0 %	67.67 m	78.0 %
Core Width	12 m		6 m		5 m		7 m		3 m		3 m	
Core Length	12 m		20 m		20 m		15 m		20 m		20 m	
Height	175 m		175 m		150 m (above base)		25 m		175 m		175 m	
Height/Width	1/6.5		1/16.5		1/17.36		1/1.7		1/32.5		1/32.5	

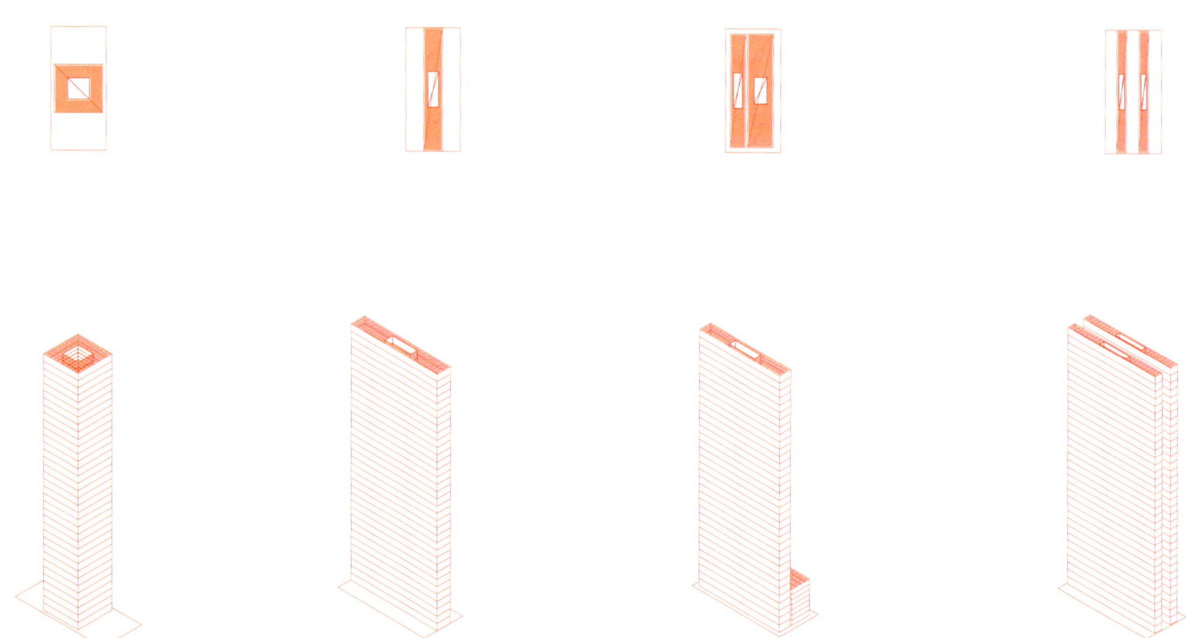

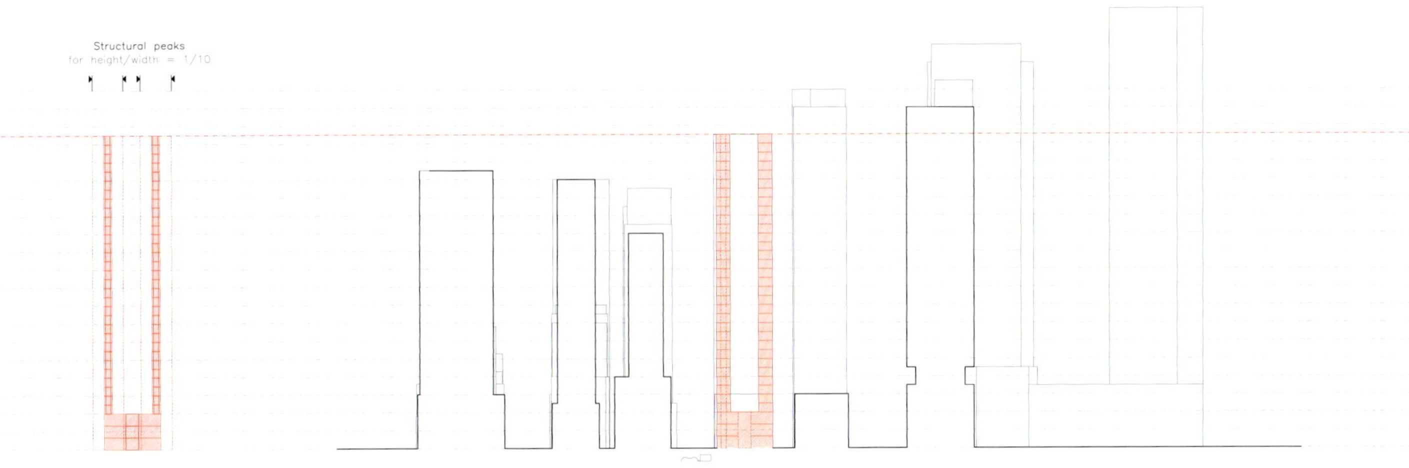

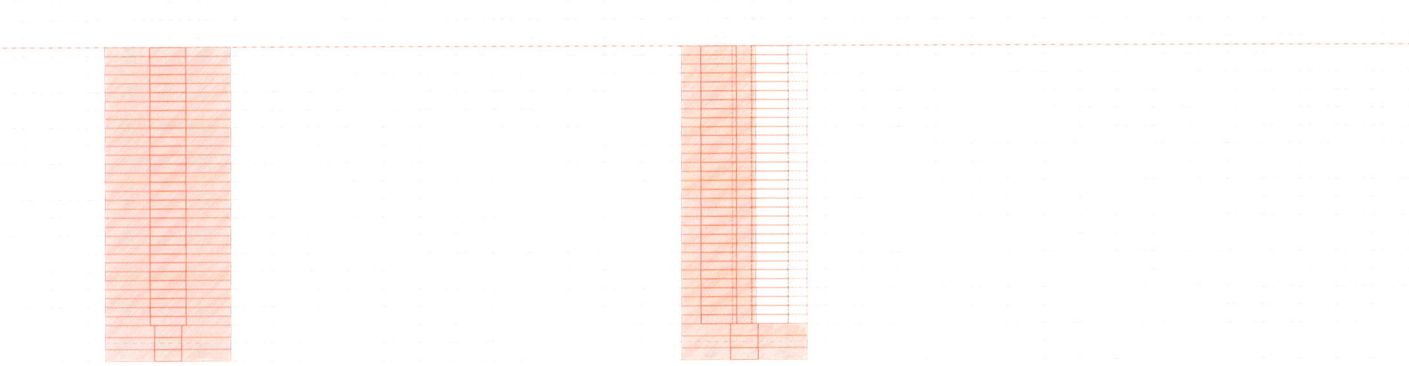

5.6.5

	Ⓐ		Ⓑ		Ⓒ	
Storeys	35		35		3	
F-F-Height	5.0 m		5.0 m		7.0 m	
Width	4.5 m	27.0 %	4.5 m	27.0 %	30.73 m	32.2 %
Length	68.9 m	73.0 %	68.9 m	73.0 %	68.9 m	67.8 %
Core Width	3.0 m		3.0 m		3 m	
Core Length	20 m		20 m		20 m	
Height	175 m		175 m		175 m	
Height/Width	1/93.6		1/93.6		1/2.4	

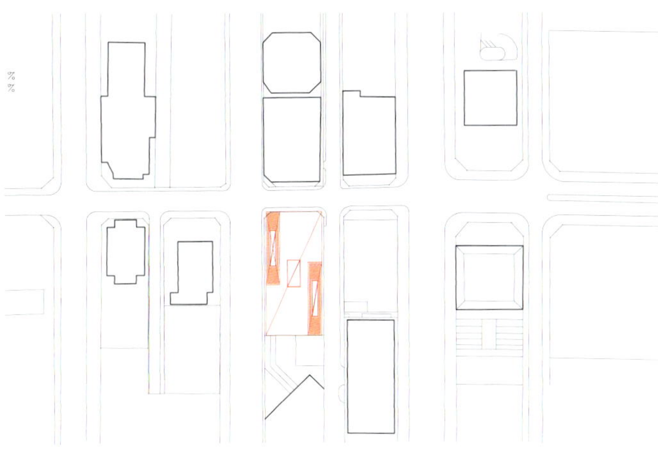

5.6.4

Ⓐ		Ⓑ		Ⓒ	
31		31		3	
5.0 m		5.0 m		7.0 m	
7.53 m	30.0 %	7.53 m	30.0 %	30.73 m	32.2 %
40.98 m	70.0 %	40.98 m	70.0 %	68.9 m	67.8 %
3 m		3 m		3 m	
20 m		20 m		20 m	
190 m		190 m		175 m	
1/25		1/25		1/2.4	

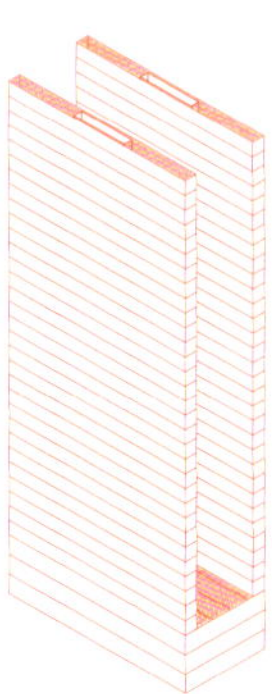

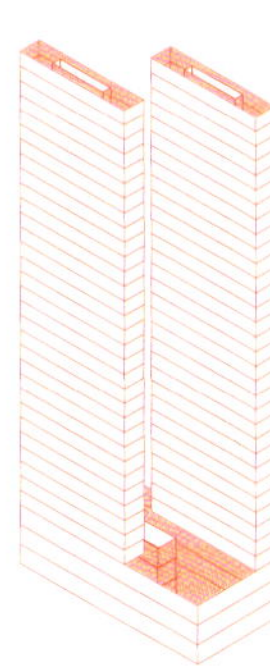

3.4 Massing Studies

The offset position of the two slabs is developed to reduce the sun exposure of the outer building surfaces.

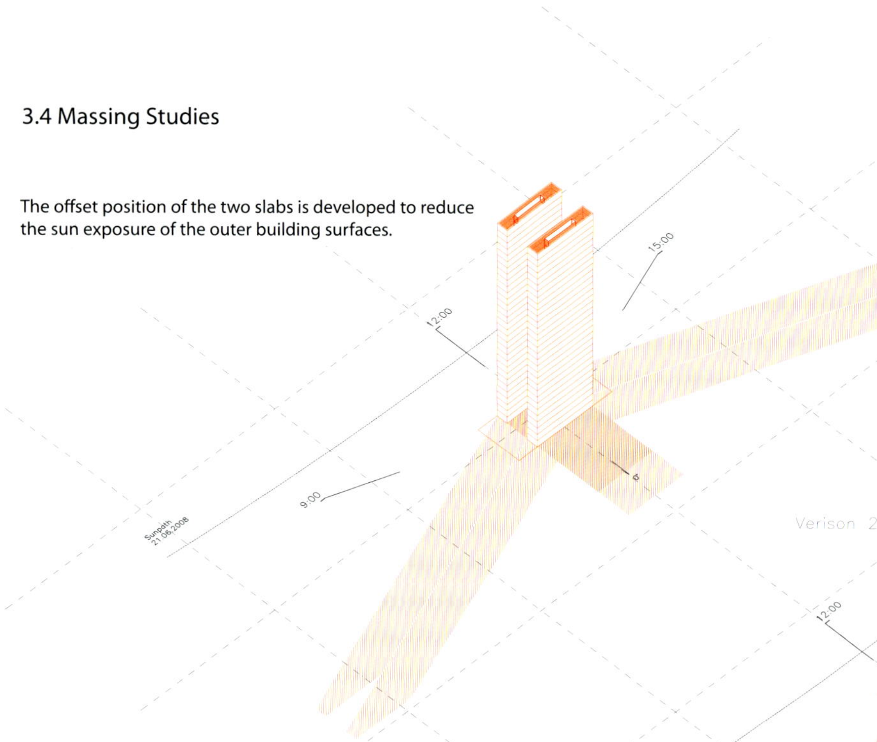

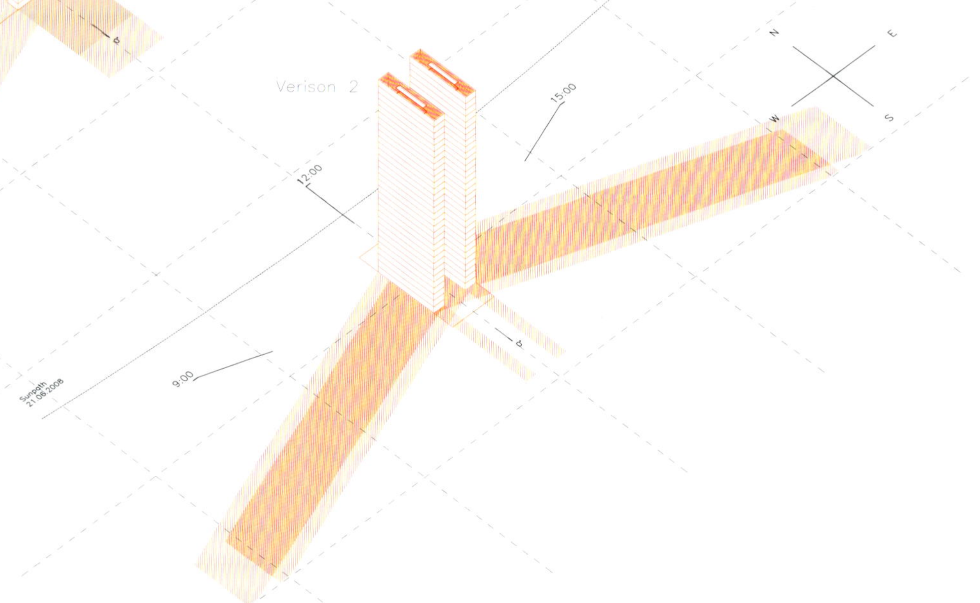

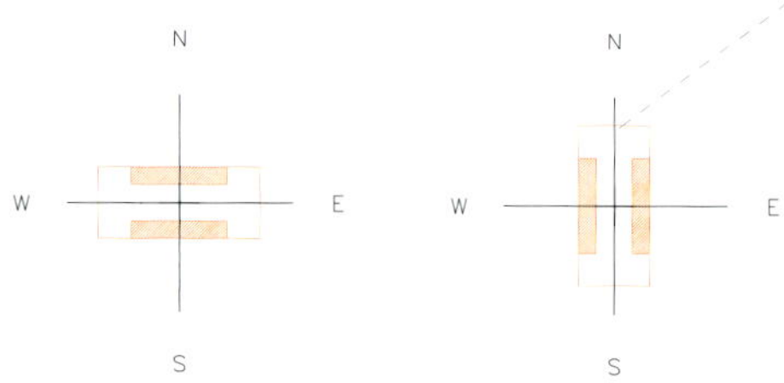

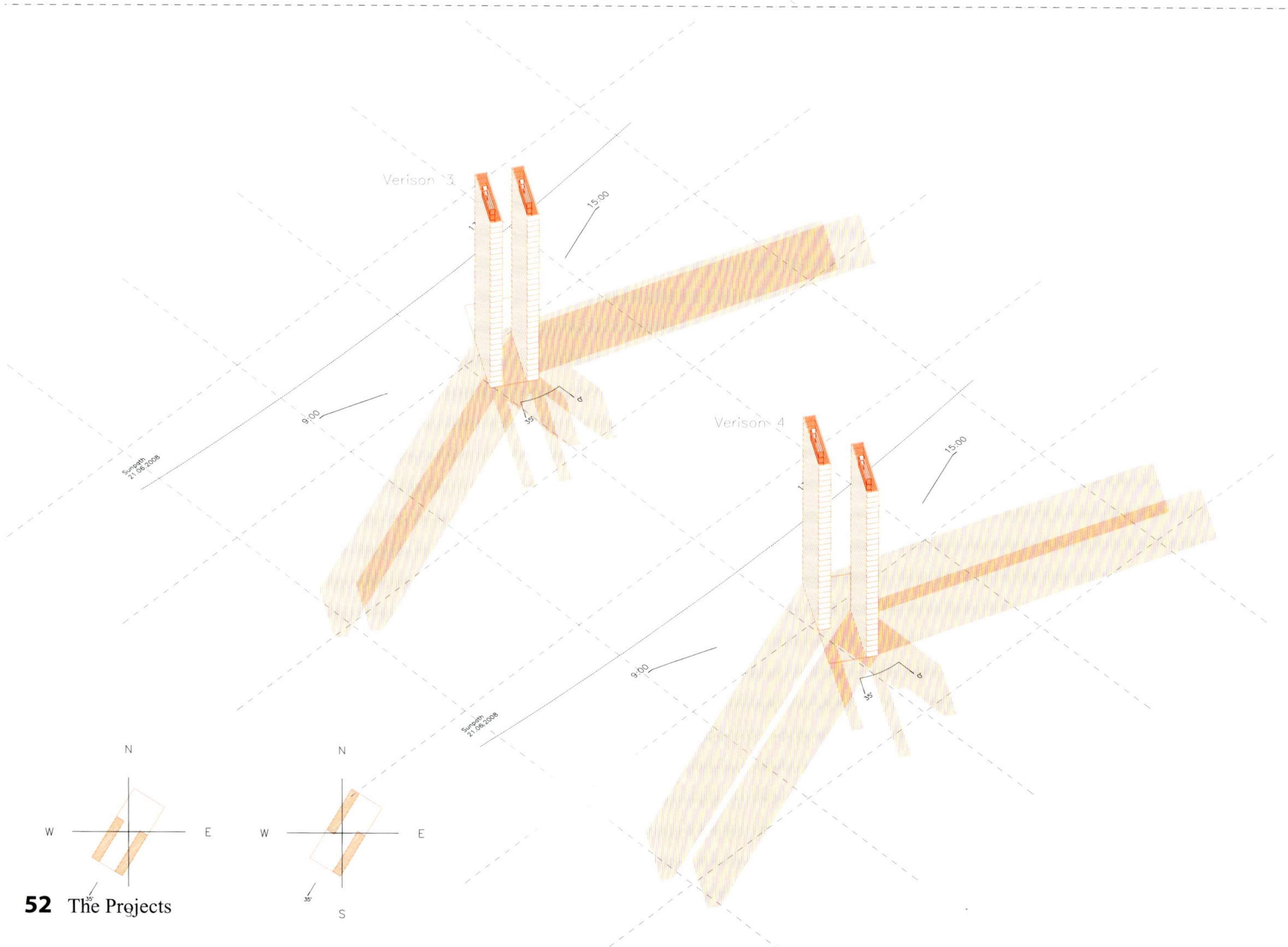

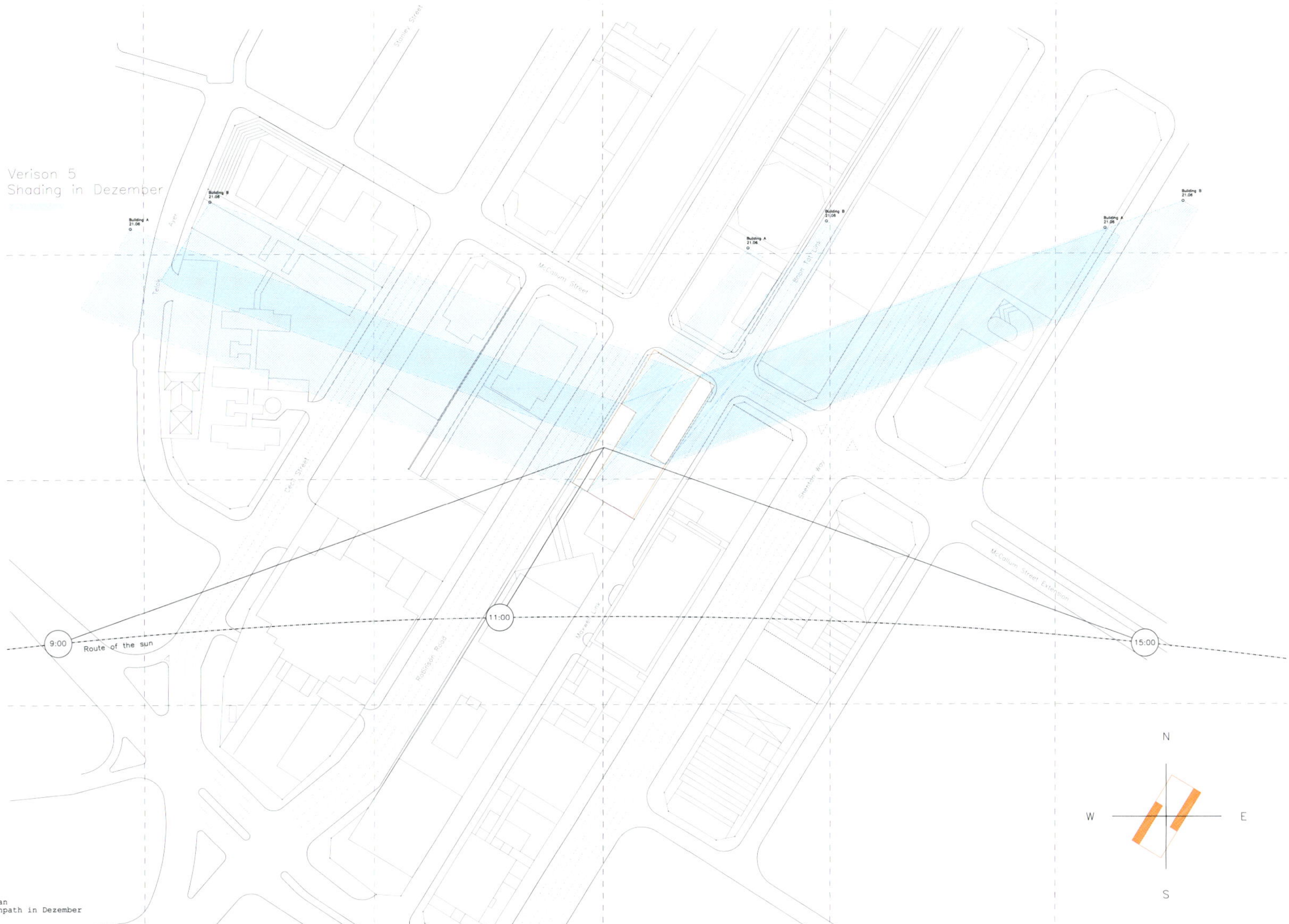
Verison 5
Shading in Dezember
Stanley Street
McCallum Street
Route of the sun
9:00
11:00
15:00
Robinson Road
McCallum Street Extension
N
W
E
S
Plan
Sunpath in Dezember

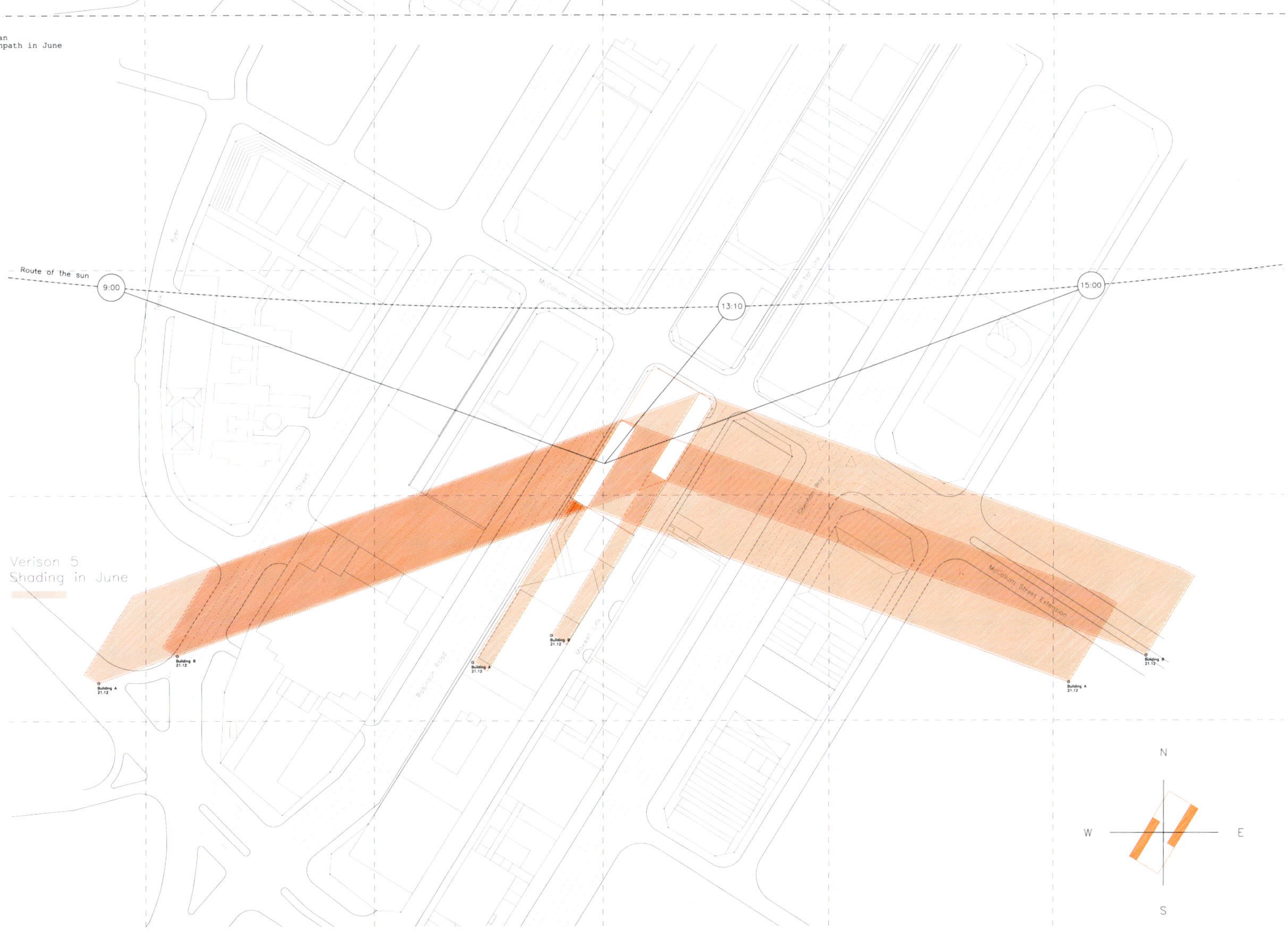
Plan
Sunpath in June
Route of the sun
9:00
13:10
15:00
Verison 5
Shading in June
McCallum Street Extension
Robinson Road
N
W
E
S

3.5 Final Massing

The drawings below show the massing concept. At the same time they illustrate the advantages as well as the problems of the extreme thinness of the building slabs.
On the one side the narrow spaces create an appreciation and awareness of the concept of living in dense heights through the constant visual confrontation with the neighbouring buildings on both sides. On the other side the section shows that the proposed dimensions are absolutely unfeasible as a structure. The two slabs could carry vertical gravity loads but any horizontal force would cause the collapse.
The dashed lines parallel to the section outline the minimum building width which could make the building structurally feasible.

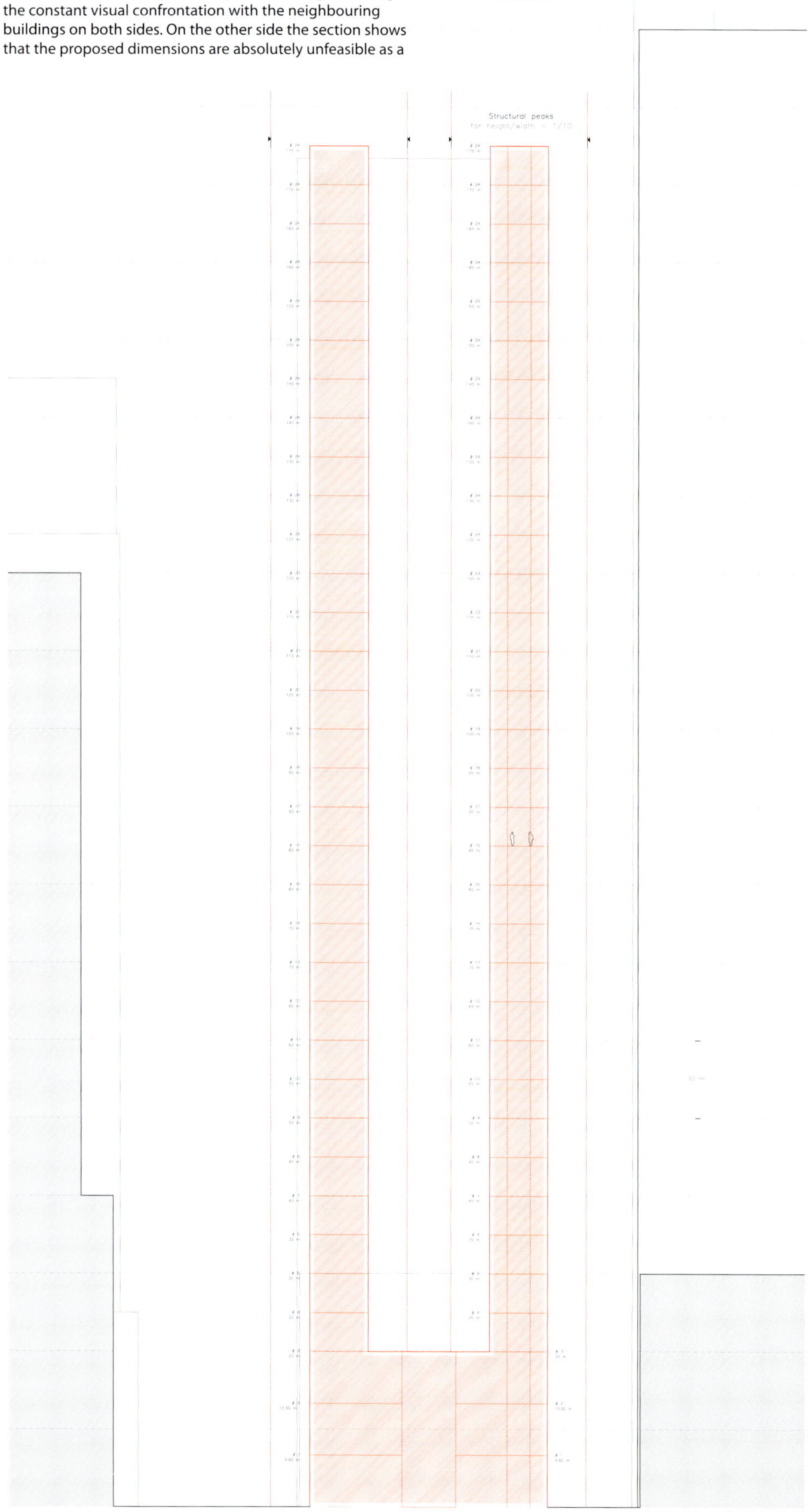

Section A-A

1

2

3

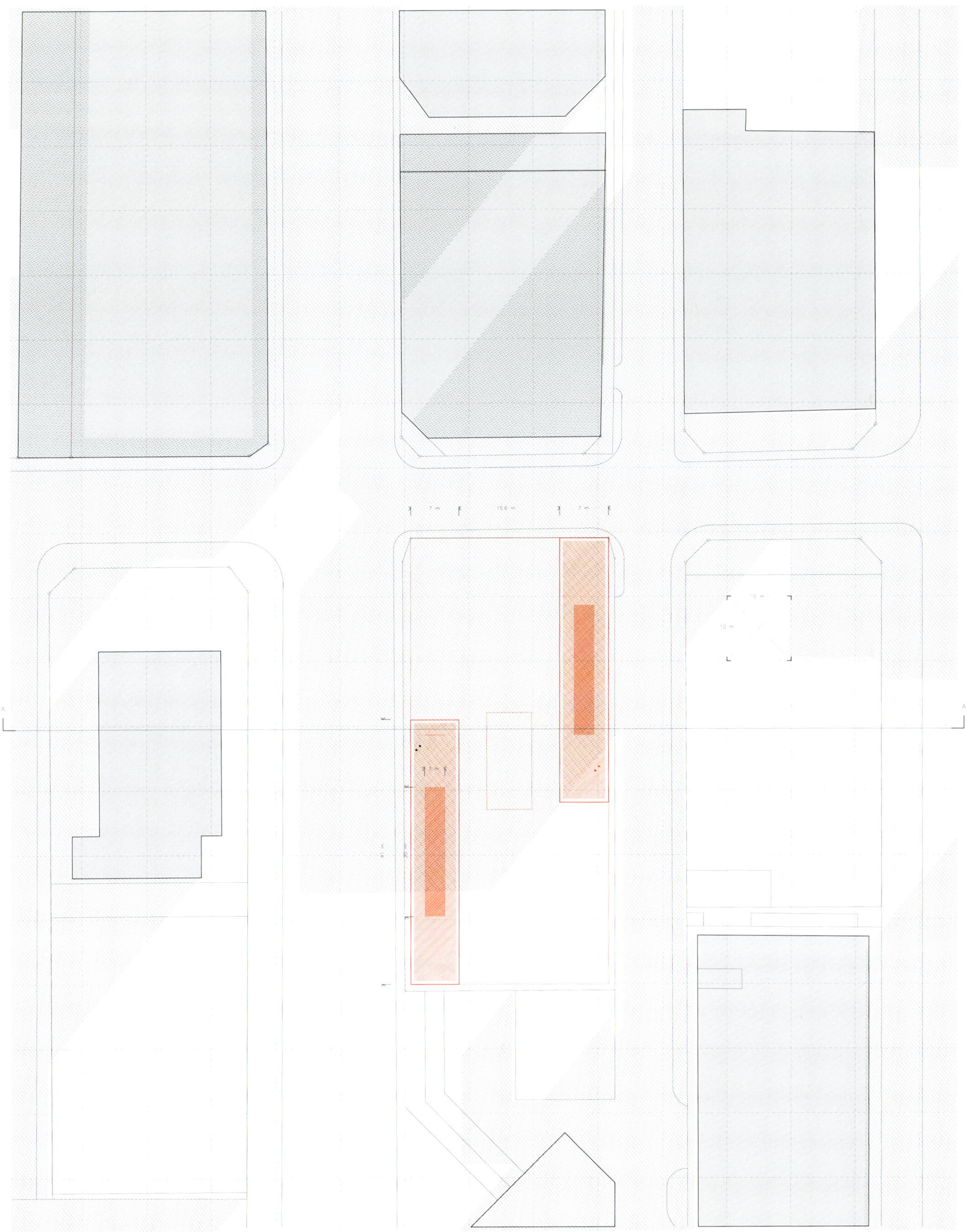

4. Final Proposal

For the final high-rise building proposal the mathematical concepts of the two previous chapters are merged together into one coherent equation. The massing interface defines the bounding box and distributes the floor space vertically while folds, embedded within the surface, provide stability and enable vertical circulation between both slabs.

Detail drawings describe the architectonic articulation to expresses the continuity of the grid derivated form the mathematical writing and the different grades of porosity embedded within that continuity. While the inner functional organs, spaces and floor layouts are organized to create a holistic system that merges the interior with the exterior and interlocks the interstitial space.

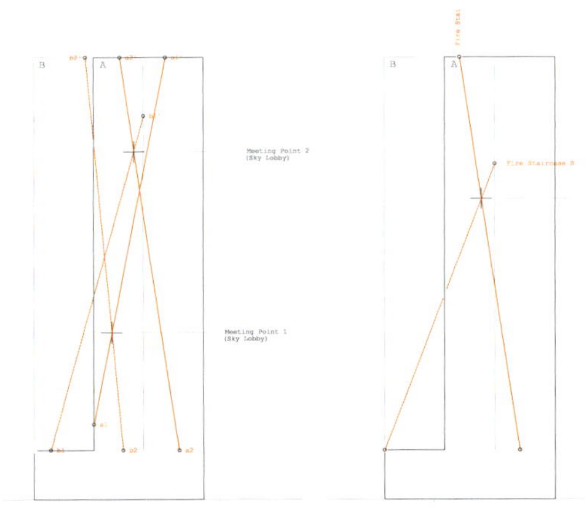

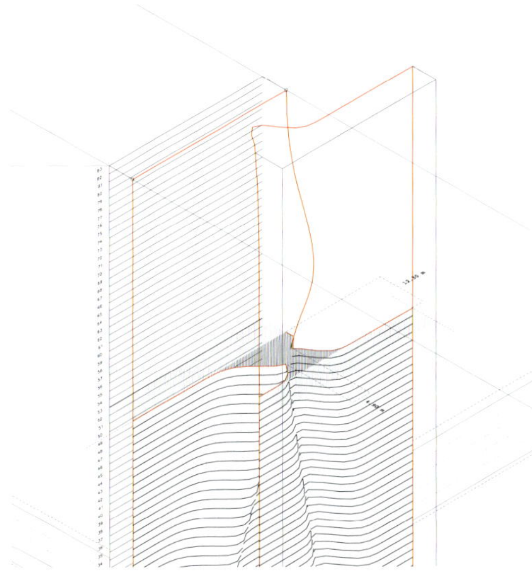

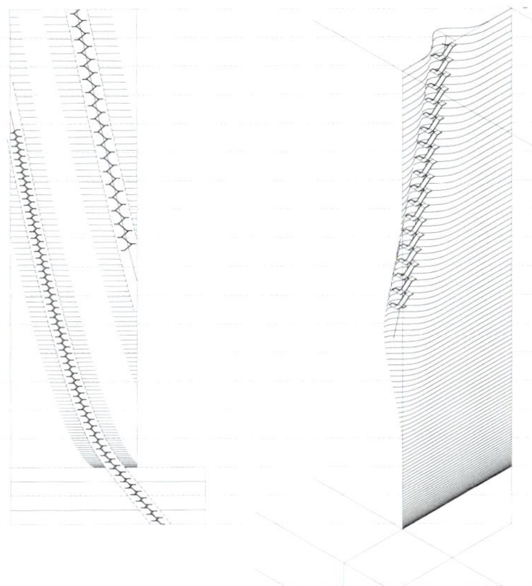

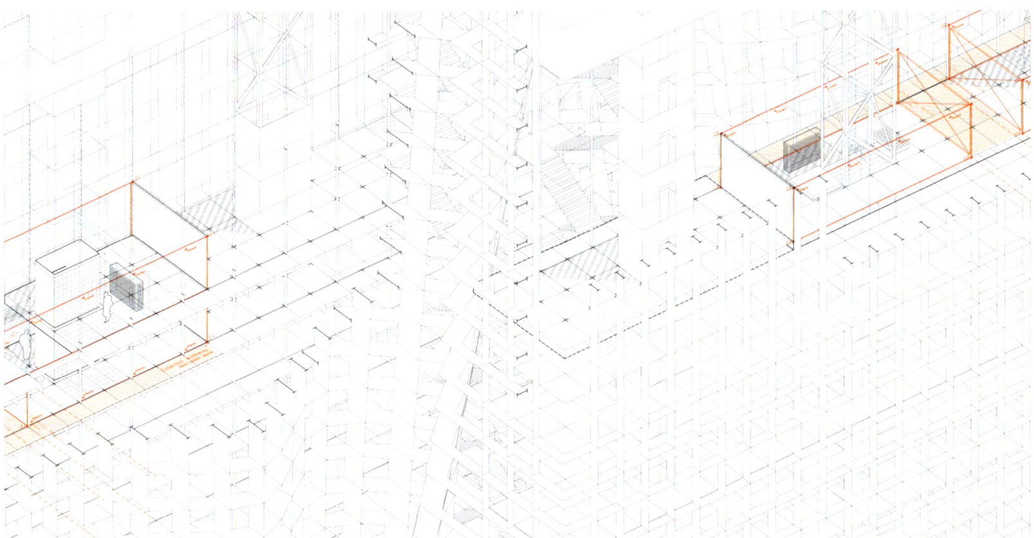

Fold Positioning

Fold Positioning

FInal Model under Construction

Circulation

Living Units

4.1 Final Massing & Distribution of Floors

In order to articulate the distribution of floor area within the boundaries of the massing additional functions are applied. The concept of a vertical distribution of horizontal elements with gradually changing intervals, as developed in the indexical model, defines the basis.

A parabolic function organises the distribution of floors and thereby controls the floor-to-floor heights while maintaining the total floor area. Different parameters control the gradual allocation of floors and simplify changes or adjustments.

The floor-to-floor heights are differentiated according to the different uses that the building has to provide. The ceiling heights for commercial area in the base are 6.66 m while the ceiling heights for office floors in the lower third of the slabs range from 3.00 m to 3.50 m and for residential floors in the upper two third the ceiling heights range from 3.50 m to 6.40 m. Last mentioned residential floors are partly equipped with mezzanine floors.

Interval Study

Parameters

$last\_in := 72 \quad height := 160 \quad in := 01, .. \; last\_in$

$last\_jn := 1 \quad width := 44 \quad jn := 01, .. \; last\_jn$

$periods\_m := 0.5$

Vertiacal Distribution of Threads

Function 1

$$v\_a_{in} := 1.4 + \left(1.7 \tan\left(\frac{in}{last\_in}\right) \cdot 0.45\,\pi - 0.2\right)$$

$$m\_final_{in,\,jn} := \sin\left(\frac{in}{last\_in} \cdot periods\_m \cdot \pi\right)^{v\_a_{in}} \cdot height$$

$$m\_final_{in,\,0} = \begin{pmatrix} 0 \\ 1.429448 \\ 3.026207 \\ 4.614935 \\ 6.177891 \\ 7.716079 \\ 9.235089 \\ 10.741528 \\ 12.241923 \\ 13.74237 \\ 15.248449 \\ 16.765232 \\ 18.297331 \\ 19.848946 \\ 21.423919 \\ 23.025777 \\ 24.657769 \\ 26.322904 \\ 28.02397 \\ 29.76356 \\ 31.54409 \\ 33.367808 \\ 35.236807 \\ 37.153028 \\ 39.118268 \\ 41.134179 \\ 43.20227 \\ 45.323899 \\ 47.500275 \\ 49.732448 \\ 52.021303 \\ 54.36755 \\ 56.77171 \\ 59.234111 \\ 61.754864 \\ 64.333855 \\ 66.970725 \\ 69.664852 \\ 72.415332 \\ 75.220955 \\ 78.080186 \\ 80.99114 \\ 83.951552 \\ 86.958753 \\ 90.009643 \\ 93.100653 \\ 96.227723 \\ 99.386259 \\ 102.571107 \\ 105.776508 \\ 108.99607 \\ 112.22272 \\ 115.448675 \\ 118.665395 \\ 121.863545 \\ 125.03296 \\ 128.162598 \\ 131.240508 \\ 134.253792 \\ 137.188573 \\ 140.029961 \\ 142.762031 \\ 145.367803 \\ 147.82923 \\ 150.127193 \\ 152.241511 \\ 154.150962 \\ 155.83332 \\ 157.265418 \\ 158.423218 \\ 159.281928 \\ 159.81614 \\ 160 \end{pmatrix}$$

Function 2

$$v\_b_{in} := 1.4$$

$$m\_final\_b_{in,\,jn} := \sin\left(\frac{in}{last\_in} \cdot periods\_m \cdot \pi\right)^{v\_b_{in}} \cdot height$$

Function 3

$$v\_c_{in} := 1$$

$$m\_final\_c_{in,\,jn} := \sin\left(\frac{in}{last\_in} \cdot periods\_m \cdot \pi\right)^{v\_c_{in}} \cdot height$$

Threads

$$iThread_{in,\,jn} := 1$$

$$jThread_{in,\,jn} := \left(\frac{jn}{last\_jn}\right) \cdot width$$

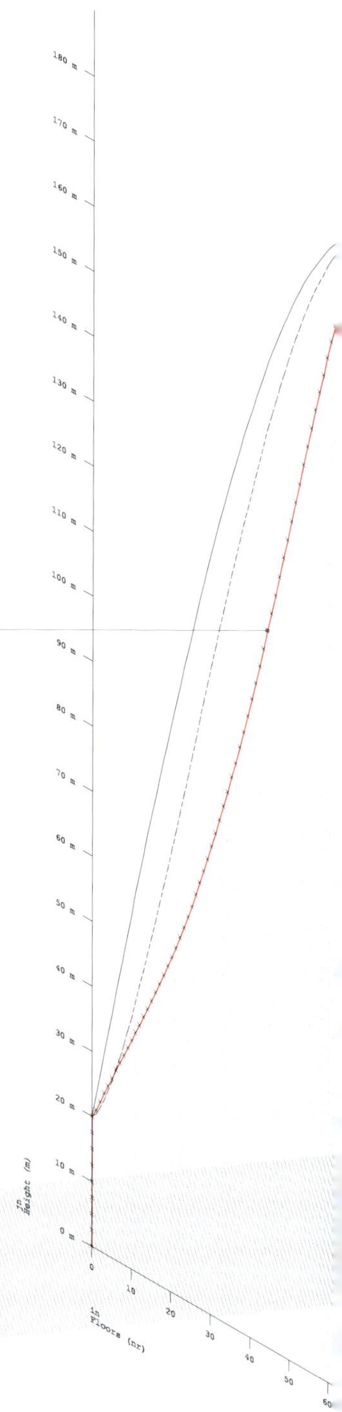

Calc | Iso

X | Di

Ucs r pl
Rot z 45
Rot x 54.75

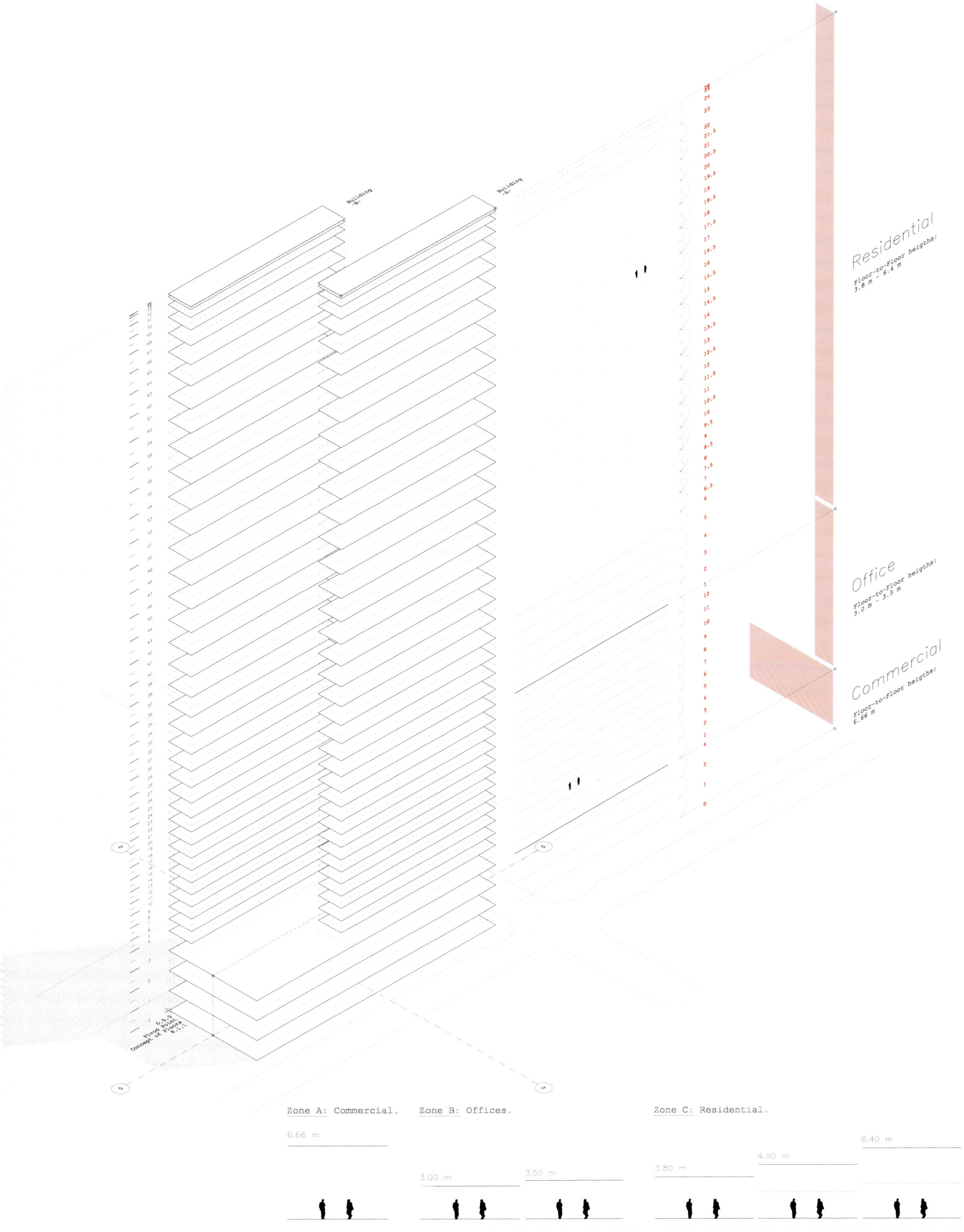

Building -B-
Building -A-
Residential
Floor-to-Floor heigths:
3.8 m - 6.4 m
Office
Floor-to-Floor heigths:
3.0 m - 3.5 m
Commercial
Floor-to-Floor heigths:
6.66 m
Zone A: Commercial.
Zone B: Offices.
Zone C: Residential.
6.66 m
3.00 m
3.50 m
3.80 m
4.90 m
6.40 m

Max Kahlen

4.2 Applying Folds & Facade Grid

As mentioned before, the thin proportions of the two slabs are structurally unfeasible. To generate enough rigidity against lateral forces the concept of the Indexical Model is used to apply folds onto the building surfaces.

These folds are situated on the two surfaces that face the interstitial space of both slabs and are used to stabilise the slabs and connect both at two points as well as the folds function as vertical circulation.

In contrast to the organisation of floors, the organisation of horizontal members within the outer surfaces is different. In this case those members represent structural elements. The form of the parabolic function that defines the vertical spacings is articulated according to the stresses that appear within the outer surface.

Furthermore the distances of the vertical structural members are simplified to one modular rhythm according to the internal plan layouts and the sizes of the functional elements.

Interval Study

Parameters $\quad last\_in := 100 \quad height := 160 \quad in := 01, .. \; last\_in$

$last\_jn := 28 \quad width := 44 \quad jn := 01, .. \; last\_jn$

$periods\_m := 0.5$

Vertiacal Distribution of Threads

Function 1 $\quad v\_a_{in} := 1.65$

$$m\_final_{in,\, jn} := \left(\frac{in}{last\_in}\right)^{v\_a_{in}} \cdot height$$

Function 2 $\quad v\_b_{in} := 1$

$$m\_final\_b_{in,\, jn} := \left(\frac{in}{last\_in}\right)^{v\_b_{in}} \cdot height$$

Threads $\quad iThread_{in,\, jn} := 1$

$$jThread_{in,\, jn} := \left(\frac{jn}{last\_jn}\right) \cdot width$$

$$m\_final_{in,\,0} = \begin{pmatrix} 0 \\ 0.08019 \\ 0.251663 \\ 0.491326 \\ 0.789803 \\ 1.141354 \\ 1.541947 \\ 1.988528 \\ 2.478669 \\ 3.010372 \\ 3.581954 \\ 4.191968 \\ 4.839149 \\ 5.522383 \\ 6.24067 \\ 6.993113 \\ 7.778896 \\ 8.597273 \\ 9.447559 \\ 10.329122 \\ 11.241376 \\ 12.183774 \\ 13.155804 \\ 14.15699 \\ 15.186879 \\ 16.245048 \\ 17.331095 \\ 18.44464 \\ 19.585323 \\ 20.752799 \\ 21.946742 \\ 23.166839 \\ 24.412792 \\ 25.684315 \\ 26.981134 \\ 28.302986 \\ 29.649618 \\ 31.020788 \\ 32.416261 \\ 33.835811 \\ 35.279219 \\ 36.746277 \\ 38.23678 \\ 39.750531 \\ 41.28734 \\ 42.847021 \\ 44.429396 \\ 46.03429 \\ 47.661535 \\ 49.310966 \\ 50.982425 \\ 52.675756 \\ 54.390807 \\ 56.127431 \\ 57.885486 \\ 59.664832 \\ 61.465331 \\ 63.286852 \\ 65.129264 \\ 66.992441 \\ 68.876258 \\ 70.780595 \\ 72.705333 \\ 74.650357 \\ 76.615553 \\ 78.60081 \\ 80.60602 \\ 82.631076 \\ 84.675874 \\ 86.740312 \\ 88.824291 \\ 90.927711 \\ 93.050477 \\ 95.192493 \\ 97.353668 \\ 99.533911 \\ 101.733131 \\ 103.951241 \\ 106.188155 \\ 108.443789 \\ 110.718058 \\ 113.010882 \\ 115.322179 \\ 117.65187 \\ 119.999878 \\ 122.366126 \\ 124.750538 \\ 127.15304 \\ 129.57356 \\ 132.012025 \\ 134.468365 \\ 136.942509 \\ 139.434389 \\ 141.943937 \\ 144.471087 \\ 147.015773 \\ 149.577929 \\ 152.157493 \\ 154.754401 \\ 157.36859 \\ 160 \end{pmatrix}$$

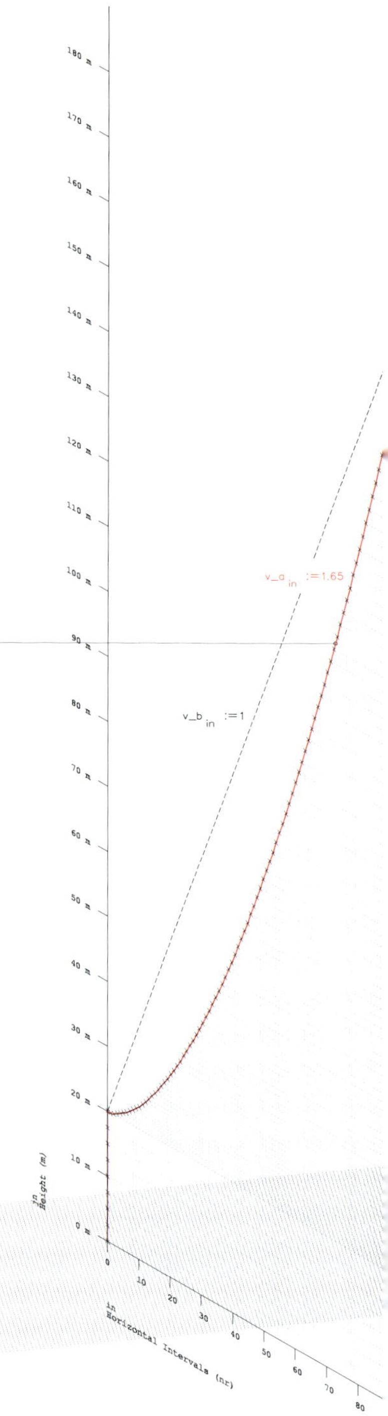

Calc Iso

X Di

Ucs r pl
Rot z *45
Rot x *54.75

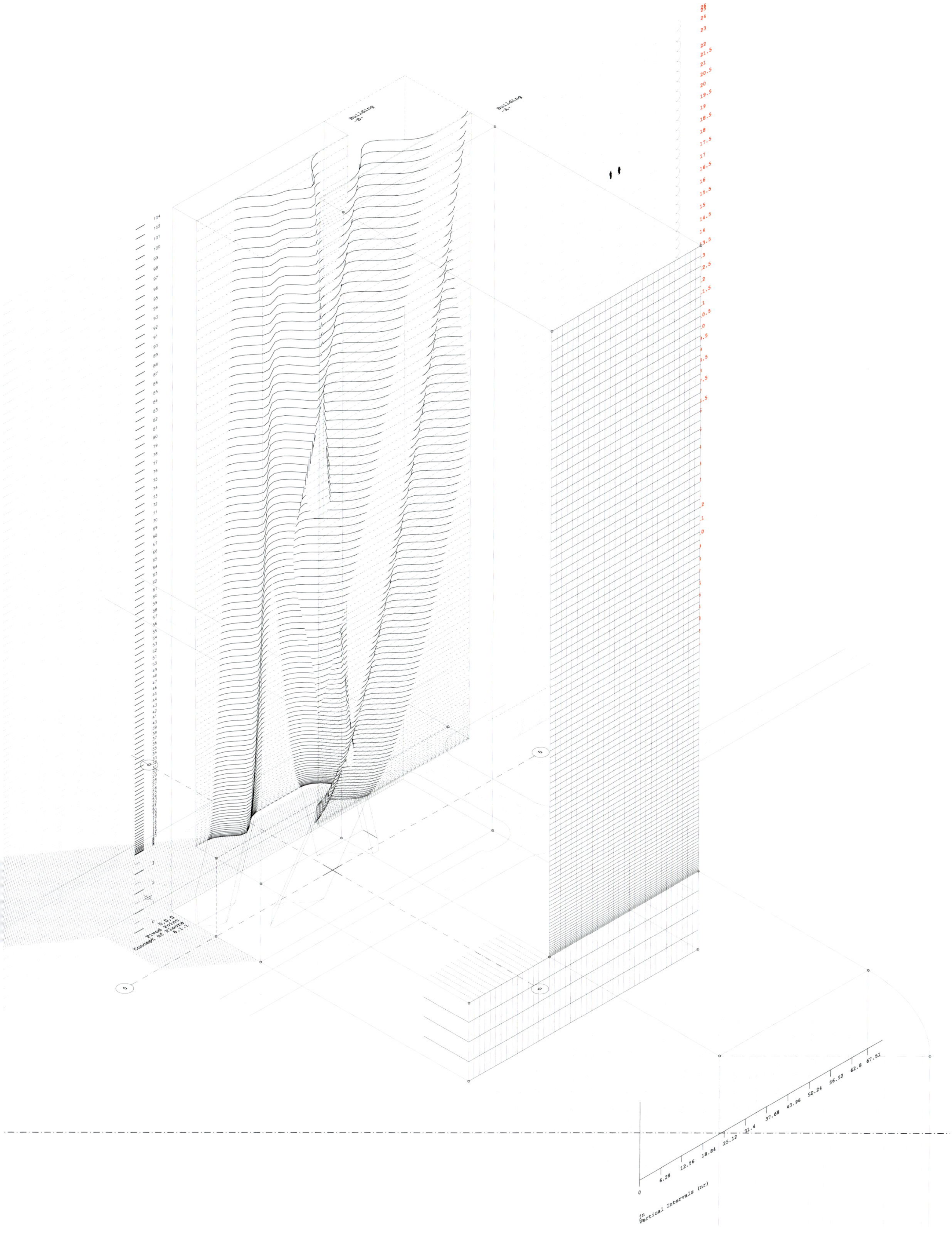
Building
-B-
Building
-A-
Vertical Intervals (m)
0
6.28
12.56
18.84
25.12
31.4
37.68
43.96
50.24
56.52
62.8
67.52

4.3 Structure & Circulation Constraints for Folds

The exact position of folds within the inner surfaces of the slabs is articulated according to two constrains.

In elevation, the points of connection (the areas where the facing folds intersect) need to be positioned to stabilize the upper ends of the slabs as well as they need to enforce the lower base where all forces meet.

In plan, the folds should maintain a maximum distance of 30 m to the opposite end of each slab, because the folds function as external fire staircases.

The lower part of the diagram below shows those constrains in red.

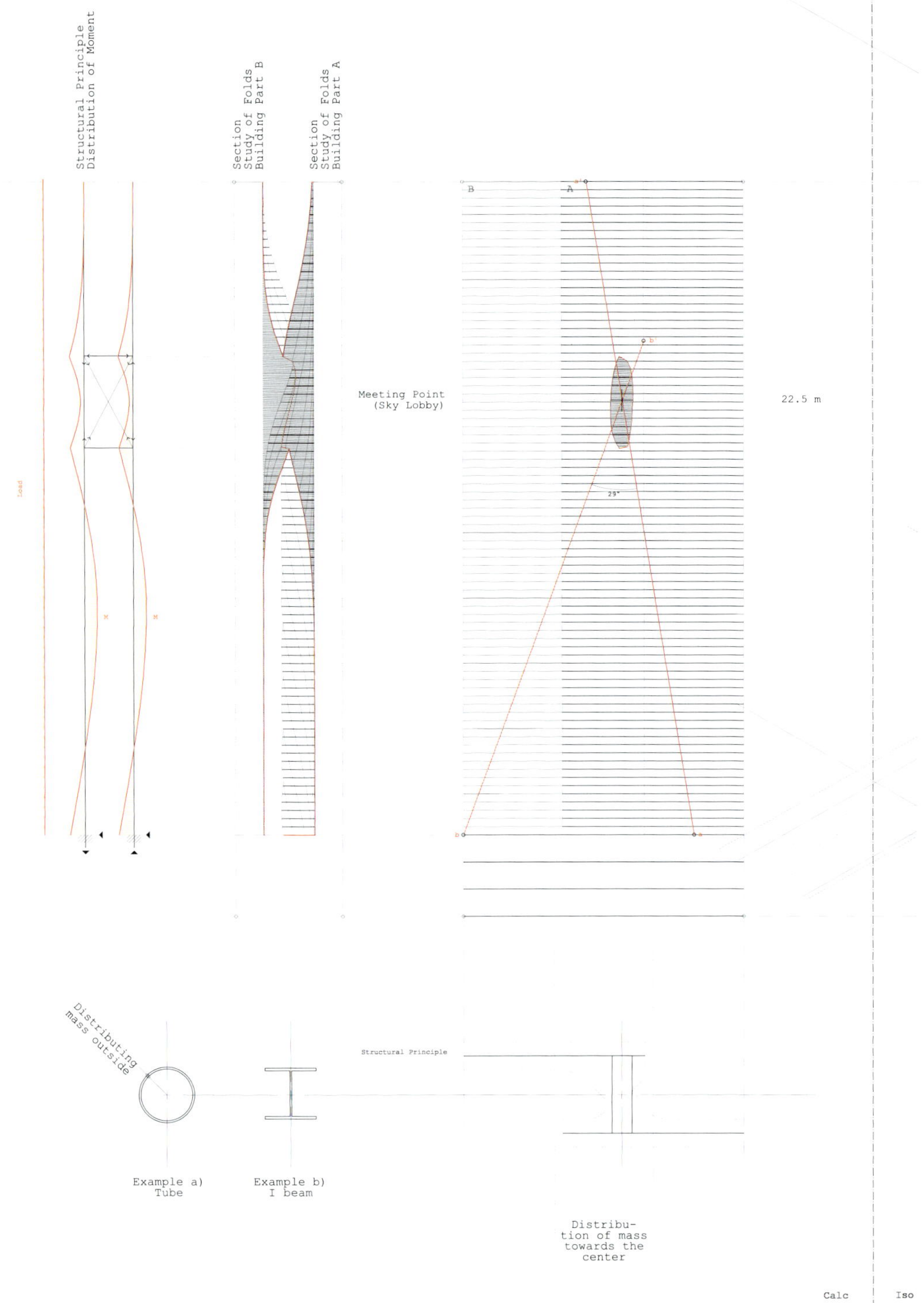

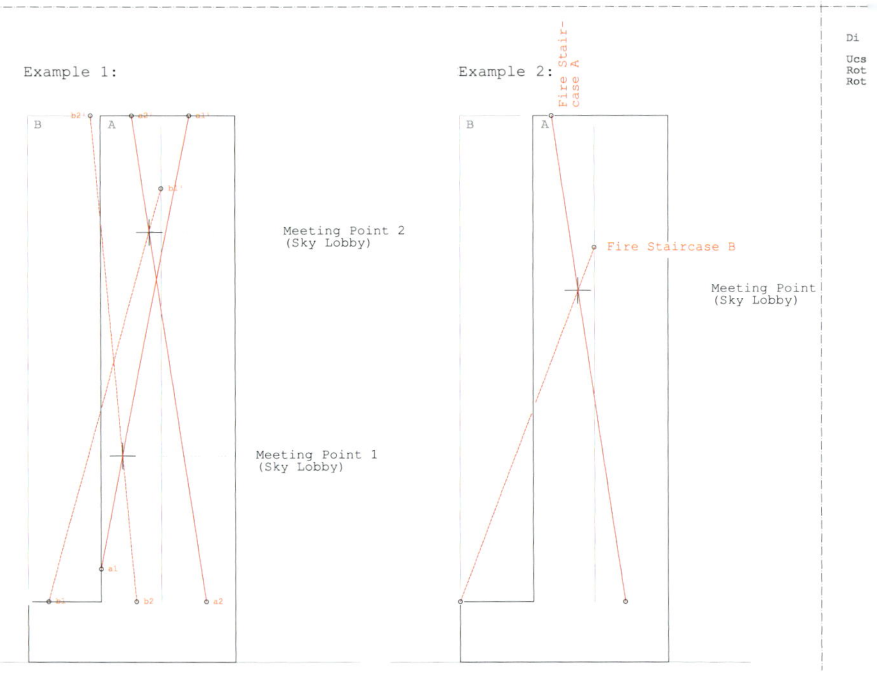

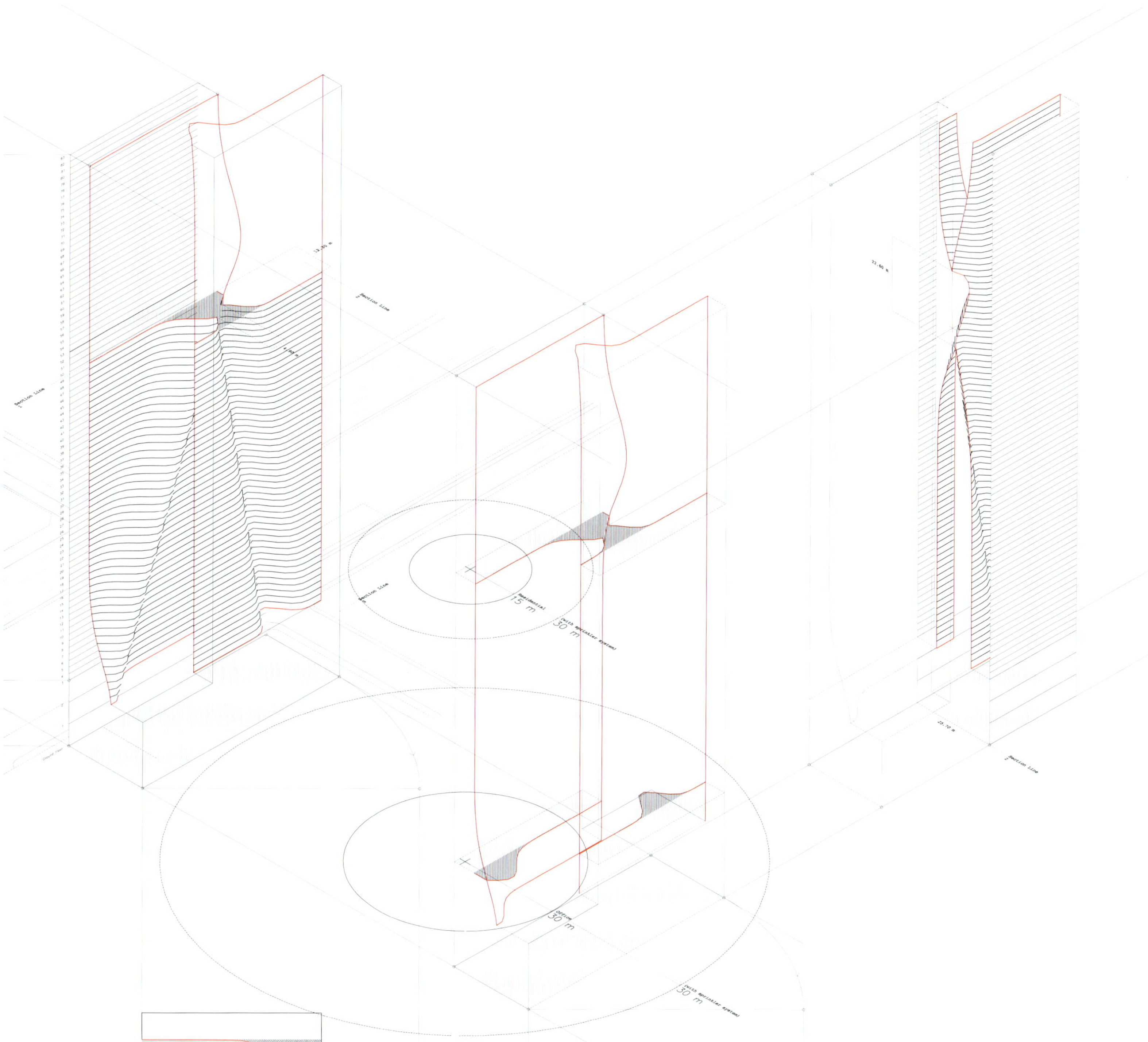

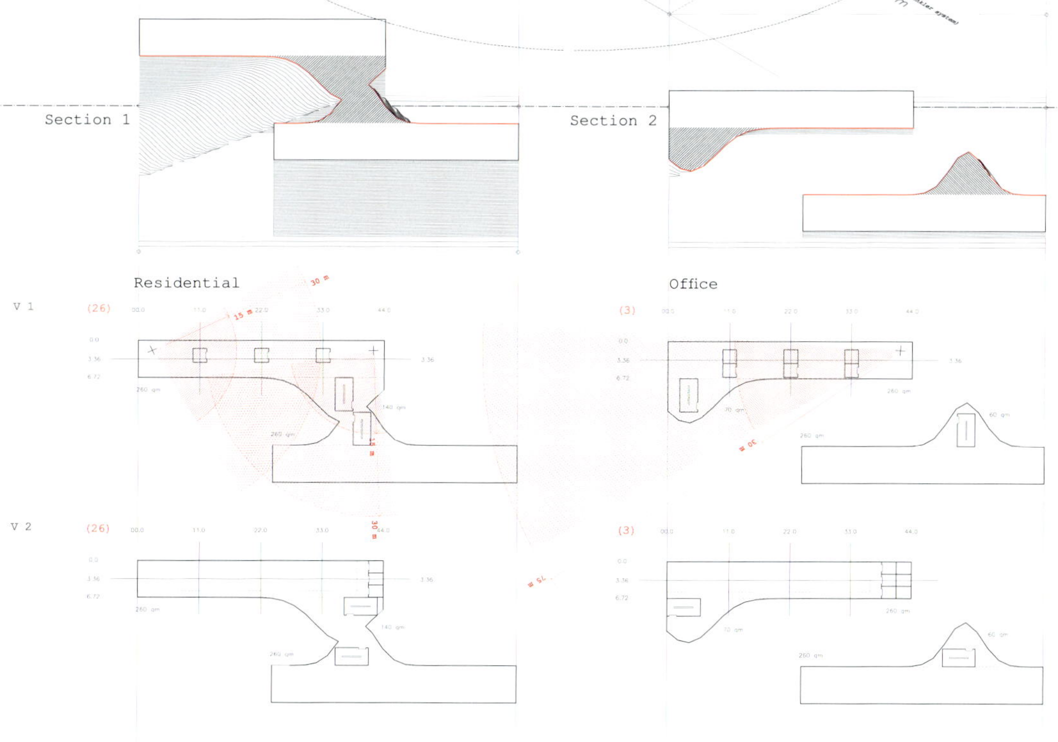

2.5 x 2.5

2.23 x 2.7

6.0 x 3.2

Firefighting Lifts:

According to (-GS) High-rise group III, includes buildings of 60m - 200m height, more than one firefighter elevators may be required. The structural components must be rated for fire resistance class R 120.

Escape Routes:

Escape routes often consist of two components. The first is where escape is only available in a single direction (for example, a romm with a single dor), and the second is where escape is available in two or more directions (for example, in a corridor leading to two alternative escape stairs).

Width Escape Routes:

(Singapore Standard)

Occupants:	Exits:
50 - 200	2
201 - 500	2

Width Escape Stairs:

(British Standard)

Occupants:	Stair Width:
100	1.00 m
120	1.10 m
130	1.20 m
140	1.30 m
150	1.40 m

4.4 Floorplates vs. Facade

The red grid represents the structural grid of the facades while the black grid indicates the floors. In elevation, the independency of these two rhythms becomes clear and expresses their programmatic autonomy.

The coexistence of both creates an interference that diversifies the rigidity of the mundane and efficient grid. Although both systems function within their own logic, they merge into one coherent matter.

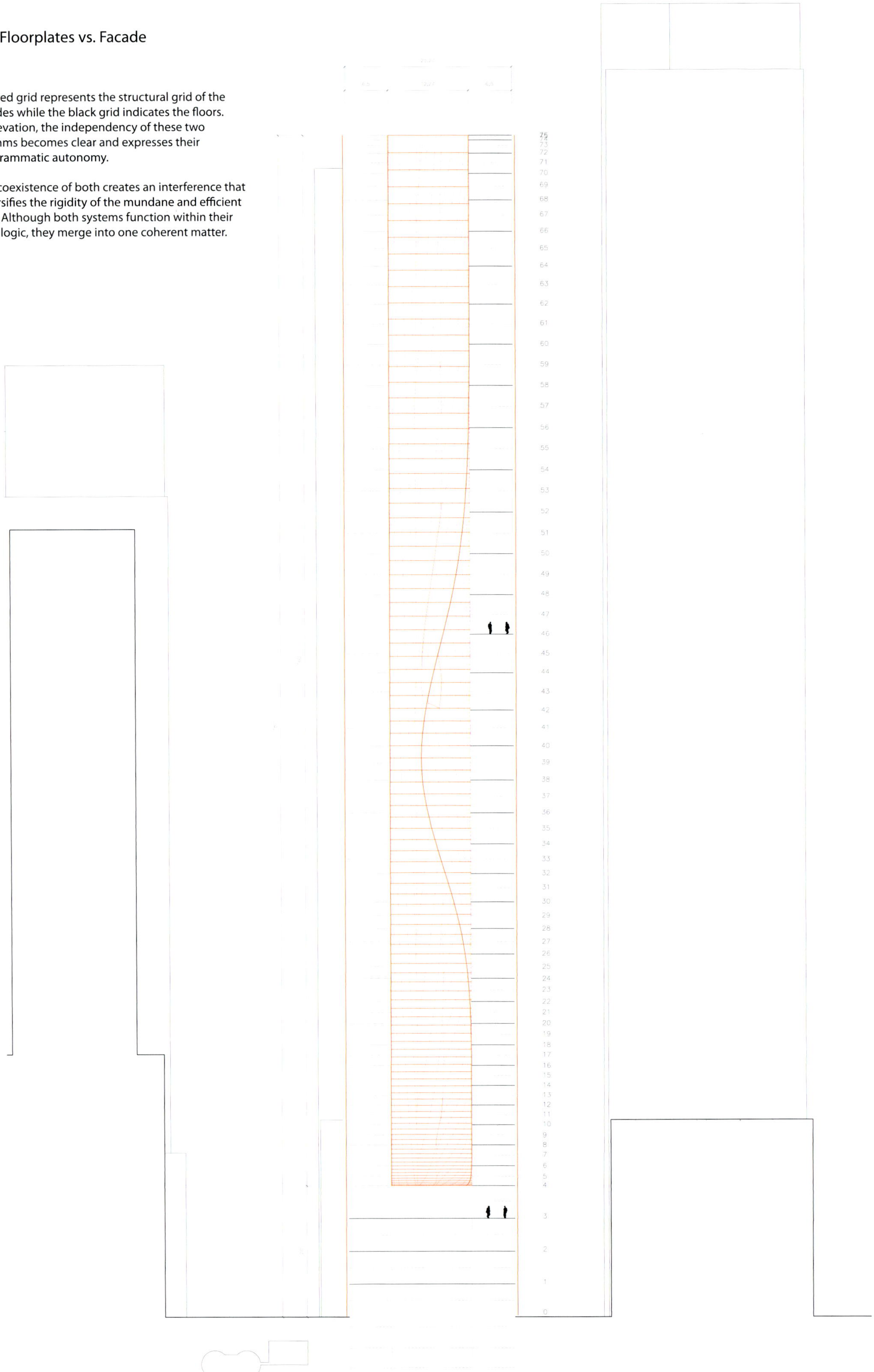

Floors	m² (total)	% of GPR	Amount of people - per floor	Amount of people - total
25	25570	100 %		
24	24998	98 %	22	22
23	24426	95,7 %	22	44
22	23854	93,5 %	22	88
21.5				
21	23282	91,2 %	22	110
20.5	–	–		
20	22710	89 %	22	132
19.5	–	–		
19	22138	86,8 %	22	154
18.5	–	–		
18	21566	84,5 %	22	176
17.5	–	–		
17	20994	82,3 %	22	198
16.5	–	–		
16	20422	80 %	22	220
15.5	–	–		
15	19850	77,8 %	22	242
14.5	–	–		
14	19278	75,5 %	22	264
13.5	–	–		
13	18706	73,3 %	22	286
12.5	–	–		
12	18134	71,1 %	22	308
11.5	–	–		
11	17562	68,8 %	22	330
10.5	–	–		
10	16990	66,6 %	22	352
9.5	–	–		
9	16418	64,3 %	22	374
8.5	–	–		
8	15846	62,1 %	22	396
7.5	–	–		
7	15274	59,8 %	22	418
6.5	–	–		
6	14702	57,6 %	22	440
5	14130	55,4 %	22	462
4	13558	0,53 %	22	484
3	12986	50,9 %	22	506
2	12414	48,8 %	22	528
1	11842	46,4 %	22	550
11	11270	44,1 %	21	571
10	10698	41,9 %	21	592
9	10126	39,7 %	21	613
8	9554	37,4 %	21	634
7	8982	35,2 %	21	655
6	8410	32,9 %	21	676
5	7838	30,7 %	21	697
4	7266	28,4 %	21	618
3	6694	26,2 %	21	639
2	6122	25,1 %	21	660
1	5550	21,7 %	21	681
4	4440	17,4 %		
2	3330	13,0 %		
1	2220	8,7 %		
0	1110	4,3 %		

Residential

Floor-to-Floor heights:
3.8 m - 6.4 m

Total Floor-Area:
14300 sqm
(54%)

Size of standard floorplate: 286 sqm
Area of elevators & stairs (20%): -60 sqm

Total useable floor area: 226 sqm

sqm per person: 13 sqm

Total amount of people per floor: 22

Total amount of residents:
550

Office

Floor-to-Floor heights:
3.0 m - 3.5 m

Total Floor-Area:
6292 sqm
(25%)

Size of standard floorplate: 286 sqm
Area of elevators & stairs (20%): -74 sqm

Total useable floor area: 210 sqm

sqm per person: 10 sqm

Total amount of people per floor: 21

Total amount of office people:
131

Commercial

Floor-to-Floor heights:
6.66 m

Total Floor-Area:
5550 sqm
(21%)

4.5 Vertical Circulation Routes

The drawings below explain how external staircases are embedded within the folds. The folds are articulated as external structure that sits outside the straight facades.

The red lines in the drawing on the right page highlight the two circulation-interchanges between the slabs.

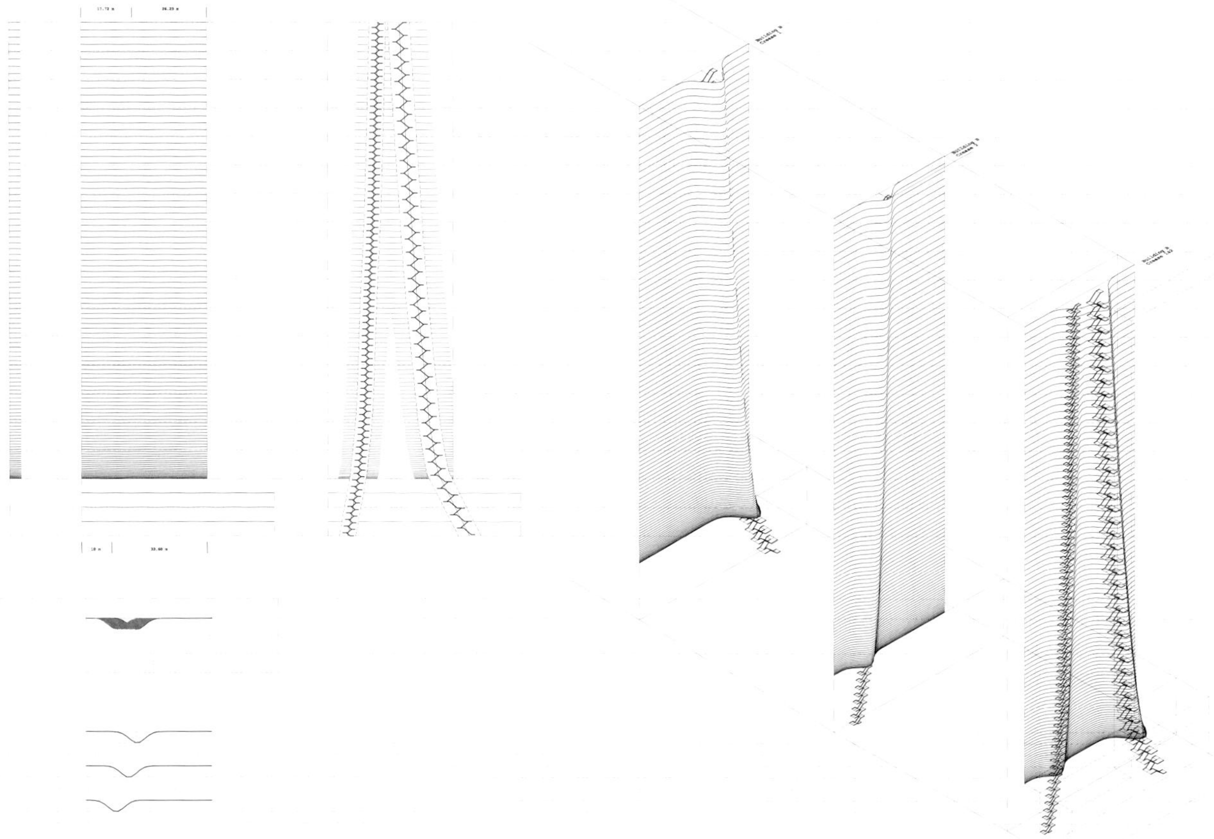

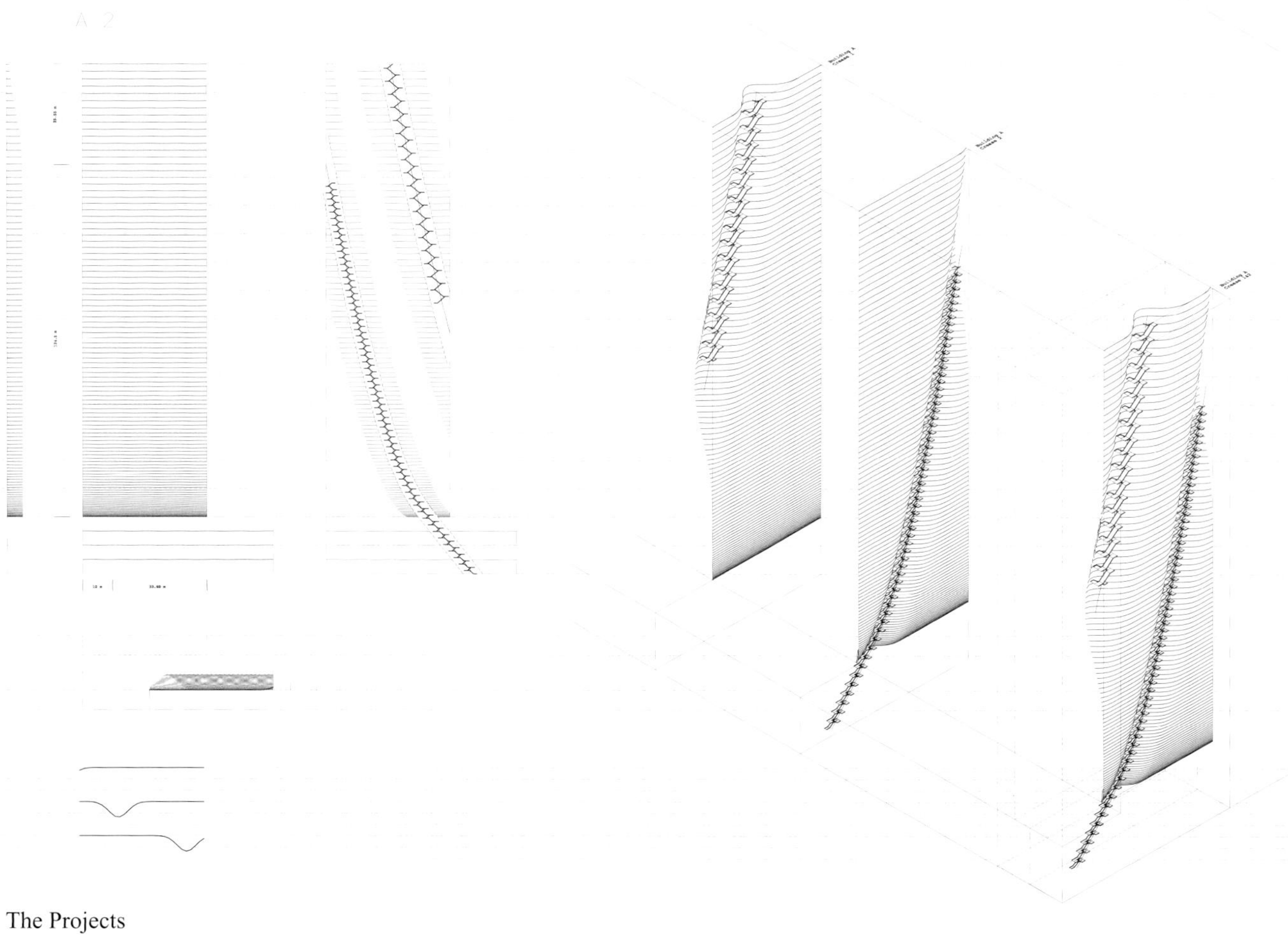

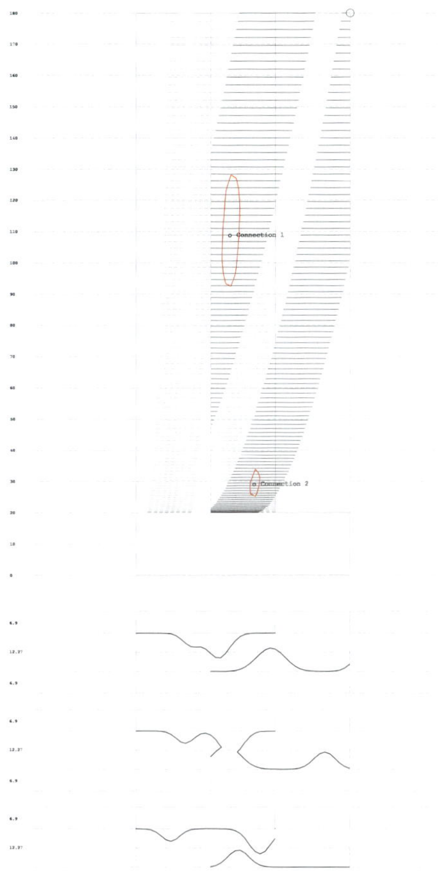

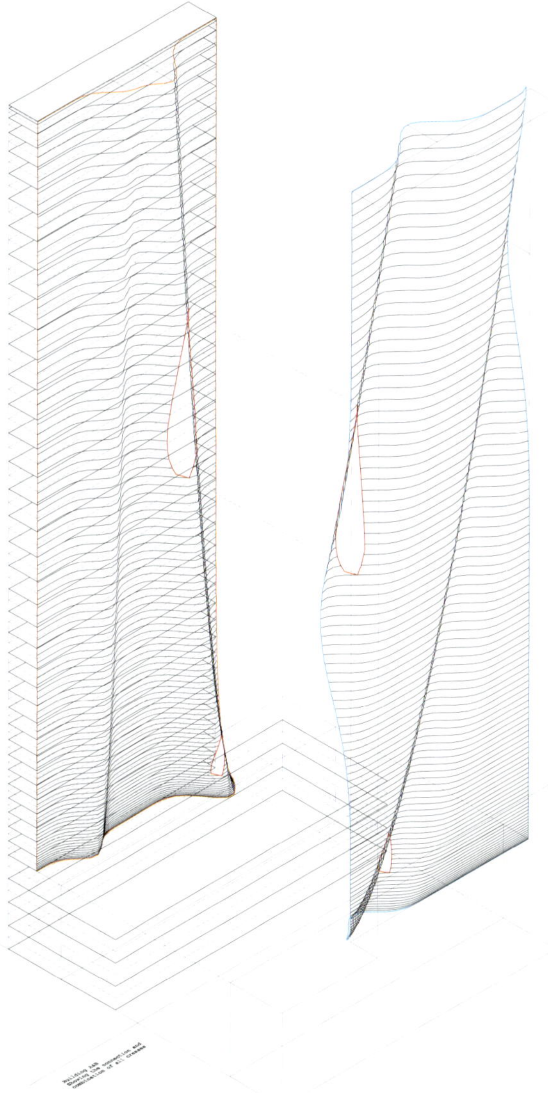

Folds, developed for circulation & structural support

Two folds run across the inner side of each building. The main fold is developed with a depth of 8.5 m and contains one (of two) fire-staircases. The other one has a depth of 5.0 m and functions partly as balconies and partly as a sub interim-storey-circulation.

The two main folds run into opposite directions and intersect each other within the upper third part of the building height. This connection has two functions: First, to produce a structural connection between both buildings in order to transfer horizontal loads. Secondly, to provide alternative escape routes in case of an emergency.

The positioning of all four folds is developed by negotiating problems of access and constrains related to fire regulations as well as structural principles and constrains caused by the extreme thinness of both buildings.

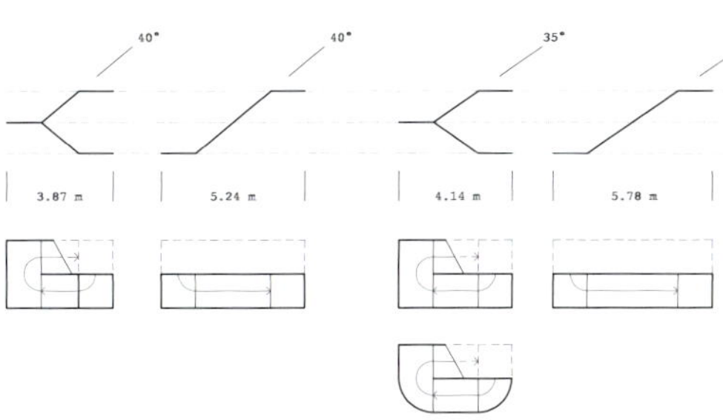

Final physical model during production.
Still, both slabs lie side by side, waiting to get bound together by the cores (standing on the right).

SANAA
Electaarchitecture

4.5 Vertical Circulation Routes

Internal and external staircases. Red indicates the two necessary fire-staircases of each building slab.

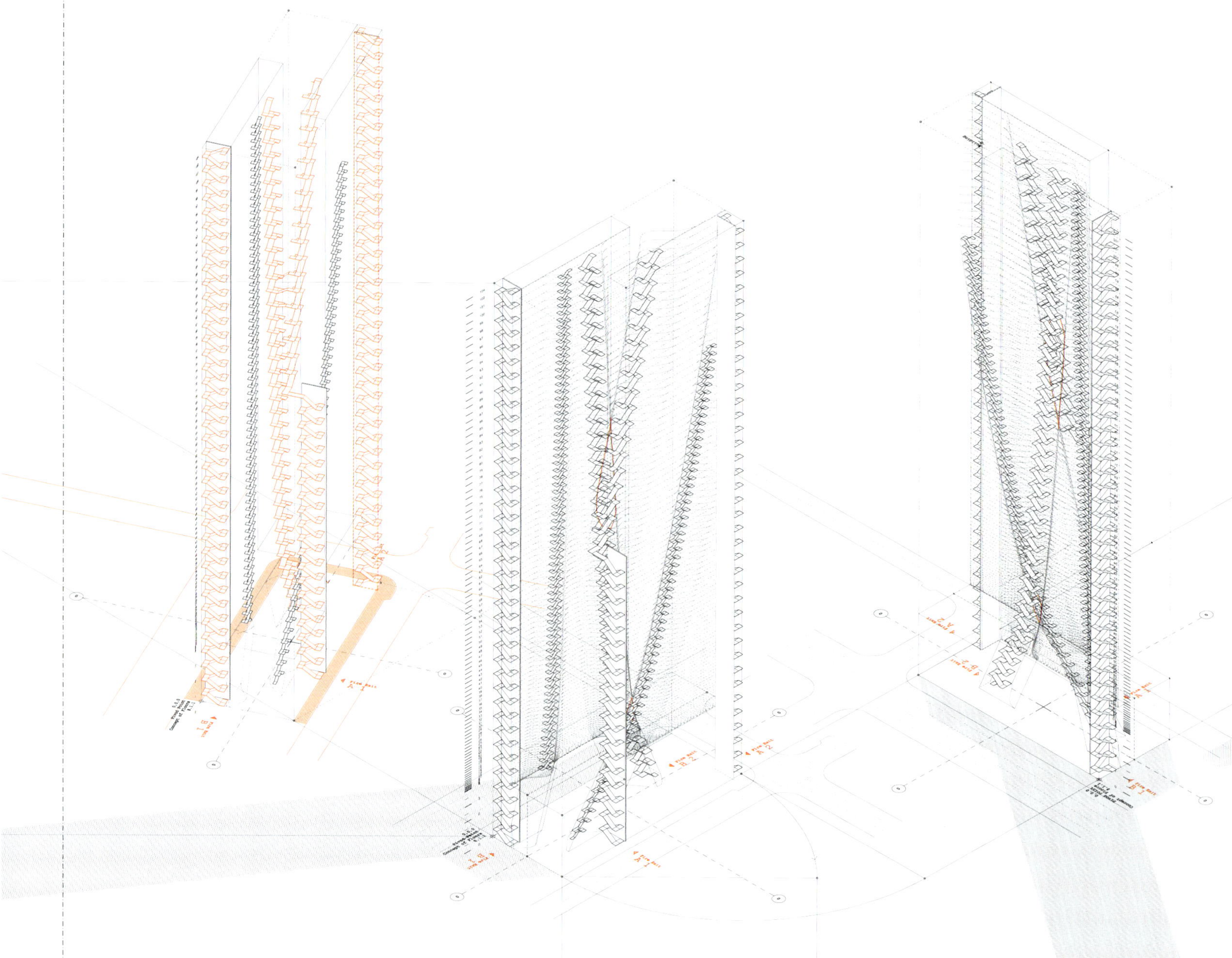

While the circulation concept for the elevator system provides an internal direct access of the individual spaces (for residential as well as offices), the concept for the external stair circulation is to create interim spaces that not only offer a functional connection between single floors but also provide terraces and platforms.

The system of staircases embedded within the folds emphasise a diverse spacial structure that is located in-between the two buildings - facing each other and thereby creating a semi-private atmosphere outside the buildings. In other words, to travel from storey 22 to storey 24 the user has to cross the outline of the building itself and is forced to experience the building from outside.

The idea is to provoke the relationship between the inside and outside of the high-rise building that underlines the extreme density and height as an urban and spacial phenomena.

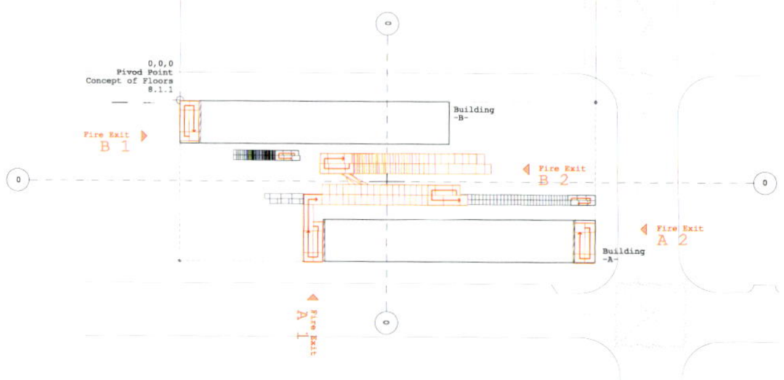

The elevator cores are spread equally over the floor plans. Red indicates elevators for residential, dark blue for offices and light blue for commercial areas. Each residential elevator services 10 - 16 units directly in order to avoid corridors within the very thin building slabs.

Therefore the residents need to operate the elevators with an individual key that allows to stop at their specific unit only.

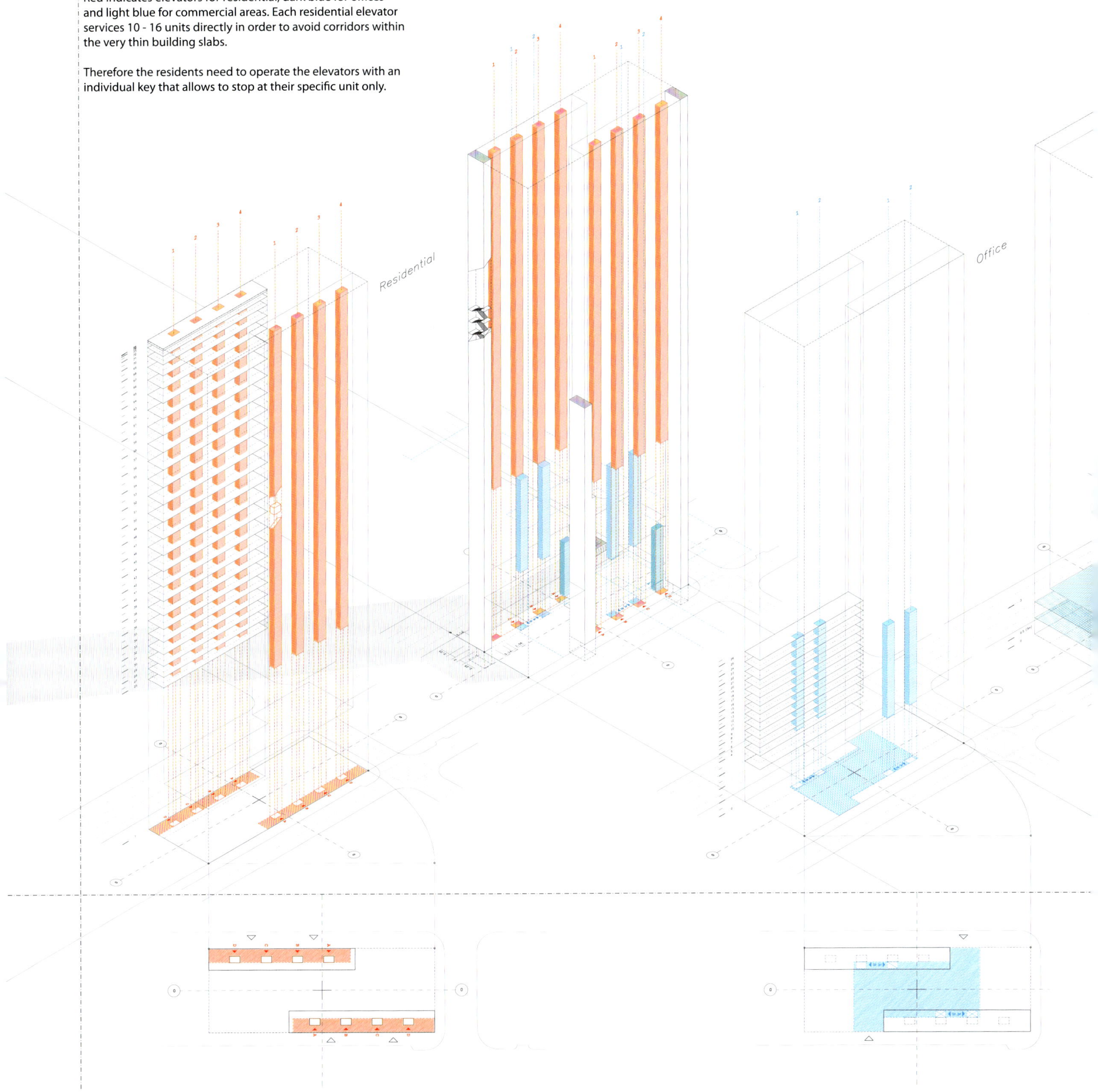

4.6 Articulation of Floor- & Facade Members

All building elements are articulated in a similar and rather crude tectonic way to express the structural purity of the grid. This expression underlines the continuity of facade and floor system and merges both into one coherent matter.

This matter should embody that it is subject to one piece of mathematical writing that manipulates the continuous grid in order to create spatial potential which will be filled with use accordingly.

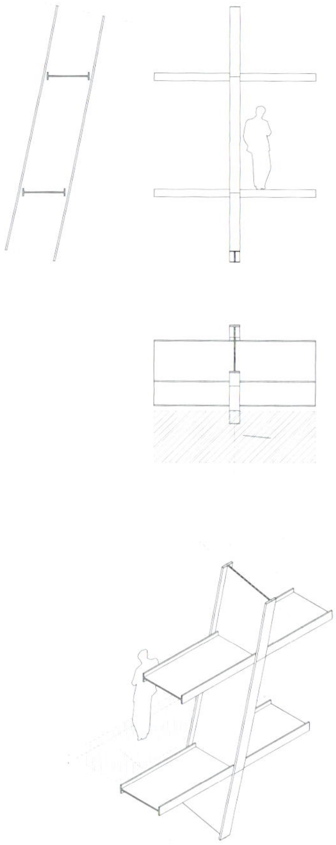

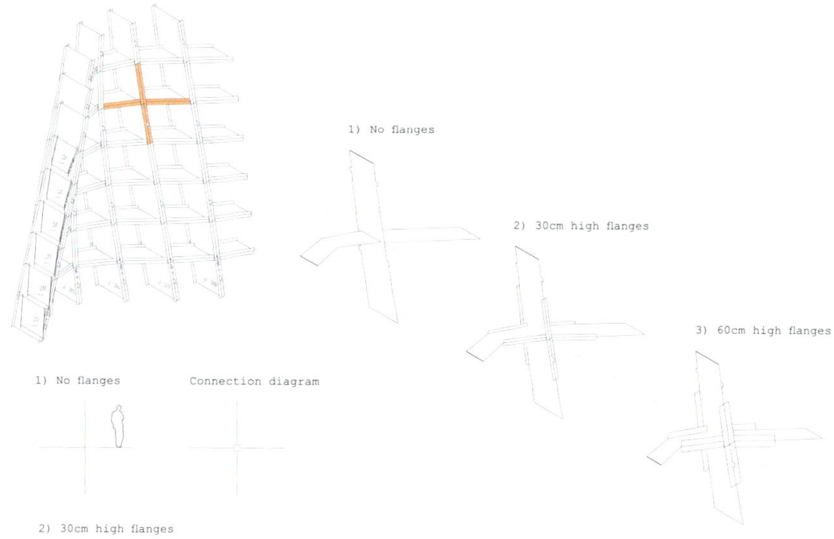

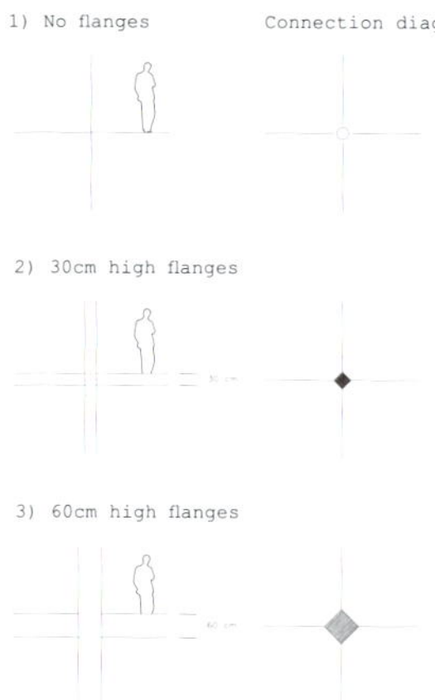

In order to stiffen the facade grid against shear forces the members need to be articulated as I-beams, while the height of the flanges essentially affect the stiffness of the overall grid.

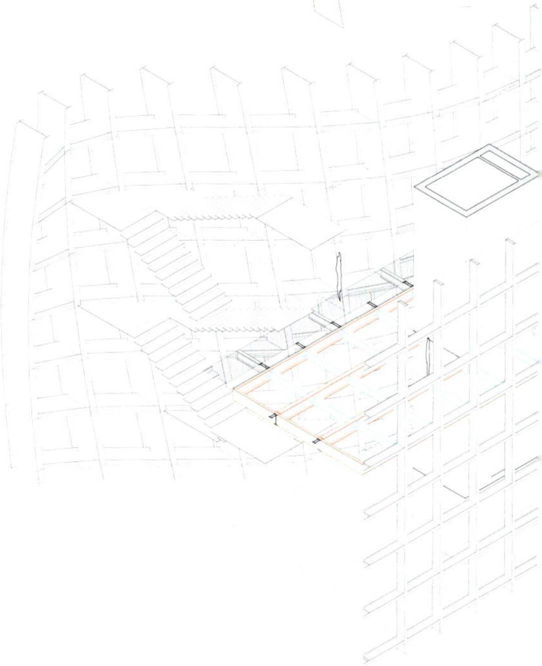

At the World Trade Center, a similar principle had been used to stiffen the connections between the floor-slabs and the vertical columns.

World Trade Center Facade Detail

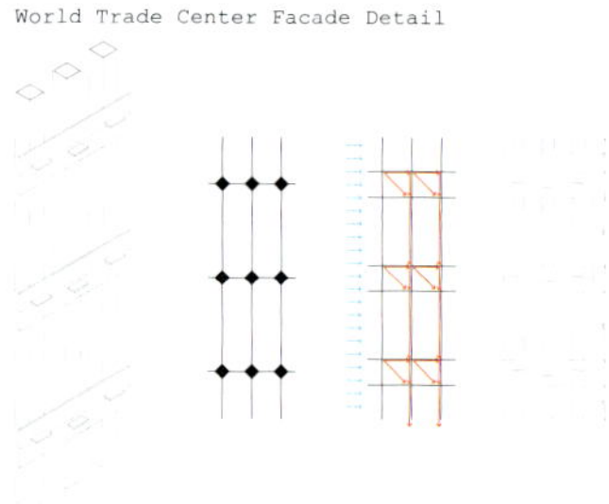

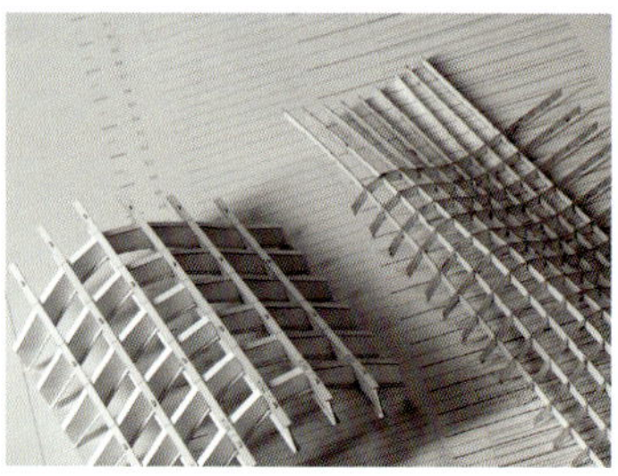

Variations of facade grid with varying flanges

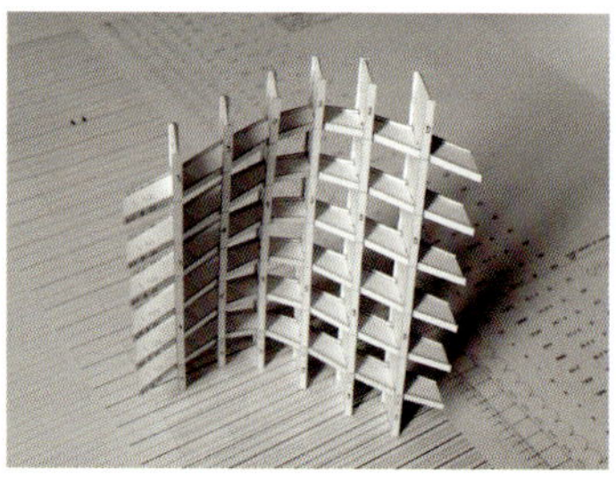

1:100 Facade MochUp

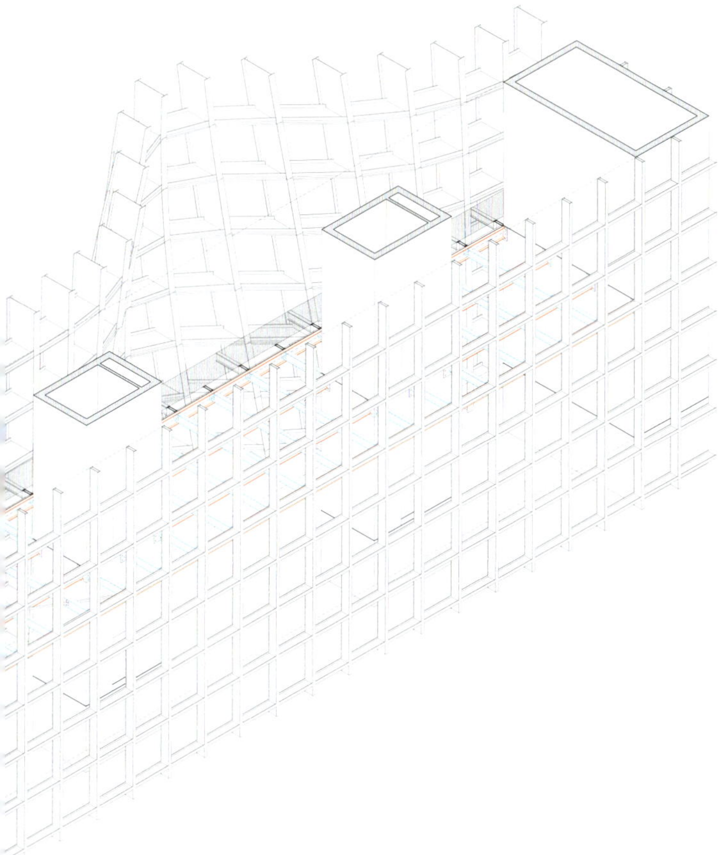

Final model.
Relation of facade structure in relation to open floor-plans.

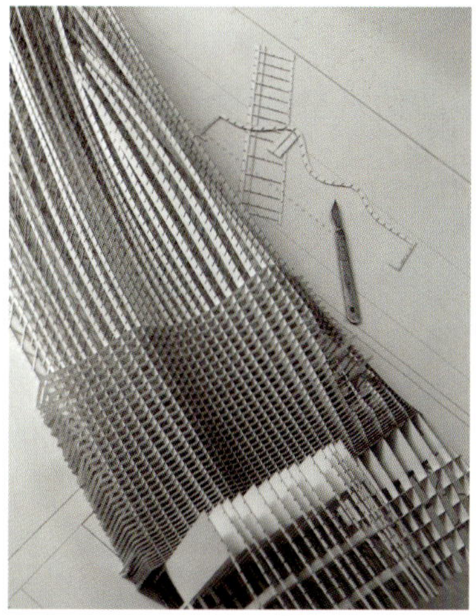

Assembly of final model.
Facade in relation to podium and vertical members.

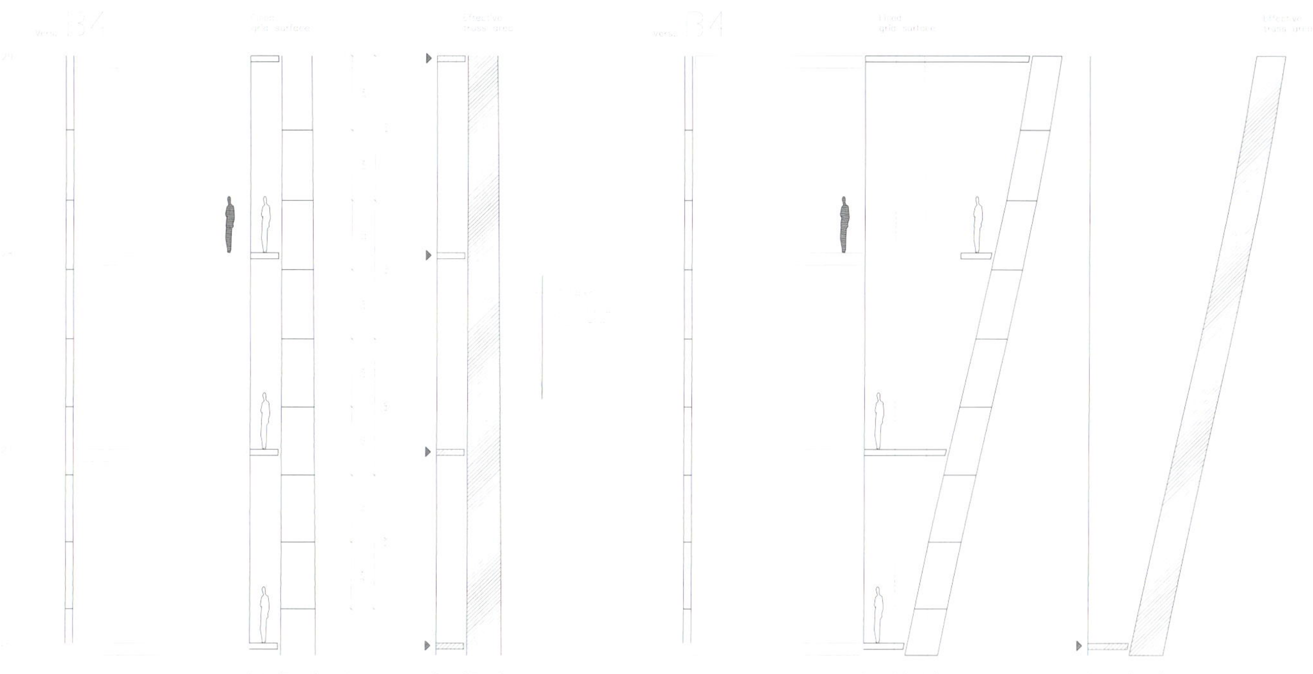

4.6 Articulation of Floor- & Facade Members

Depending on the viewpoint, the overall structure seams almost transparent and the different rhythms of floors and facades can be read clearly, while, the more this viewpoint changes, the outer surfaces appear almost solid.

The three diagrams on the right indicate in red the section of the inner surface-grids that vary in thickness from bottom to top and that merge in the middle.

Parameters:

$last\_in := 100$ $in := 01, .. last\_in$ $height := 160$

$last\_jn := 1$ $jn := 01, .. last\_jn$ $width := 1$

Vertical Distribution of Threads:

$v\_a_{in} := 1.65$

Threads:

$$iThread\_tap1_{in,\,jn} := 0$$

$$jThread\_tap1_{in,\,jn} := \left(\frac{jn}{last\_jn}\right) \cdot width$$

$$vThread\_tap1_{in,\,jn} := \left(\frac{in}{last\_in}\right)^{v\_a_{in}} \cdot height$$

$$iThread\_tap2_{in,\,jn} := 1.5 - 0.6 \sin\left(\frac{in}{last\_in} 0.5\,\pi\right)^{4}$$

$iThread\_tap2_{in,\,1} =$

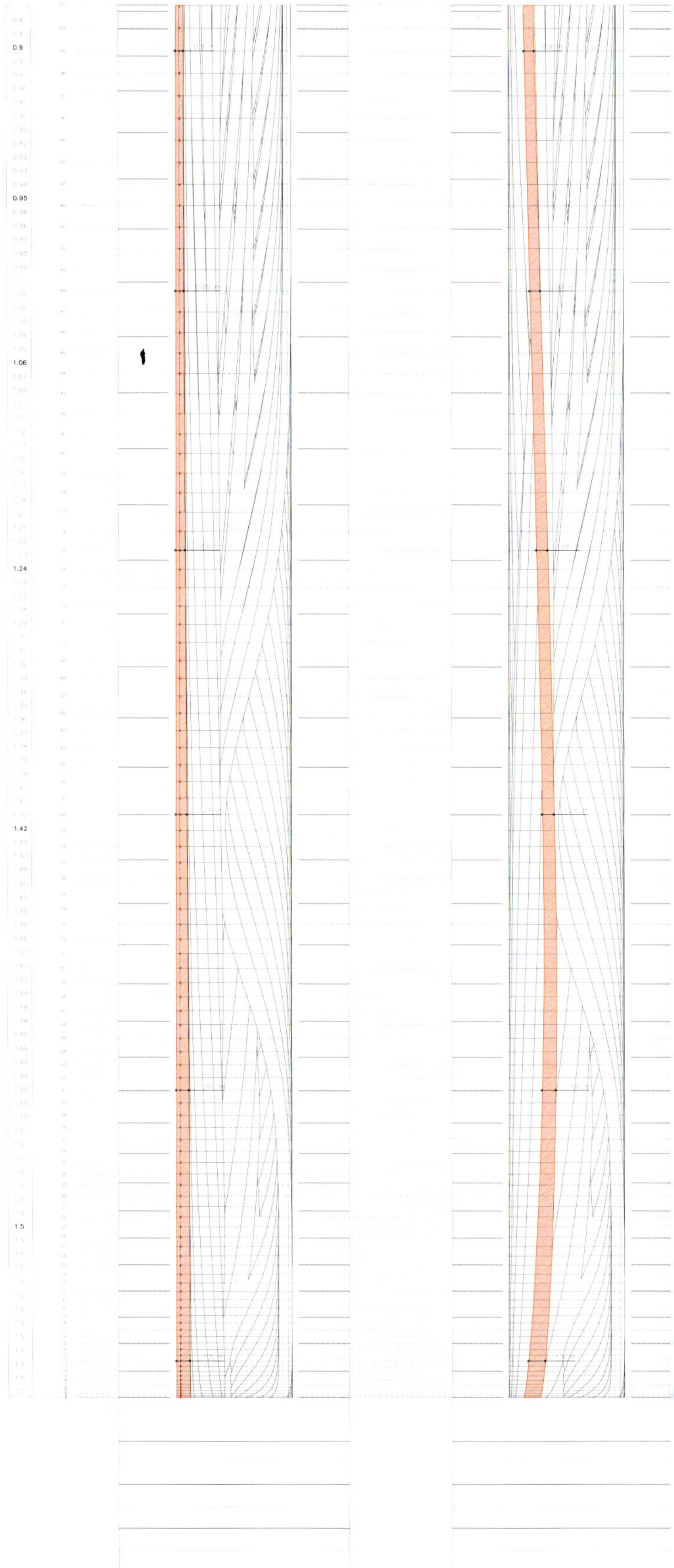

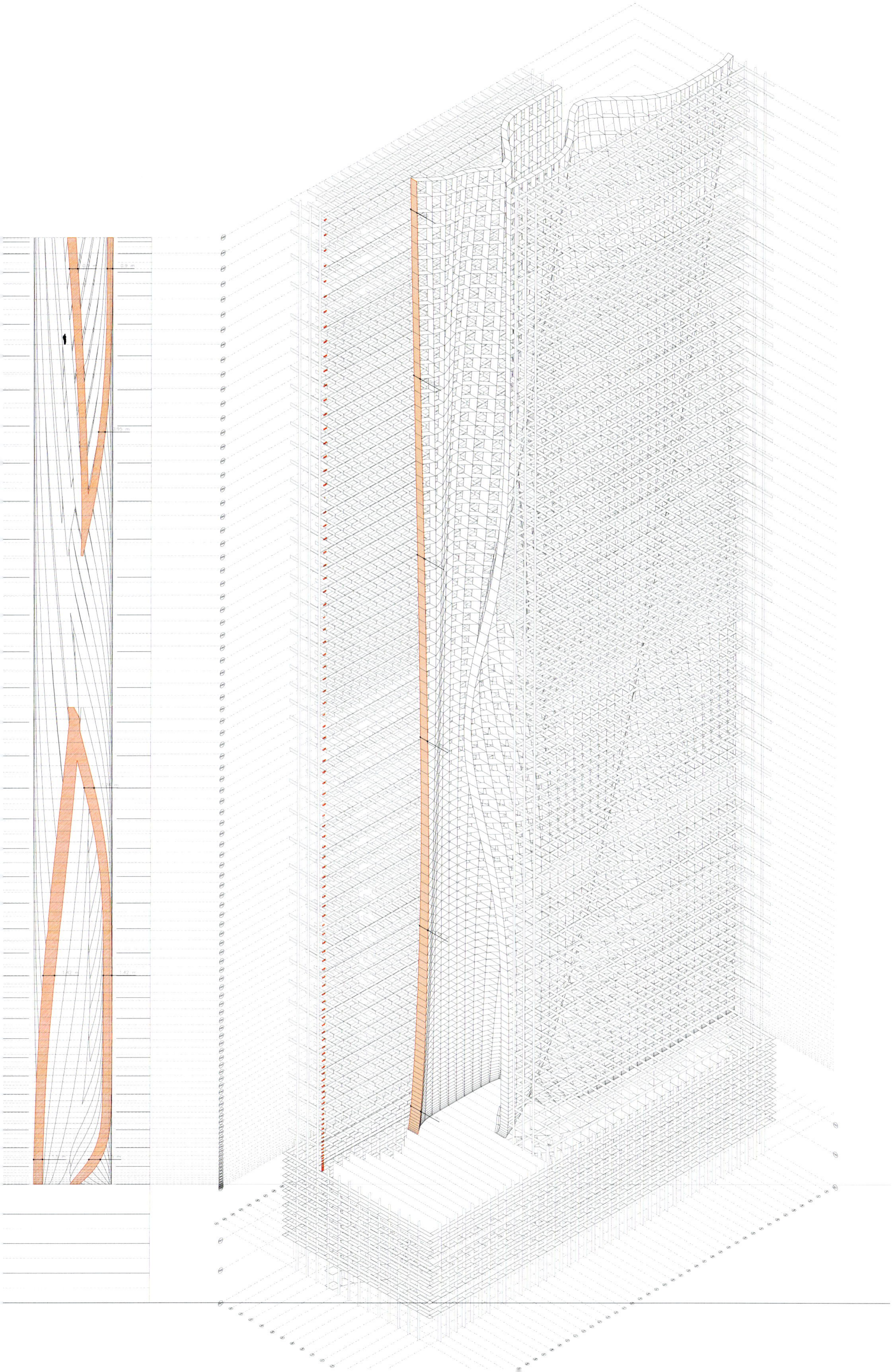

Max Kahlen

4.7 Apartment Units

The drawing below shows how residential units are embedded within the final building articulation.

This articulation blurs the boundary of the efficient straight grid on the outer facades and the excessive redundant forest of structure on the inner faces.

In some areas the facade-structure can be inhabited and used as terraces, in other areas this structure becomes a simple curtain-wall-facade or sun shelter. The surfaces remain continuous but create various grades of porosity that activate the interstitial space between the two building slabs and interconnect the interior with the exterior. In areas where folds are positioned, the facade-structure merges from an internal grid into an external skeleton and creates an surreal lanscape based on functionality and the rigidity of the grid.

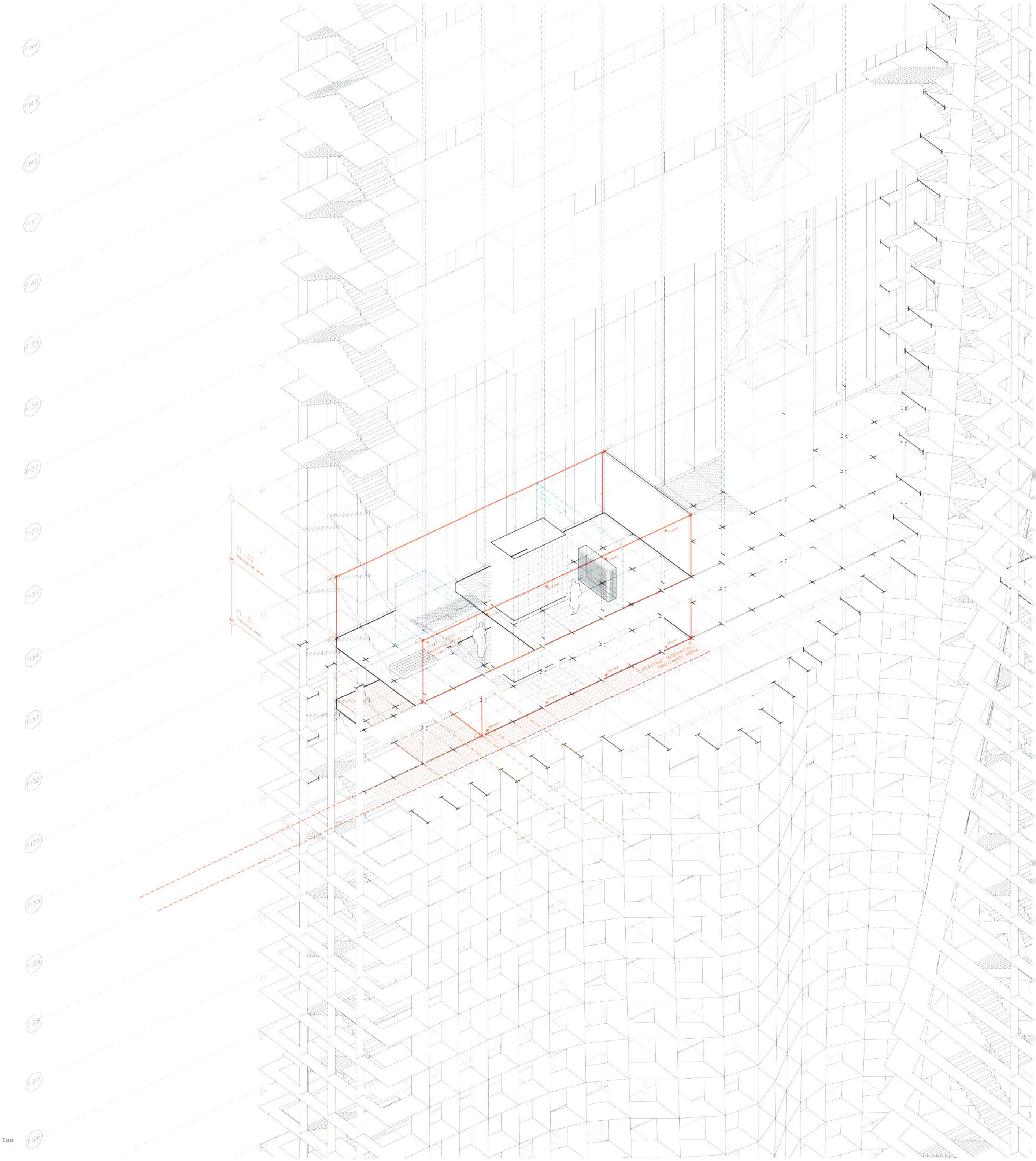

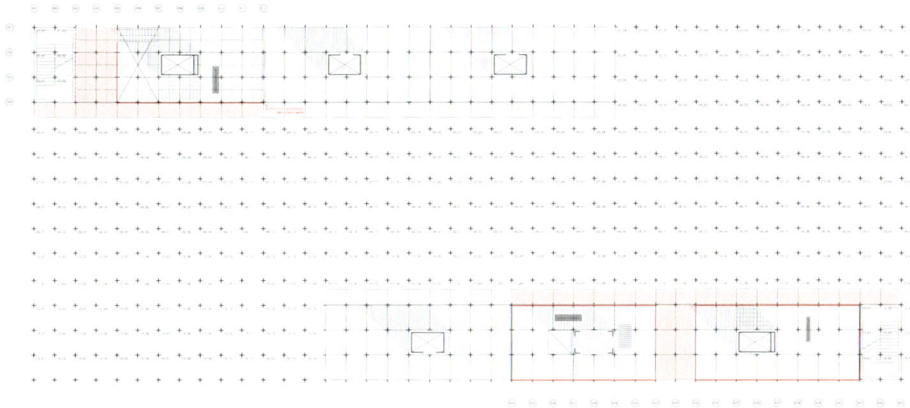

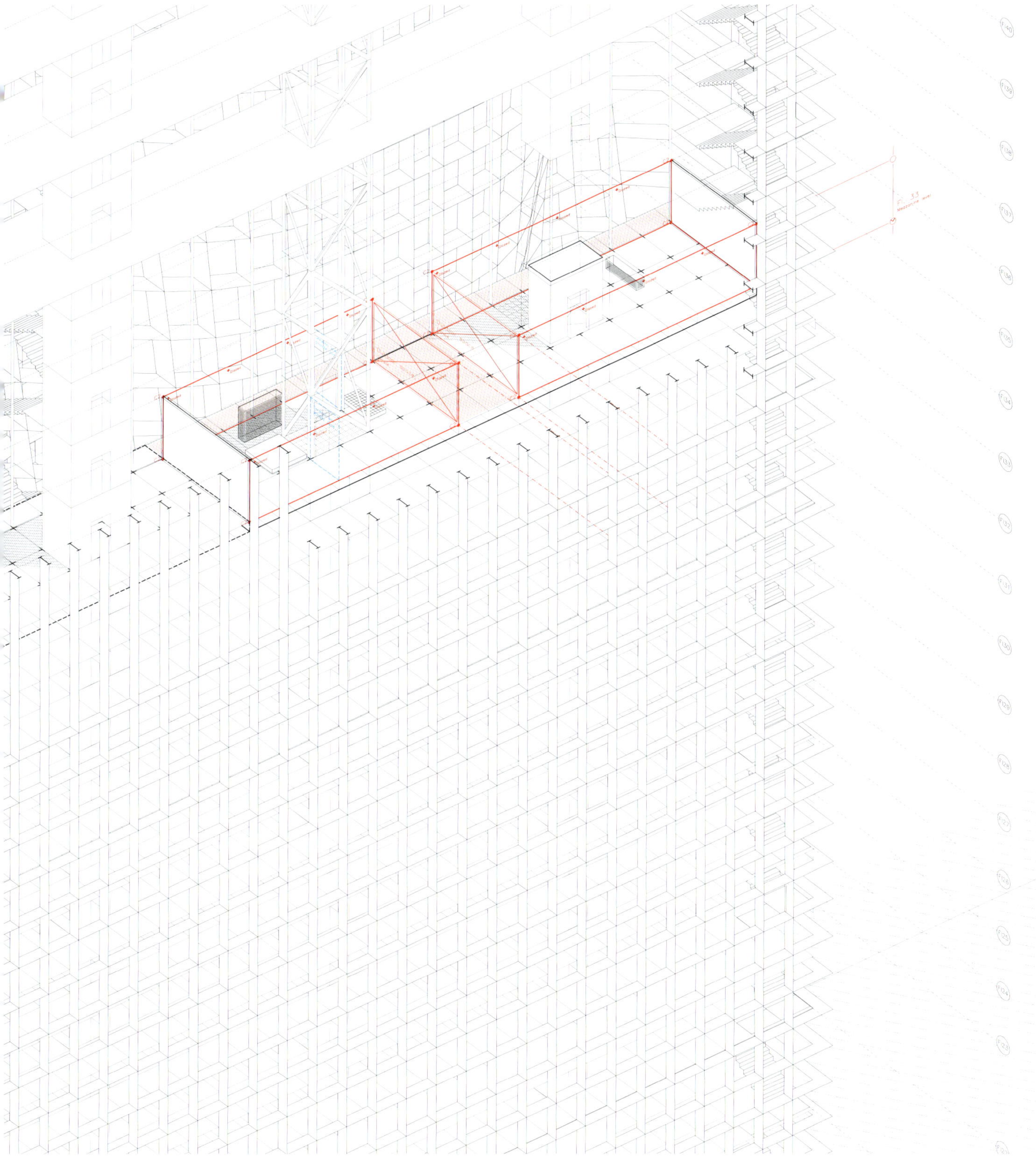

4.7 Apartment Units

The residential units are arranged around the elevator cores which function as main entrance. Inner courtyards punch through the whole building slabs and help to organize the layout of the units. Those private courtyards, marked in red, create the connection to the external walkways that connect to the external staircases.

Each unit is laid out as a continuum that circulates around the elevator cores to which all functional areas such as bathrooms and kitchens are attached.

The sizes of units range from 50 m^2 to 250 m^2.

The drawing below shows a unit of 65 m^{2}with a mezzanine floor, seen from front side and back side.

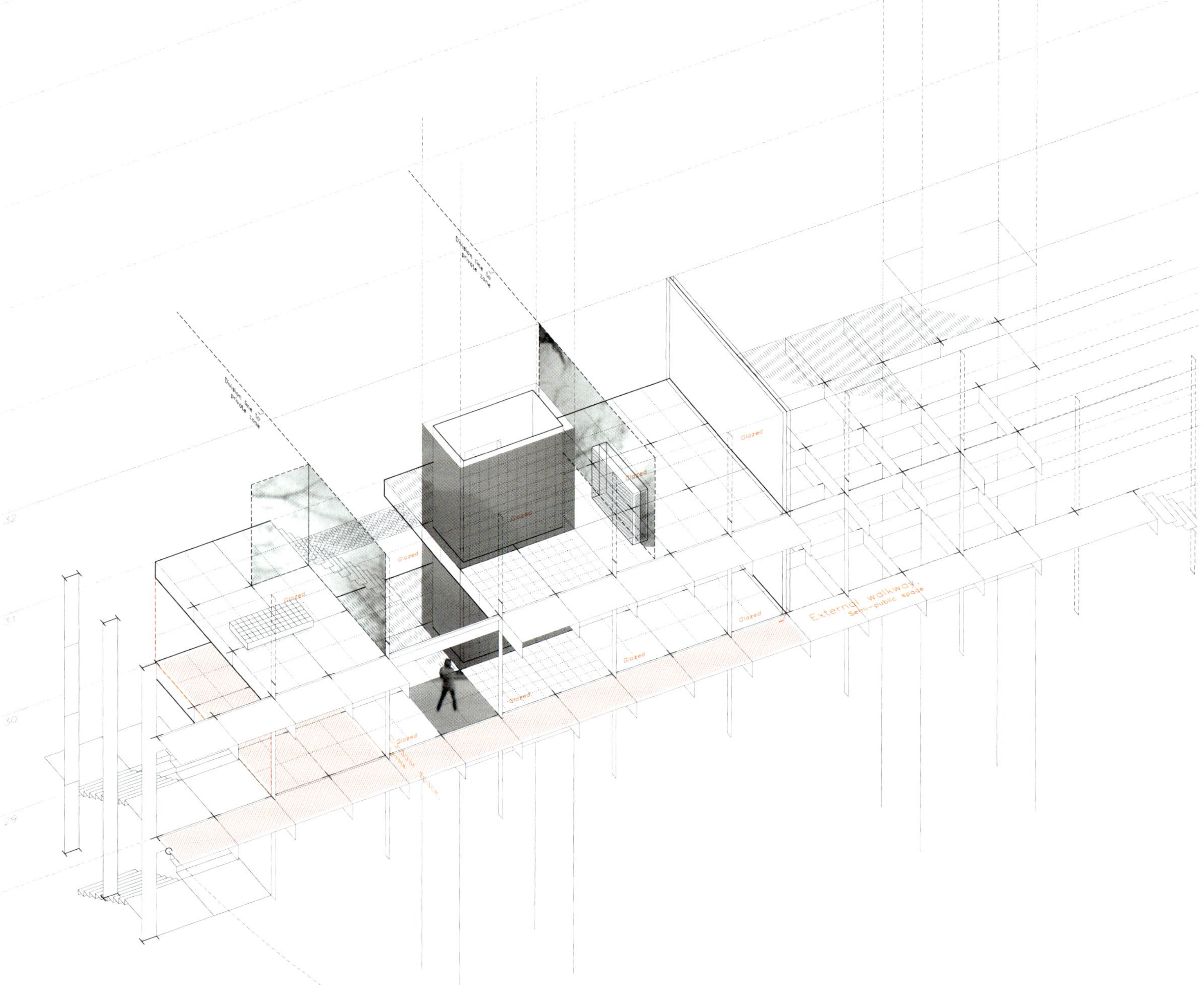

Max Kahlen

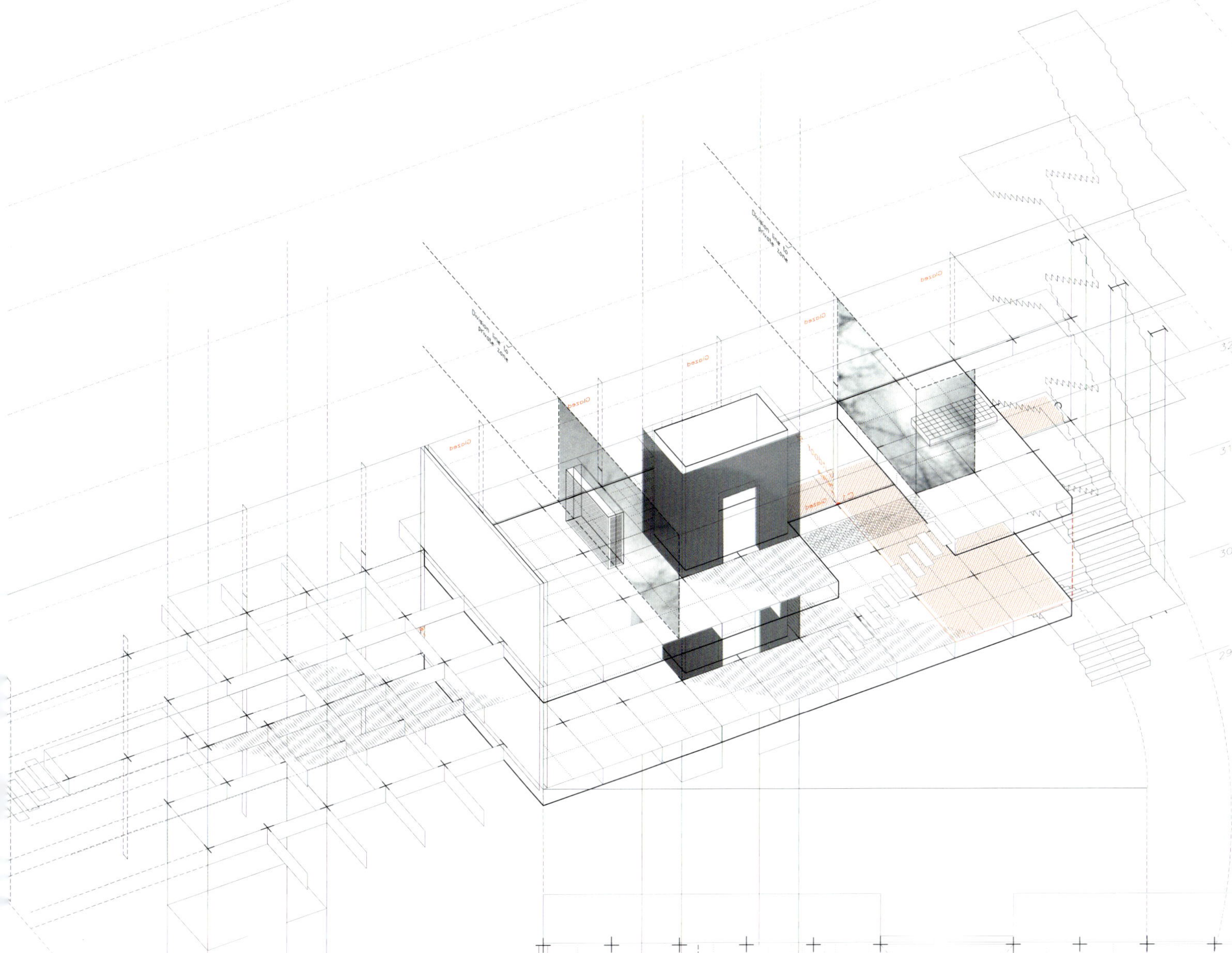

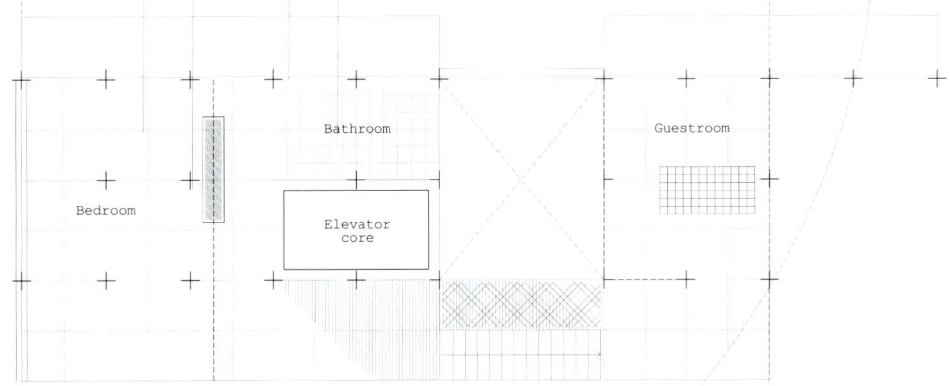
Bathroom
Guestroom
Bedroom
Elevator
core

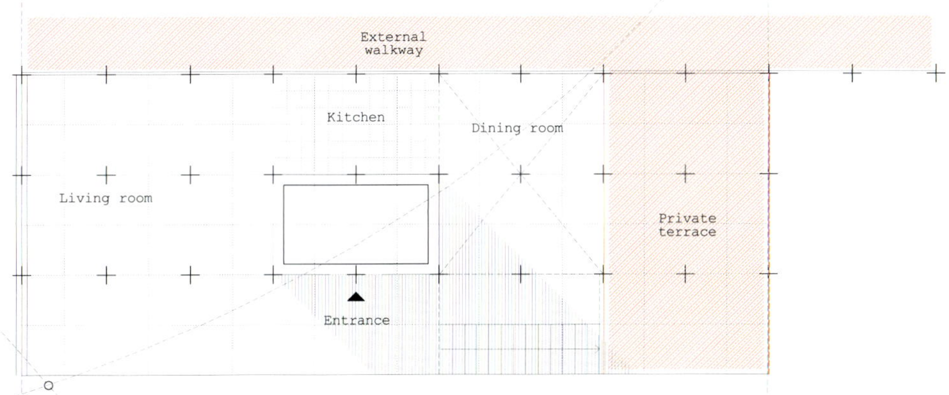
External
walkway
Kitchen
Dining room
Living room
Private
terrace
Entrance

4.7 Apartment Units

The drawing below shows a unit with one floor, seen from front side and back side.

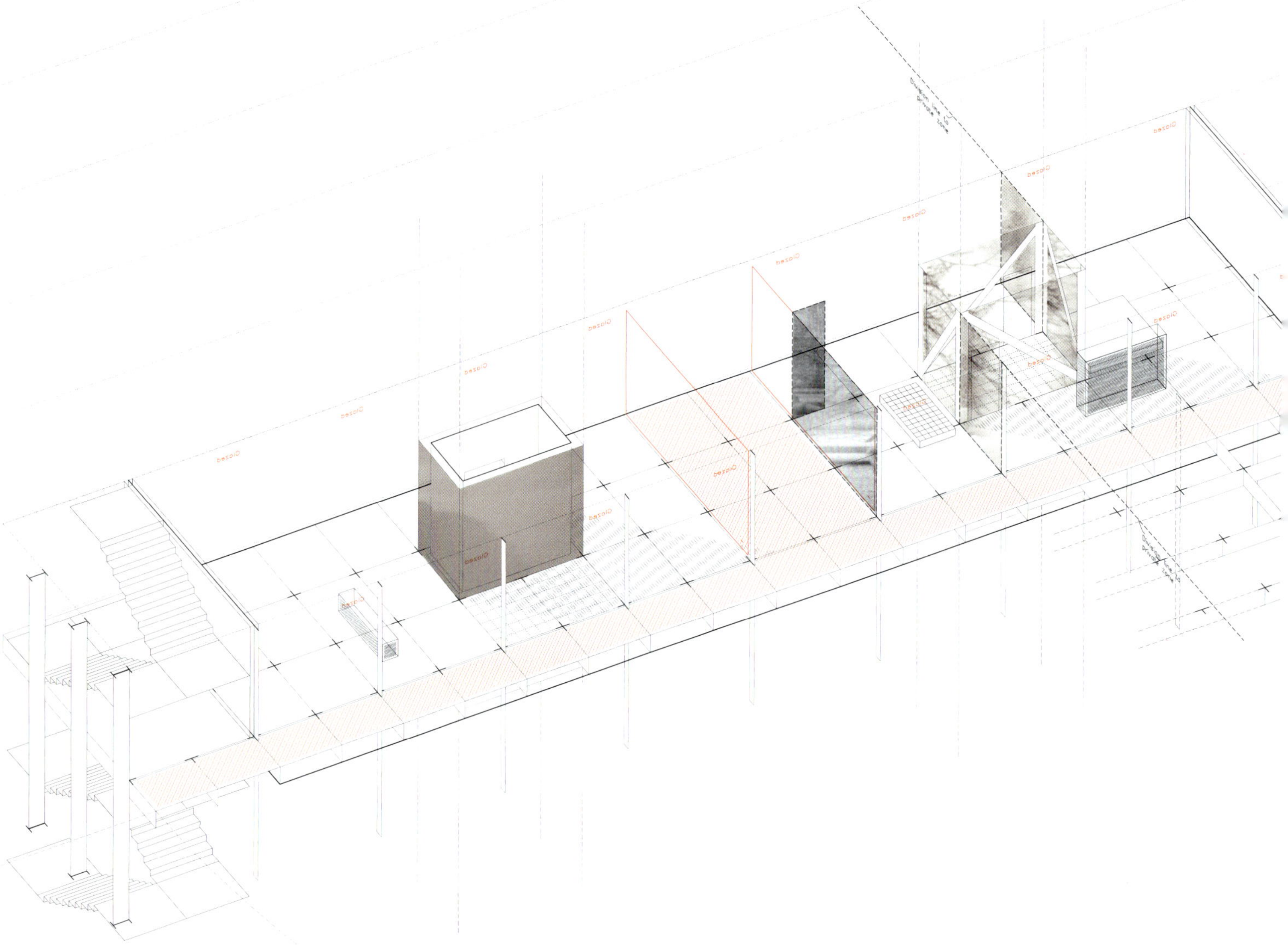

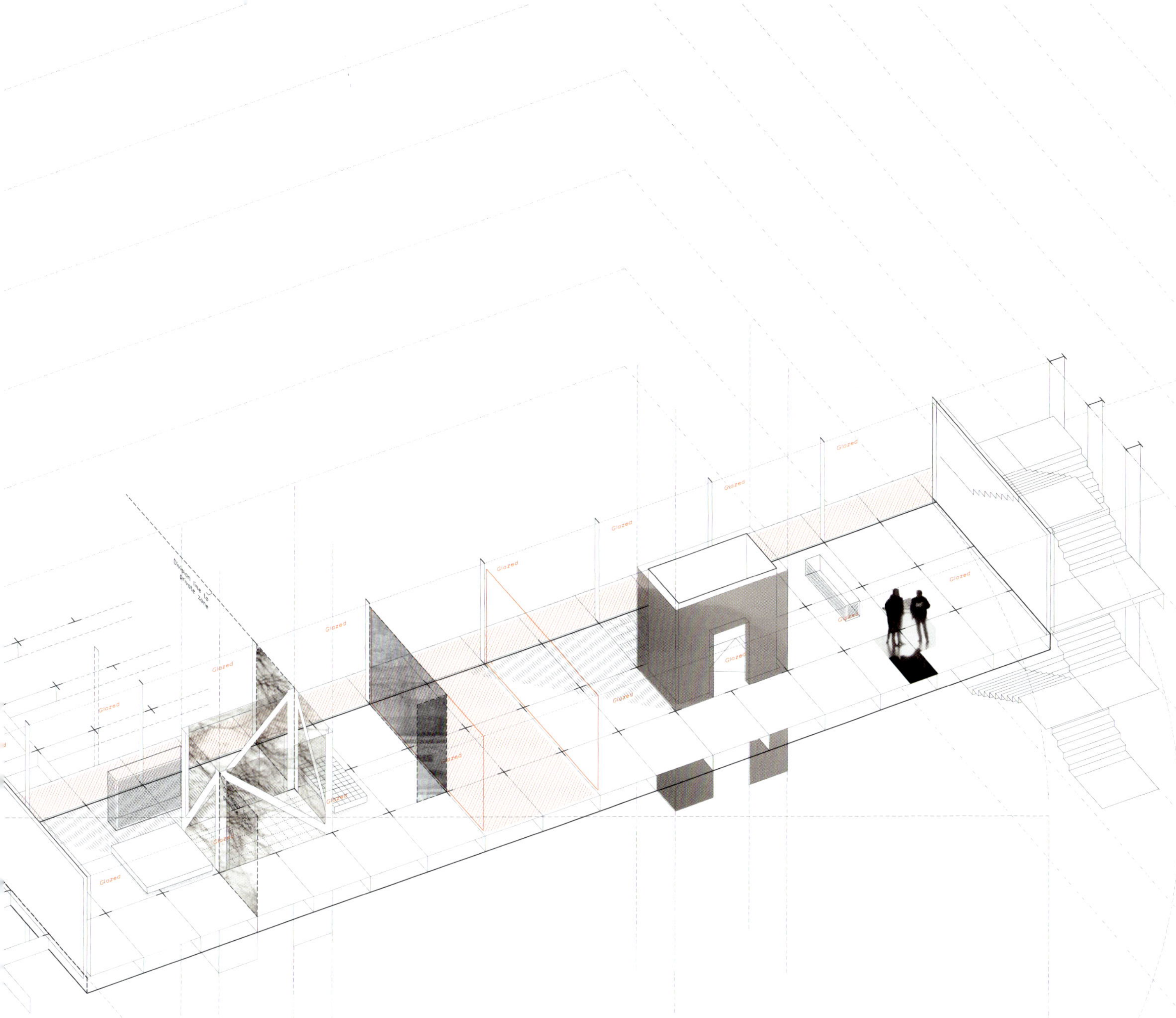

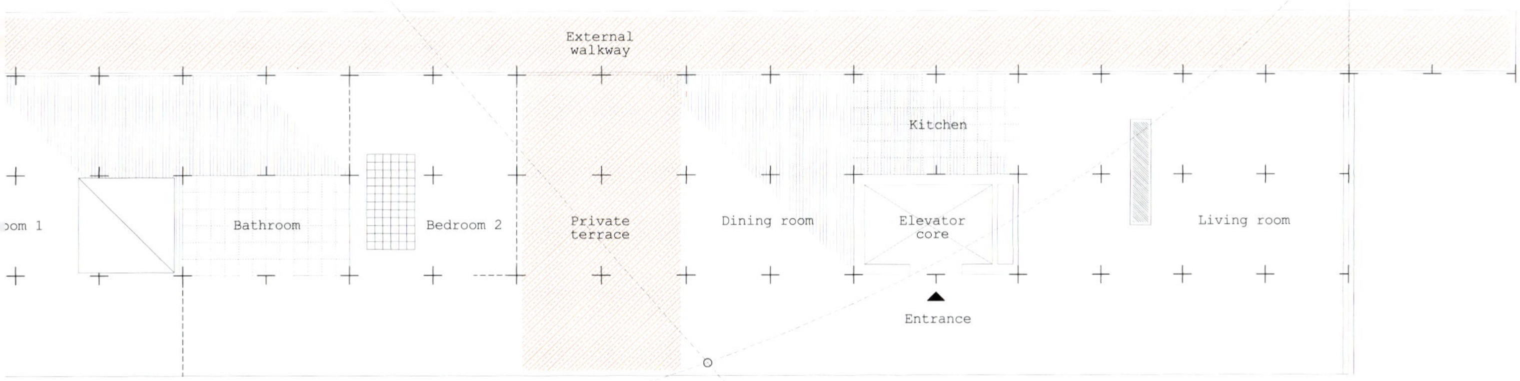
External walkway
Kitchen
Bathroom
Bedroom 2
Private terrace
Dining room
Elevator core
Living room
Entrance

4.8 Final Views

The drawing shows a perspective view of one elevation and indicates the transparency of the structure which exposes the vertical circulation and the inner spaces.

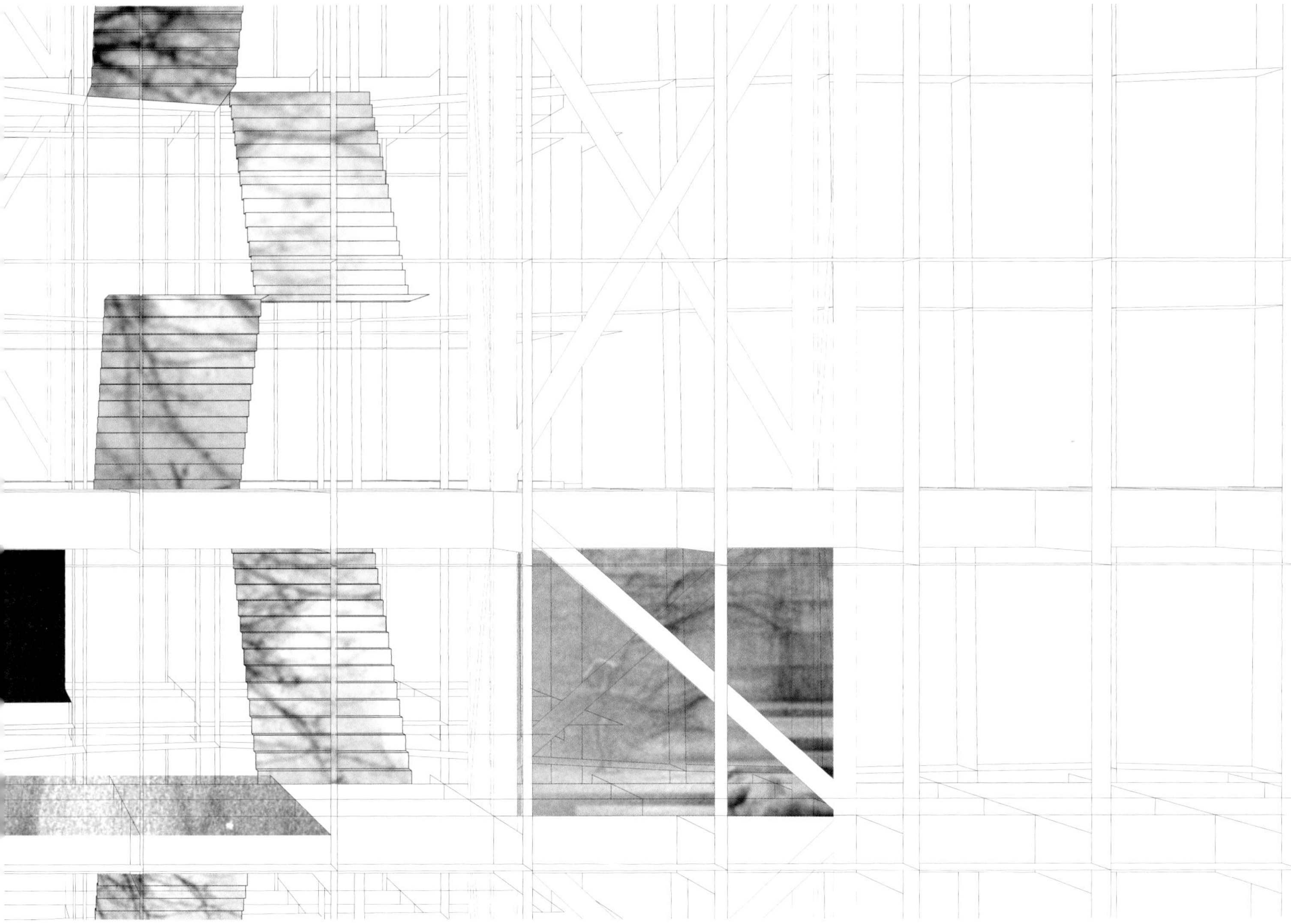

4.8 Final Views

In the end, the project is still about the madness of repetition.

The outer facades reflect the monotony of the neighbouring buildings while the inner facades divert and gain in porosity to interlock the interstitial space with the interiority of the slabs.

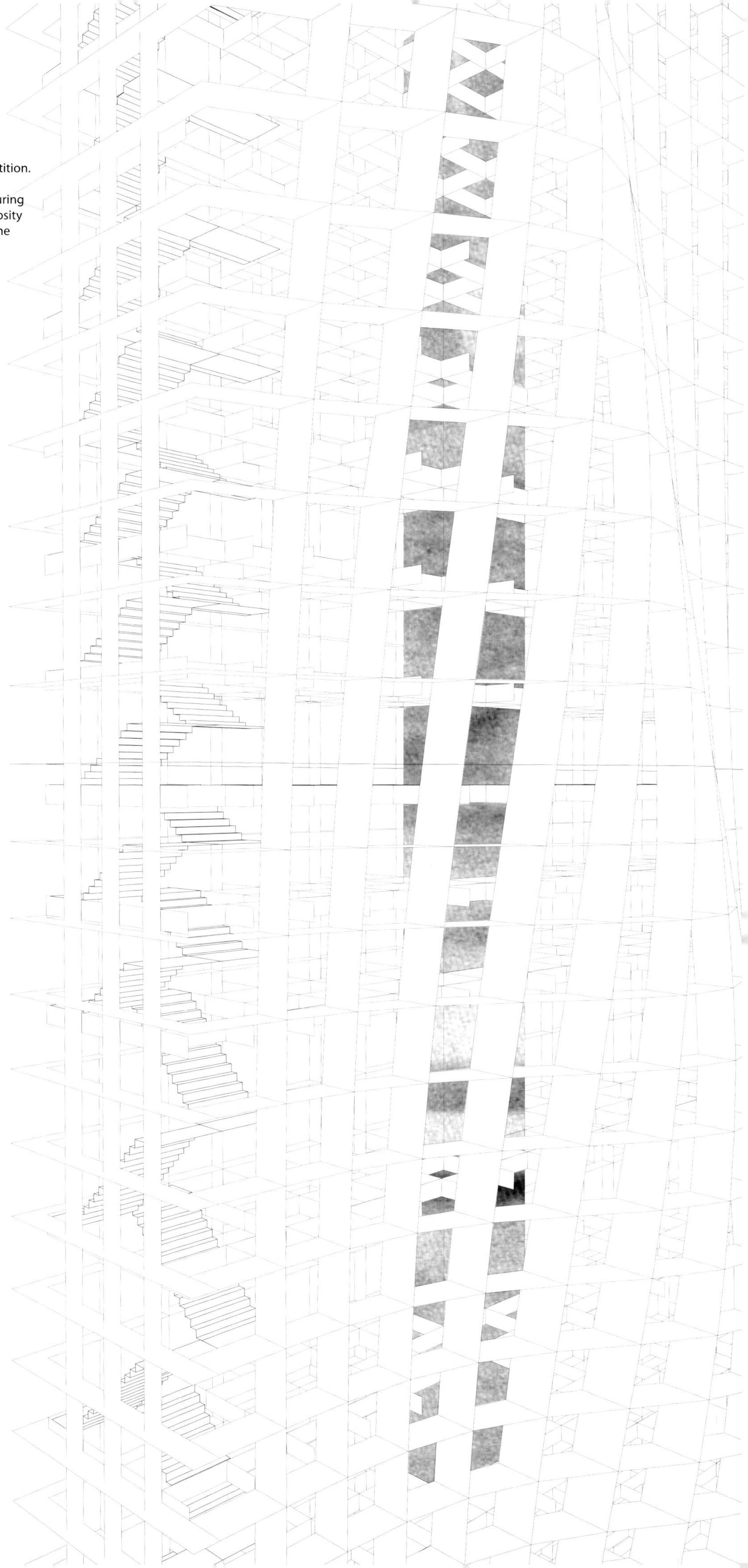

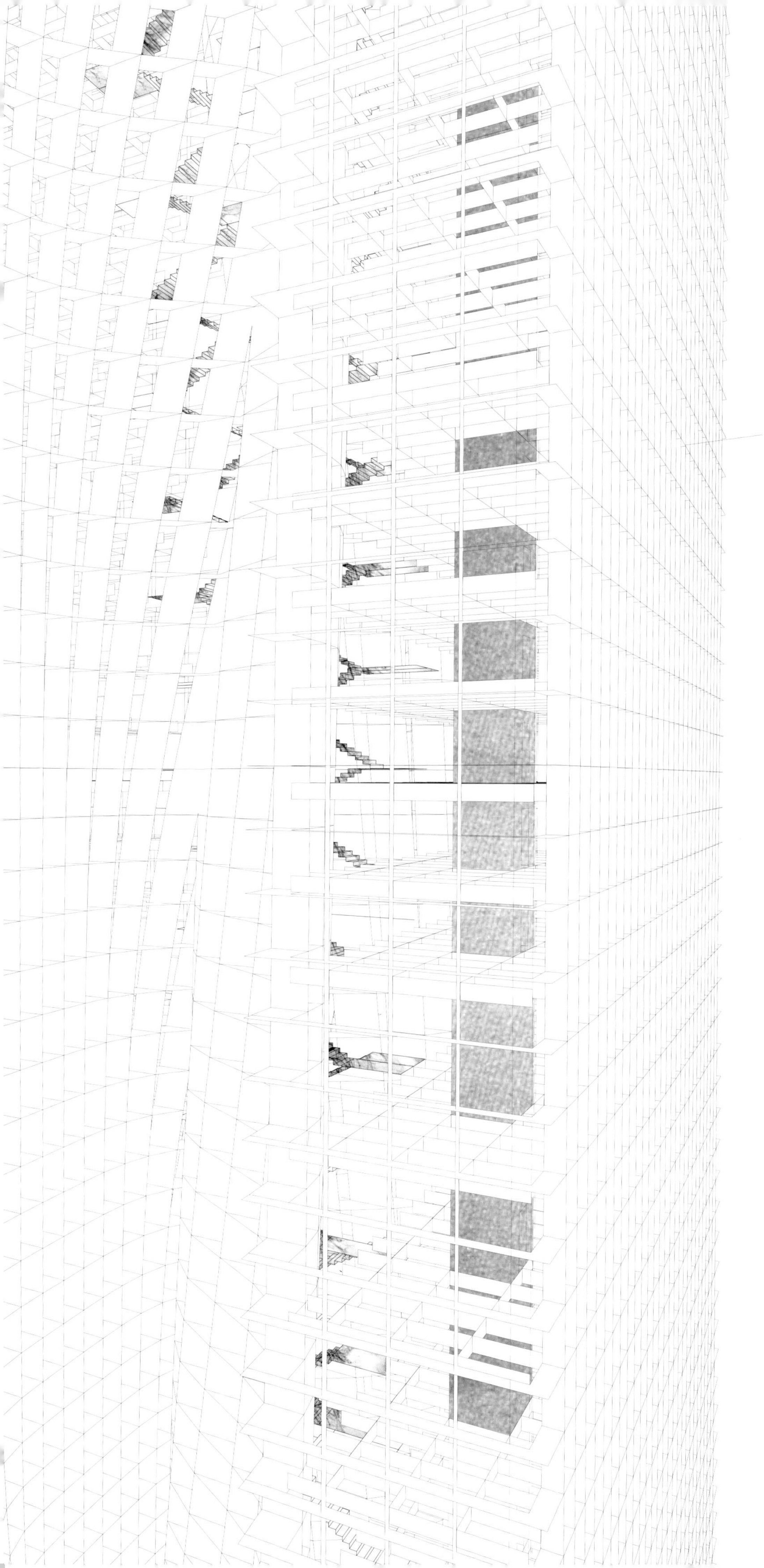

4.8 Final Views

4.9 Final Model

4.8 Final Views

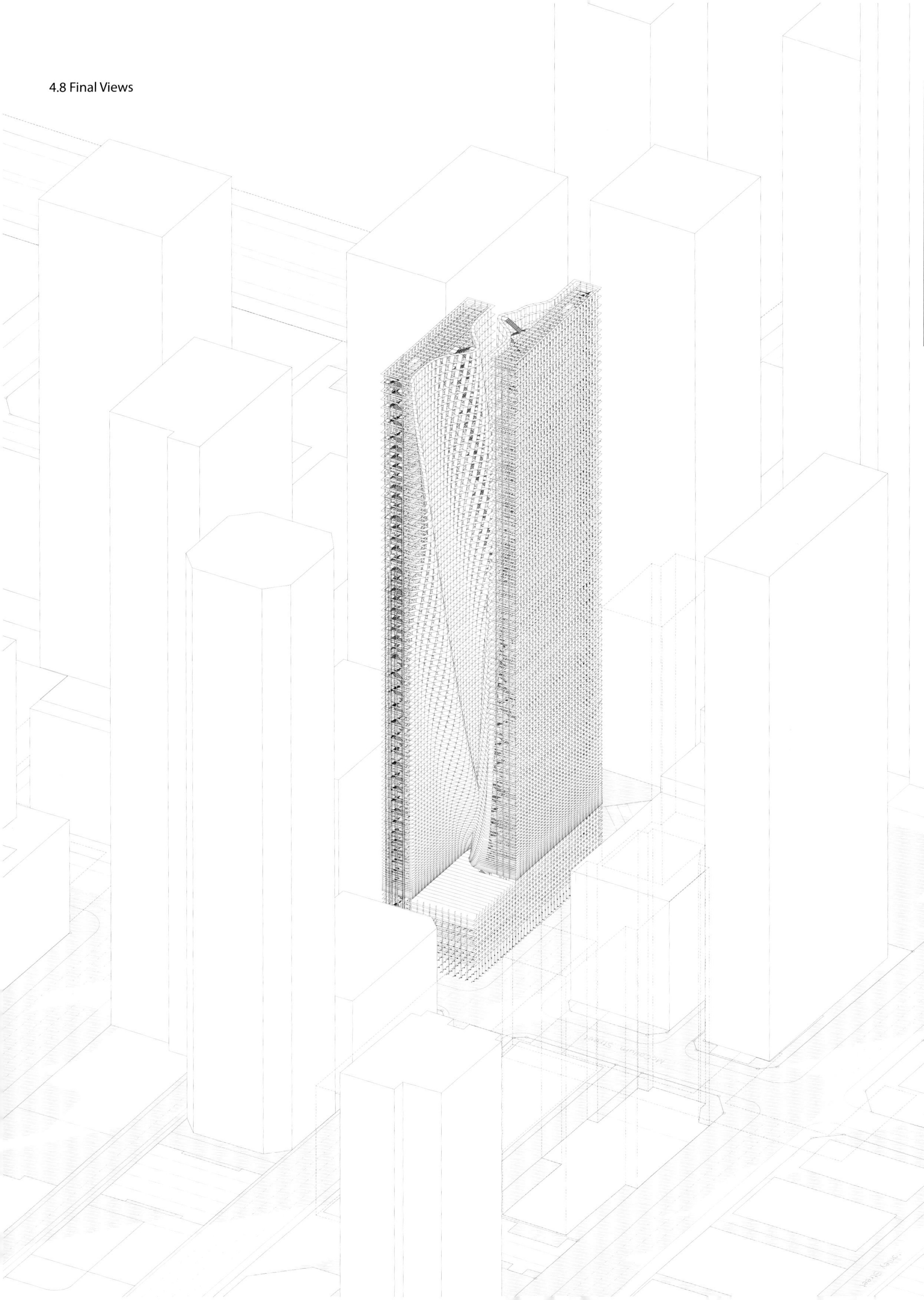

4.9 Final Model

Thanks to

Dad
George
Isabel
Pil
Erlend
Gernot & Anna
Bibi

Photos by Soenne, Aachen

The Holy City and its Discontents

Deena Fakhro
Diploma unit 6

AA Diploma Honours, 2008
Shortlisted for RIBA Silver Medal, 2008
RIBA Part II High Pass, 2008
Technical Design Thesis High Pass, 2008
Beijing Biennale "(Im)material Processes," 2008

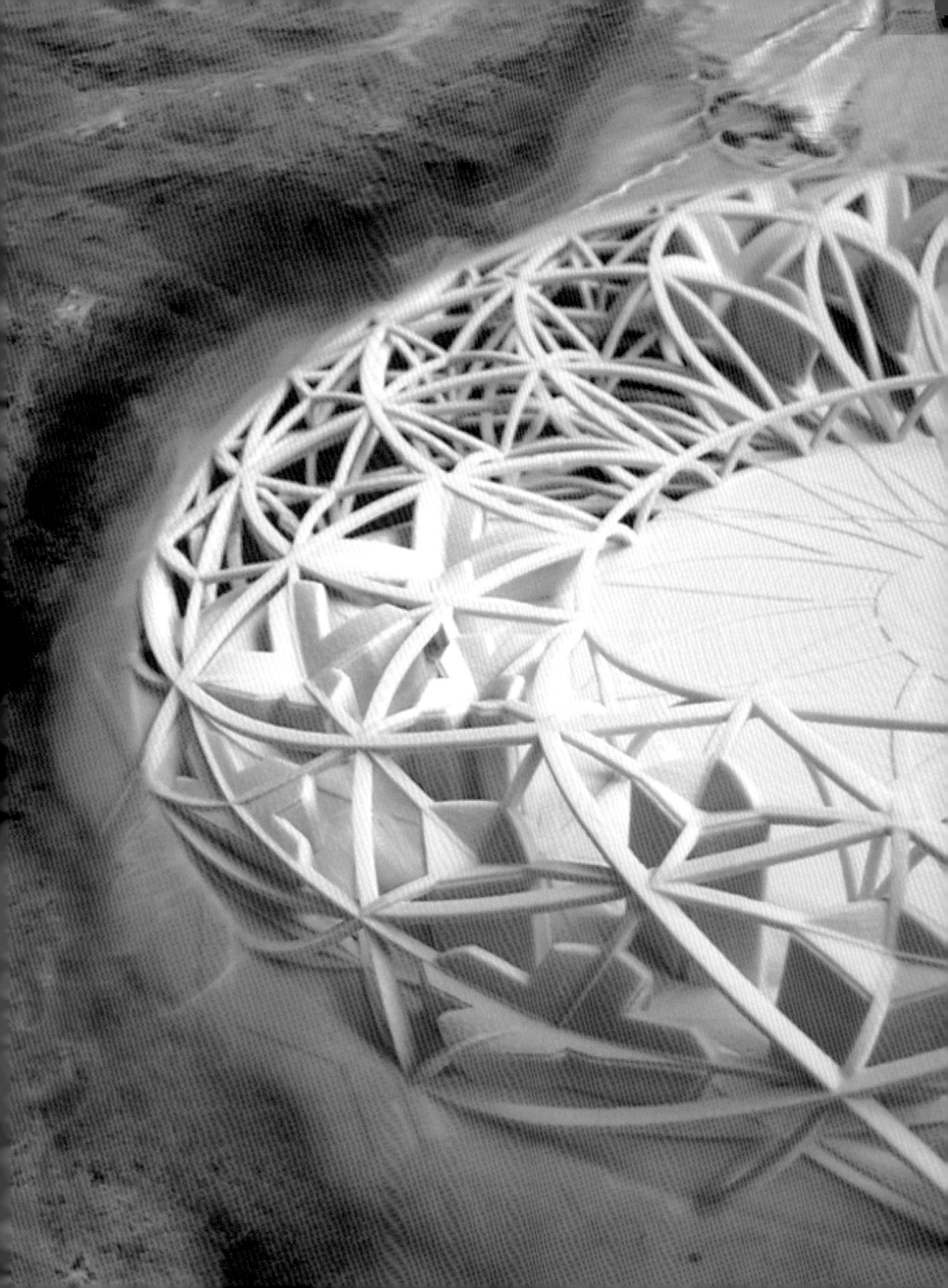

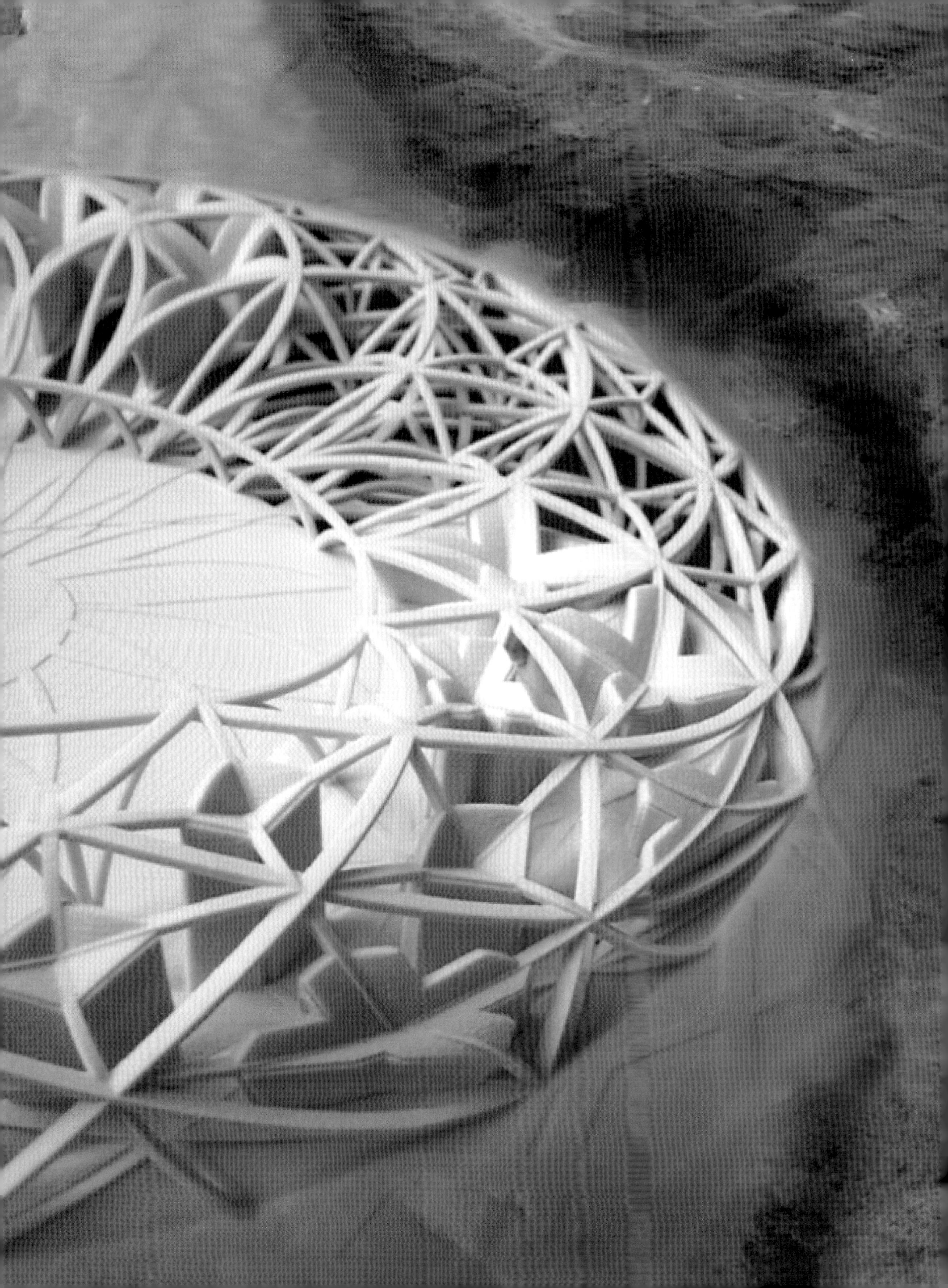

Deena Fakhro

AA Dipl. (Hons), ARB/RIBA Part II

Deena studied at the Architectural Association where she received the AA Diploma with Honours in 2008. During her studies she worked in offices in Bahrain and Chicago on several masterplanning and urban design projects worldwide.

Currently, she is working at Acme in London while running workshops for AA Diploma 6.

Table of Contents

1. Introduction

Typological Reasoning:

At the core of Diploma 6's agenda is the rethinking of the urban plan through elemental building types. Employed by generations of architects to shape the city, type is adopted as a pliant component with the ability to seed, regulate and differentiate a larger urban fabric. Through typological changes to a dominant type, alternative ideas of the city are proposed that address issues of political, economic, social and environmental renewal.

Divided into three parts or briefs, diploma 6 projects culminate in a series of typological guidelines, an urban plan and a projective idea for the city. The research-based first part is analytical and critical, where typological conflicts and contentions within a chosen site are identified. The second part is generative; based on the notion of working through typological series and precedents, it encourages experimental and evolutionary differentiation of the dominant type in reaction to the critical agenda of brief one. The third part is conjectural; while using the changed or differentiated dominant type to resolve the outlined conflicts through a resilient urban plan, it also projects an alternative idea for the city.

Aerotropolis:

Airports have grown into massive amenity-ridden and integrated transport, commercial, economic, and social hubs. No longer disdainfully cast to city-peripheries, the mammoth type now engulfs its own cities feeding off its sprawling infrastructural tentacles. As the brief for Dip 6 07-08, the emerging aerotropolis is dissected, reconfigured and reimagined as a dominant type with the potential to project new ideas for a twenty-first century city.

The Holy City and its Discontents:

The Hajj pilgrimage in the Holy City of Makkah is the world's largest gathering of people in one place, at one time, for one purpose. Centred around the sacred Kaaba, this ancient ritual is the once-in-a-lifetime religious duty for each of the world's 1.2 billion Muslims. And so, with the mass accessibility of air-travel, the annual pilgrimage has exponentially mushroomed into a colossal scale that demands infrastructural miracles of the humble and comparatively tiny Holy City. Where once a year, virtually overnight, a relentless 3 million pilgrim-tsunami momentarily floods it to breaking point. Thus, the extraordinary Hajj anomaly challenges the city and its airport in completely unparalleled ways, where despite military style planning and a strict cap to control pilgrim numbers (through a rigorous allocated pilgrim quota to each country), deadly logistical nightmares and stampedes keep recurring annually, with thousands of casualties.

Although Saudi Authorities have invested over 14 billion dollars in the past 25 years on short-term improvements to deal with the brief surges, the seemingly insurmountable task has bred a hopeless "Just-Live-With-It" attitude among them. Especially because for the majority of the year, as the Holy City stagnates during the stretched off-peak season, the fleetingly critical Hajj infrastructure remains mostly redundant and underwhelmed. The maintenance of this robust infrastructure then is a costly financial burden that constantly requires new economic generators for subsidy. Since the 1970s the Kingdom of Saudi Arabia has willingly pumped massive amounts of oil revenues into the Holy City, fulfilling its duty as custodian of the two Holy Mosques. Yet as the bleak and unsustainable future of oil is upon us, a "new era of visionary expansion" is dawning. Following the example set forth by its neighbouring Gulf States, Saudi Arabia is moving "towards a knowledge economy," facilitated by the rapid proliferation of multi-billion dollar Knowledge Economic Cities.

Makkah is now faced with a intertwined twin crisis triggered by the aviation revolution: the logistical nightmare of the Hajj and its resulting financial disaster. In light of these two crisis, the project proposes strategies on both a regional and architectural scale. As a response to the logistical bottlenecks of the Hajj, a polynodal Gateway strategy is explored that restores historic multidirectional entries into the Holy City. While on the type scale, the airport is opportunistically combined with a knowledge industry in the form of an Islamic Mosque-based University – enabling a switch between pilgrims surges and student populations that eliminates redundancy.

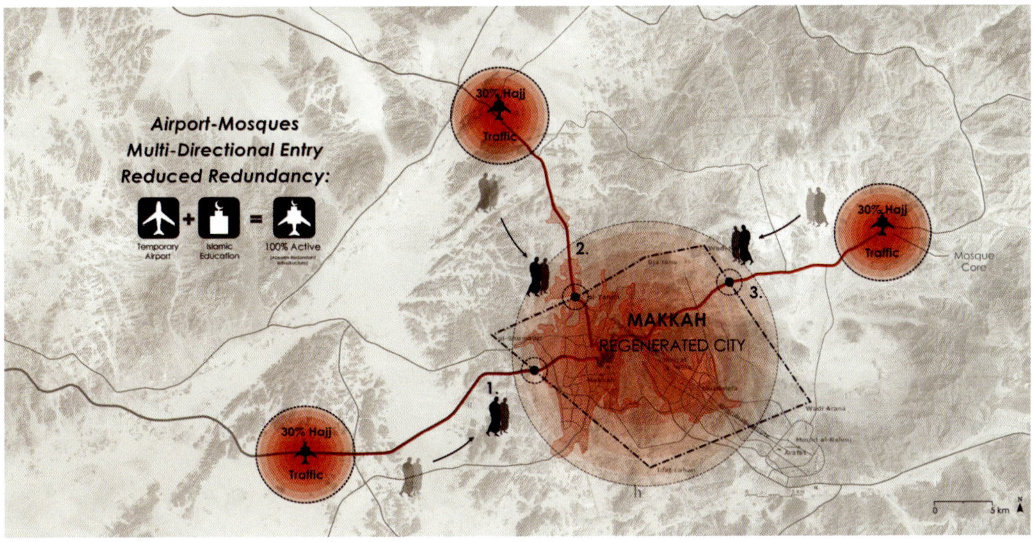

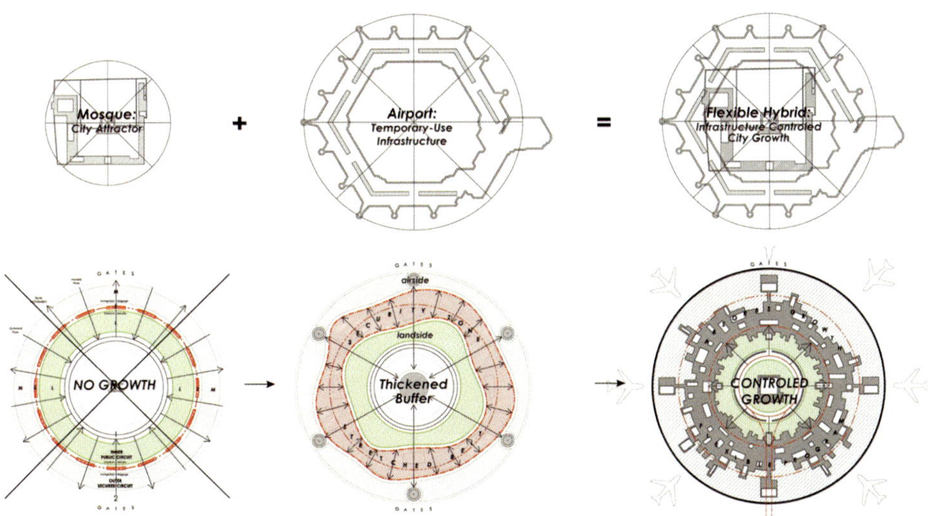

Makkah: Extreme Pilgrim Surges

Regional Strategy: Polynodal Gateways

Typological Change: Airport-Mosques

1. Introduction

An Ancient Pilgrimage Destination: Makkah Al-Mukarama

One of the five pilars of Islam, the obligatory Hajj pilgrimage in the Holy City of Makkah is centred around the Sacred Kaaba, or House of God, and has existed as a global pilgrimage destination even before Islamic times.

1. Region: Hejaz - Makkah/Madina Provinces

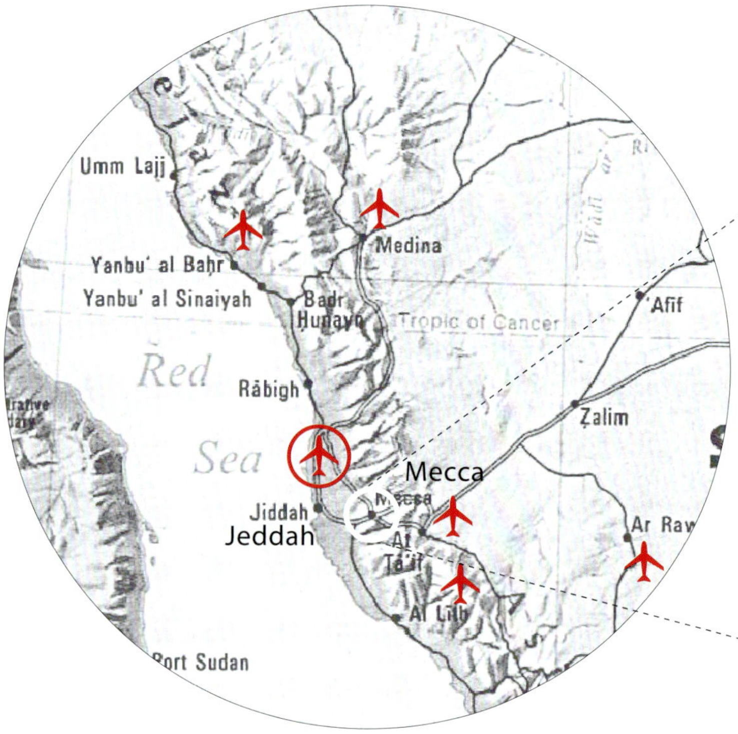

2. City: Makkah Al-Mukarramah

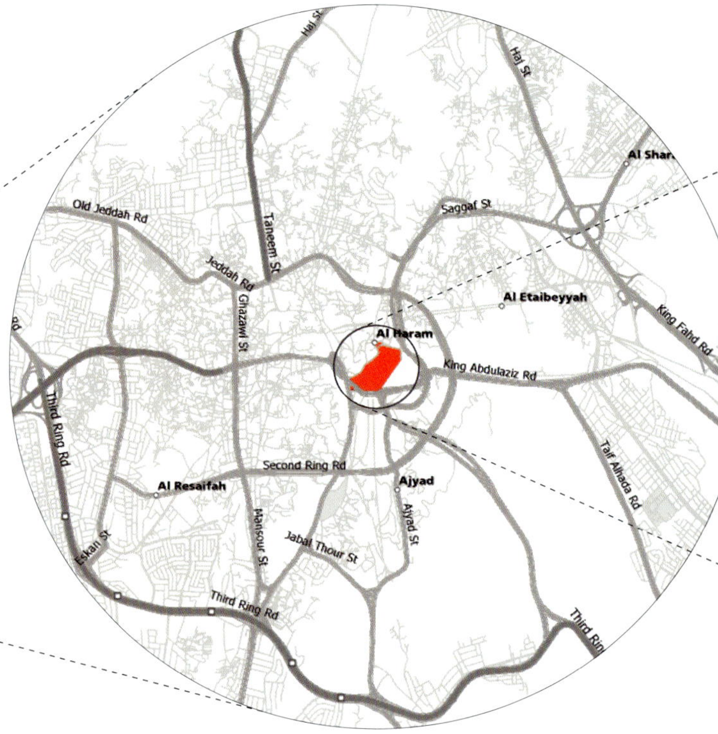

International Airports

1. King Abdul Aziz Int'l Airport

View within the Haram Mosque, with the Holy Ka'aba at the center

Building: Masjid Al-Haram

4. Ritual: The Hajj

Deena Fakhro

Aviation Revolution: Exponential Growth of Pilgrimage

No changes in transport have affected the pilgrimage to the extent that the aviation revolution has. Airports and cheap flights have not only exponentially increased Makkah's accessibility to the point that pilgrim numbers have to be strictly capped annually, but it has also cut the time of the pilgrimage to an absolute minimum. While in the pre-aviation days the pilgrimage was a time consuming ordeal one planned a whole lifetime for, today it can virtually be completed within a mere 10 days to a month. This new found mass accessibility has severely upset the harmony the Holy City had established with its pilgrimage over the past fourteen hundred years.

Umrah Pilgrims 1986-2004:

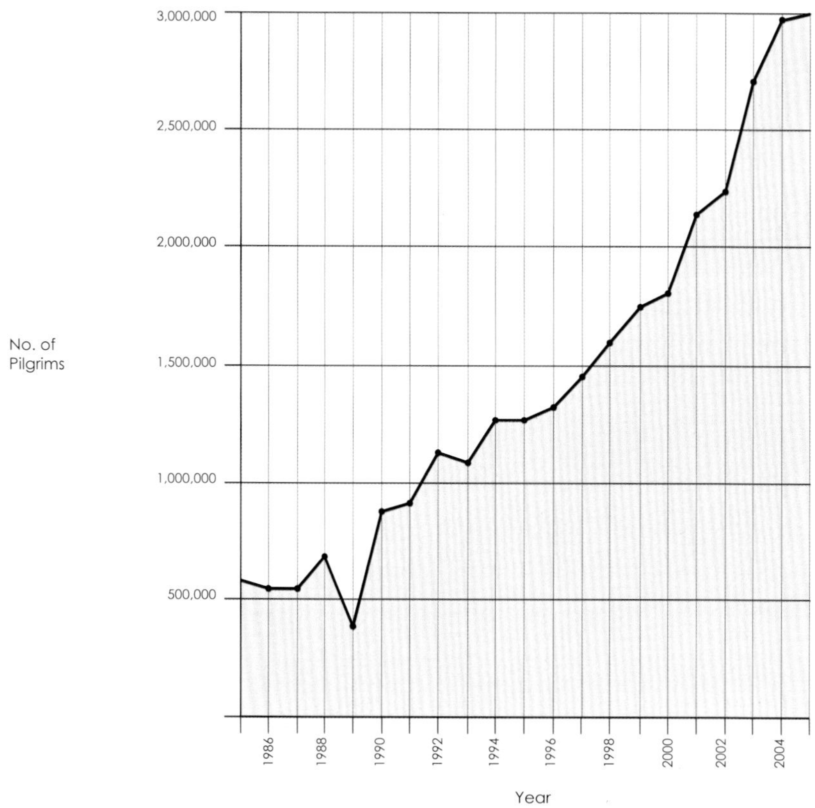

Scale Comparison:

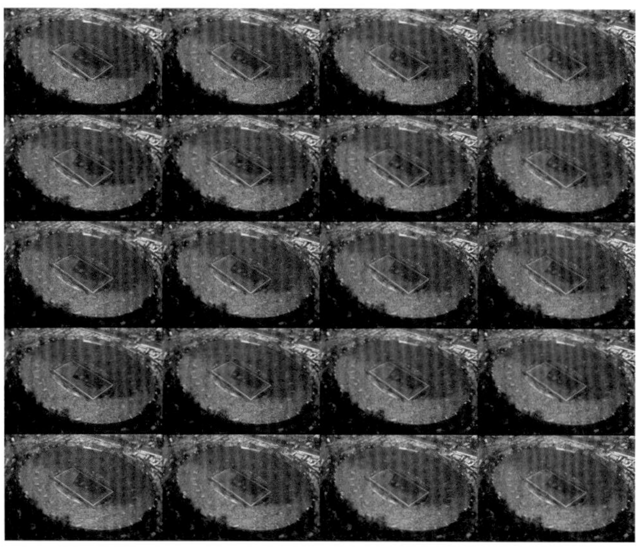

x 20 stadiums

"By comparison consider a dozen Super Bowl games getting out at the same time and place, everyone all heading in the same direction..."
- David E. Long

=

x 1 annual Pilgrimage

1 stadium = 98,000 people

1 Hajj = 2,000,000+ Pilgrims

Hajj Pilgrims 1970-2004:

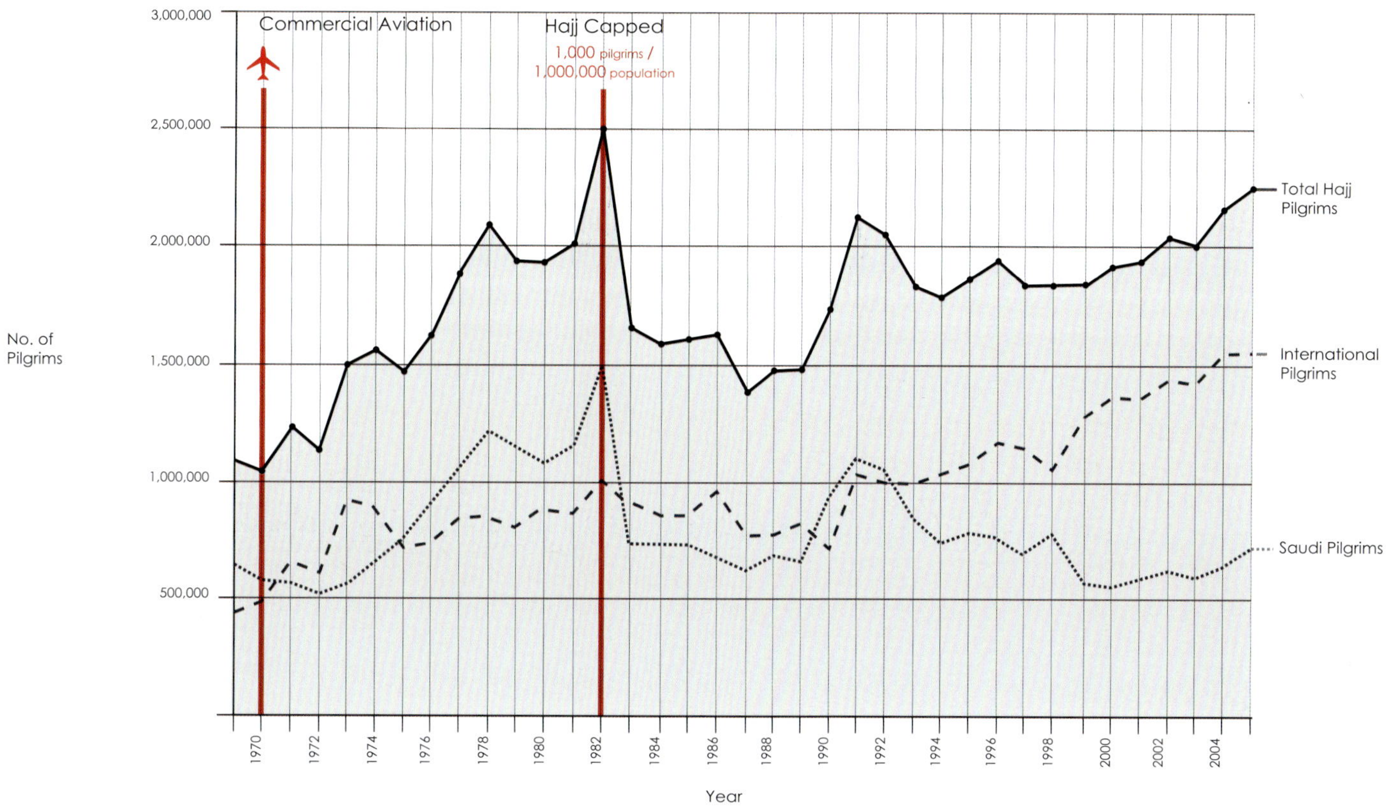

Extreme Annual Pilgrim Surge:

Logistical Chaos:

During the Hajj month, the Holy City comes to a complete standstill as pilgrims flood its streets. The Hajj's logistical problems are further worsened by the fact that the pilgrimage is composed of time-sensitive, dispersed rituals forcing 2.5 million people to do the exact same thing at the same time. The resulting extreme congestion must be well coordinated and maneuvered to avoid devastating and deadly results.

WATCH One-Minute World News

Last Updated: Thursday, 12 January 2006, 17:21 GMT

E-mail this to a friend   Printable version

Hundreds killed in Hajj stampede

At least 345 Muslim pilgrims have died in a crush during the stone-throwing ritual at the Hajj pilgrim... Arabia, officials

Hundreds of pilgri... been injured. A BI... correspondent at... Mina saw dozens... lined up on the g...

Pilgrims Squat All Around as Airport Chaos Continues

K.S. Ramkumar & Hasan Hatrash, Arab News

JEDDAH, 22 October 2007 — Chaotic conditions at the North Terminal of King Abdul Aziz International Airport which began with the rush of Eid...

Blame Game as Crunch Continues at Jeddah Airport

Ghada Aboud & Mahmoud Ahmad, Arab News

JEDDAH, 22 November 2004 — Blame for the crunch at King Abdul Aziz International Airport in recent days is flying almost as much as the jetliners that land and take off from the overcrowded facility. The problem has evoked different reactions from members of the travel trade. Instead of addressing the issues and trying to solve them, airlines, travel agencies, and Haj and Umrah companies seem to be blaming one another for their lack of organizational and communication skills.

...r waiting for airline ...ll at the North ...ziz International ...day. (AN photo by

a. Mina-Makkah Highways:

Pilgrims walking 8 km from Mina back to Makkah as part of the 5 step rituals that take place over a distance of 43 km. The streets and major highways are flooded with people, with little room for vehicular traffic.

b. Mina "Tented City":

A temporarily active city made of permanent tents and an intensely robust infrastructural system, virtually deserted 51 weeks of the year, is for 1 week a bustling city crowded with millions of pilgrims.

a. b. c. d.

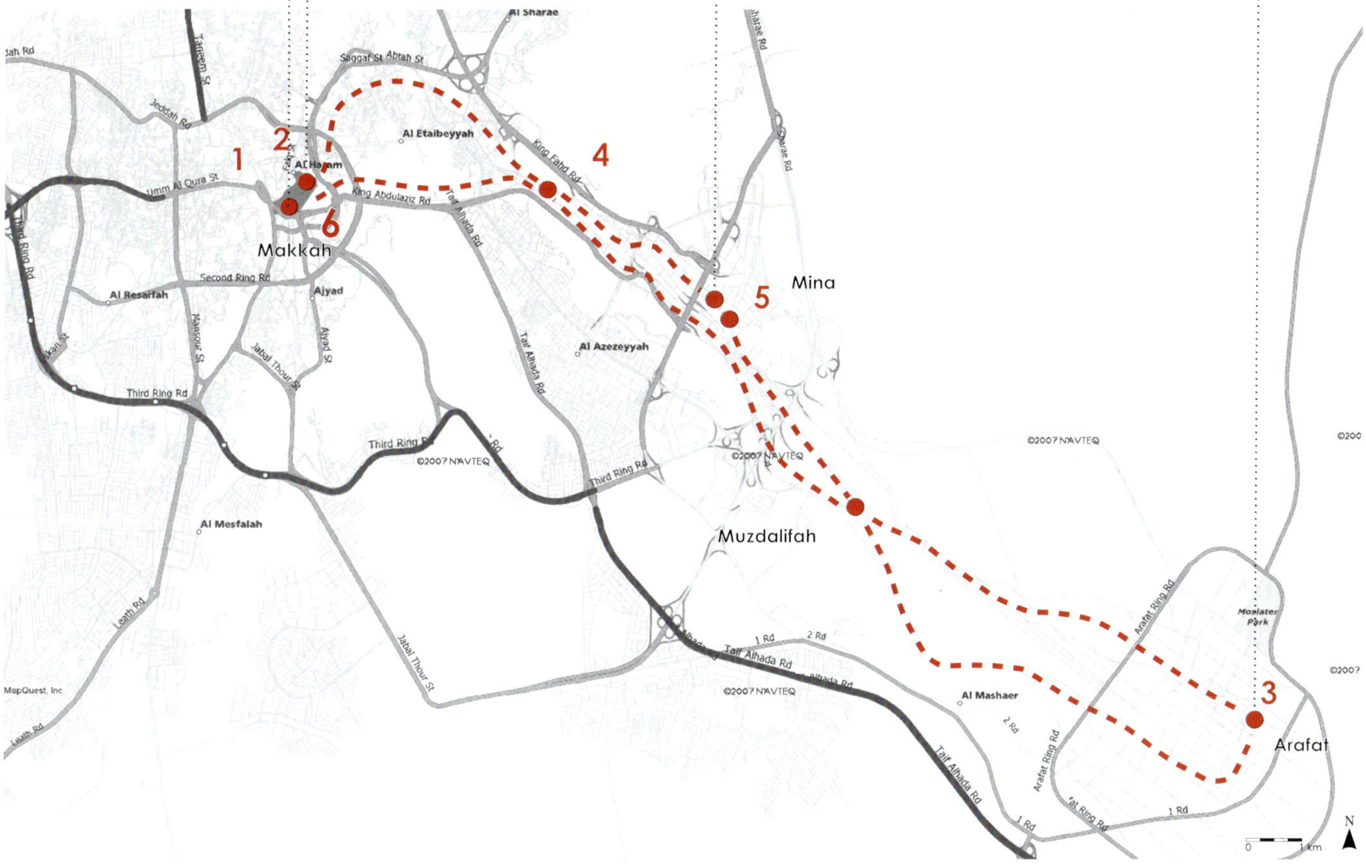

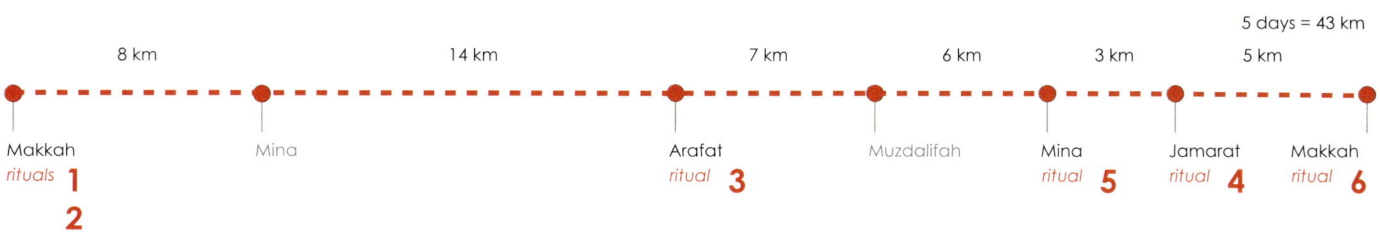

The Hajj: A Series of Dispersed Rituals

1. Tawaf: Circling the Ka'aba 7 times (a.)

2. : Saiy: Walking between Safa and Marwah hills (b.)

*Transit_Mina: Temporary tent city on the way to Arafat

3. Wukuf: Prayer and meditation on Arafat hill (c.)

*Transit_Muzdalifah: Prayer and meditation on Arafat hill

4. Jamarat: Symbolic stonning of the devil (d.)

5. Sacrifice: Of an slaughtered animal to the poor

6. Final Tawaf: Circling Kaaba before leaving Makkah

2. The Hajj Pilgrimage: Twin Crisis

2.1 A Logistical Nightmare!
2.2 The Pilgrims Are Our Crops

The Aviation Revolution:

Everyday the world's Muslims turn in unison to face Makkah in prayer. And once a lifetime, those who are able, will perform the pilgrimage to the Holy City. Makkah receives over 16,000,000 pilgrims annually, all with the desire to glorify God within the world's most sacred mosque, at its core. Historically sustaining its service-based economy, these pilgrims are vital to Makkah's survival, where any changes to transportation have critical and severe repercussions. For over two millennia the Holy City had existed in relative harmony with its pilgrimage. Yet despite wars, political unrest, and natural disasters over the centuries, nothing has permanently and irreversibly transformed the city to the extent of the aviation revolution.

A Logistical Nightmare!

Where the journey to Makkah was once long and arduous, an endeavour one meticulously planned and prepared a lifetime for, the airport has made it into an almost immediate, overnight experience. The speed, efficiency and accessibility of air travel has triggered an insatiable expansion-mode at Jeddah's Airport. Where the purpose built Hajj terminal there operates only during the six week Hajj season, attempting to process a chaotic 100,000 pilgrims per day at its peak. Buckling under this immense pressure, it was clear from the onset of the 1980s SOM-designed terminal that it had failed miserably to address the idea of surge distribution on a regional scale. Instead, the terminal's innovative naturally lit and ventilated tensile structure fatefully focuses on energy efficiency to tackle its seasonal redundancy.

Proudly boasting the world's largest roof structure, the Hajj terminal has made the fatal error of believing that bigger equals better. Receiving 90% of Hajj traffic, the terminal channels 1.6 million pilgrims within a 73 km infrastructural corridor – the Jeddah Makkah Highway that connects the terminal to Makkah's Haram core. Typically taking 40 minutes to traverse, during the Hajj the crippled highway link becomes a frustrating 8 hour deadlock. By maintaining a single, albeit massive, global entry point the terminal has secured a far less flattering title – that of the world's largest and most insufferable bottleneck.

Yet in stark contrast, traditionally pilgrims approached Makkah from multidirectional routes where the often gruelling desert-journey was intrinsic to the Hajj experience. The pilgrimage culminated when the Holy City, veiled by its protective wall of mountains, was finally revealed at gateway laden valleys. Here, long awaited and anticipated arrival was the absolute climax of the Hajj. With the airport however, the provocative power of these monumental gateway entries have been destroyed in the name of efficiency - but one that fails to even deliver.

The Pilgrims Are Our Crops

Despite the exponential growth that the jet age has brought upon the Holy City, the cruel irony is that the only condition worse than this overactivity is its subsequent underactivity. Pre-aviation, pilgrims constantly trickled in to and out of Makkah, sometimes staying for years at a time while earning a return journey home. Lacking in any natural resources or industries, the city thrived on the Hajj commerce and service economy that the pilgrims sustained, as reflected in the famous Makkan proverb: "We sow no wheat or sorghum, the pilgrims are our crops." With the speed of air travel and increasingly stringent baggage restrictions, pilgrims today are spending a fraction of the time and money during a sharply curtailed Hajj season. The stagnating Holy City is in economic decline, and growing ever more reliant on external subsidy.

The pilgrimage's seasonality, combined with the costly logistical demands of the massive surges, means that the city and its robust infrastructural system constantly oscillate between total desolation and crushing congestion – a condition financially disastrous to sustain. As Saudi Arabia enters the post-oil era, an emerging "Knowledge Economy" is forcefully being sewn to inevitably contribute towards subsidizing the Hajj. A movement sweeping the Arabian Gulf, the first multi-billion dollar "Education Cities" already operate in Doha and Dubai. Based on Western university campus models, these tabula rasa projects even adopt prestigious Western educational branding, such as Cornell and Georgetown University. Following suit, Saudi Arabia is building four economic cities within Makkah's vicinity, including Madinah Knowledge Economic City and King Abdulla University of Science and Technology. With another one for Makkah in planning, this regenerative project will undoubtedly introduce new infrastructural crisis while completely ignoring the existing burden of redundant hajj infrastructure.

So what is the alternative? Returning to culturally rooted precedents, the Islamic University is a far less invasive model. As the world's oldest formal universities, the archetypal Al Azhar University (Cairo, Egypt) and University of Al Karaouine (Fez, Morocco) grew through a natural and gradual accumulation of individual schools around a vital mosque core. Unlike the imported Education City model, the flexible mosque, able to absorb varied programs from schools to markets to political platforms, has persisted as the Islamic Community's most effective social and economic generator.

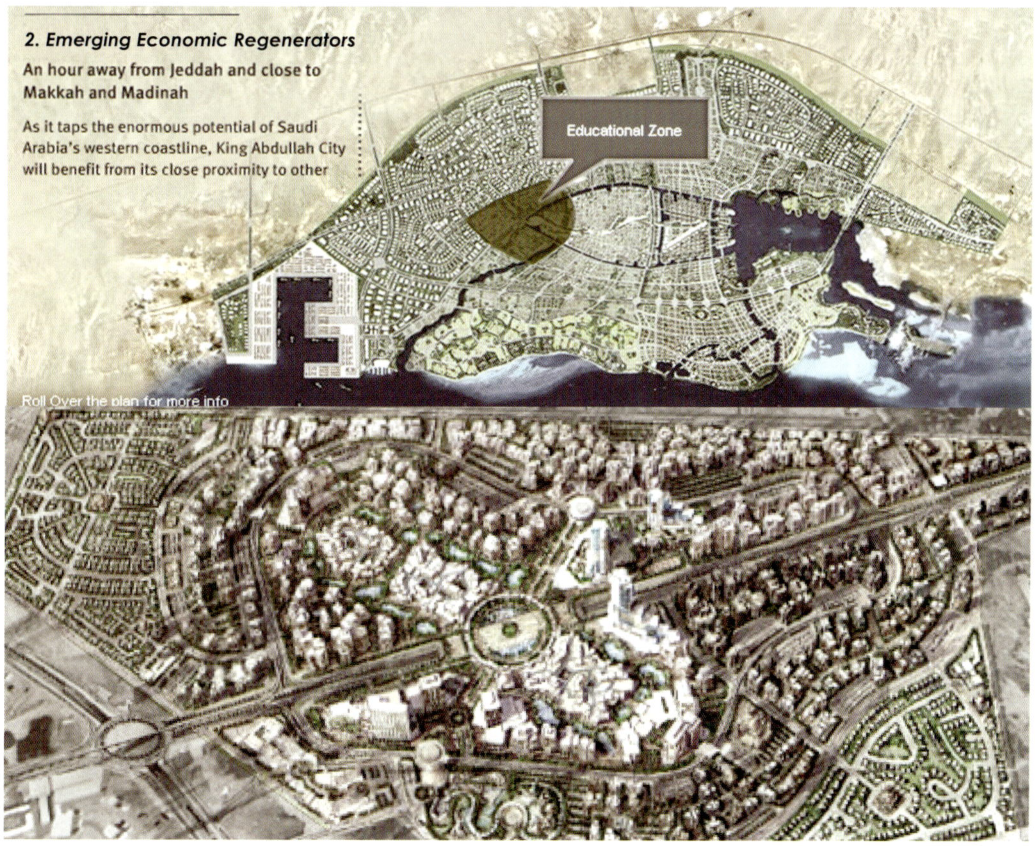

Temporarily Active Hajj Terminal

Proposed Knowledge Economic Cities

2.1 A Logistical Nightmare!

Current Single Corridor vs Historic Gateways:

Hisorically Hajj pilgrims slowly trickled into Makkah from multidirectional routes, on foot, camel or by sea. These ancient routes were also lined with multifunctional stations and outposts, that prepared year round for pilgrims while also becoming Hajj-independent markets and farms.

In total contrast, today's Hajj Terminal Airport is both singular and monoprogrammatic. By maintaining one entry in to Makkah, receiving 90% of Hajj traffic, it has created a massive bottleneck. By focusing on passive energy efficiency to tackle its off-peak season, the stagnating terminal misses an opportunity to take advantage of the redundant infrastructure.

Current: Single Entry and Temporary Monoprogrammatic Terminal

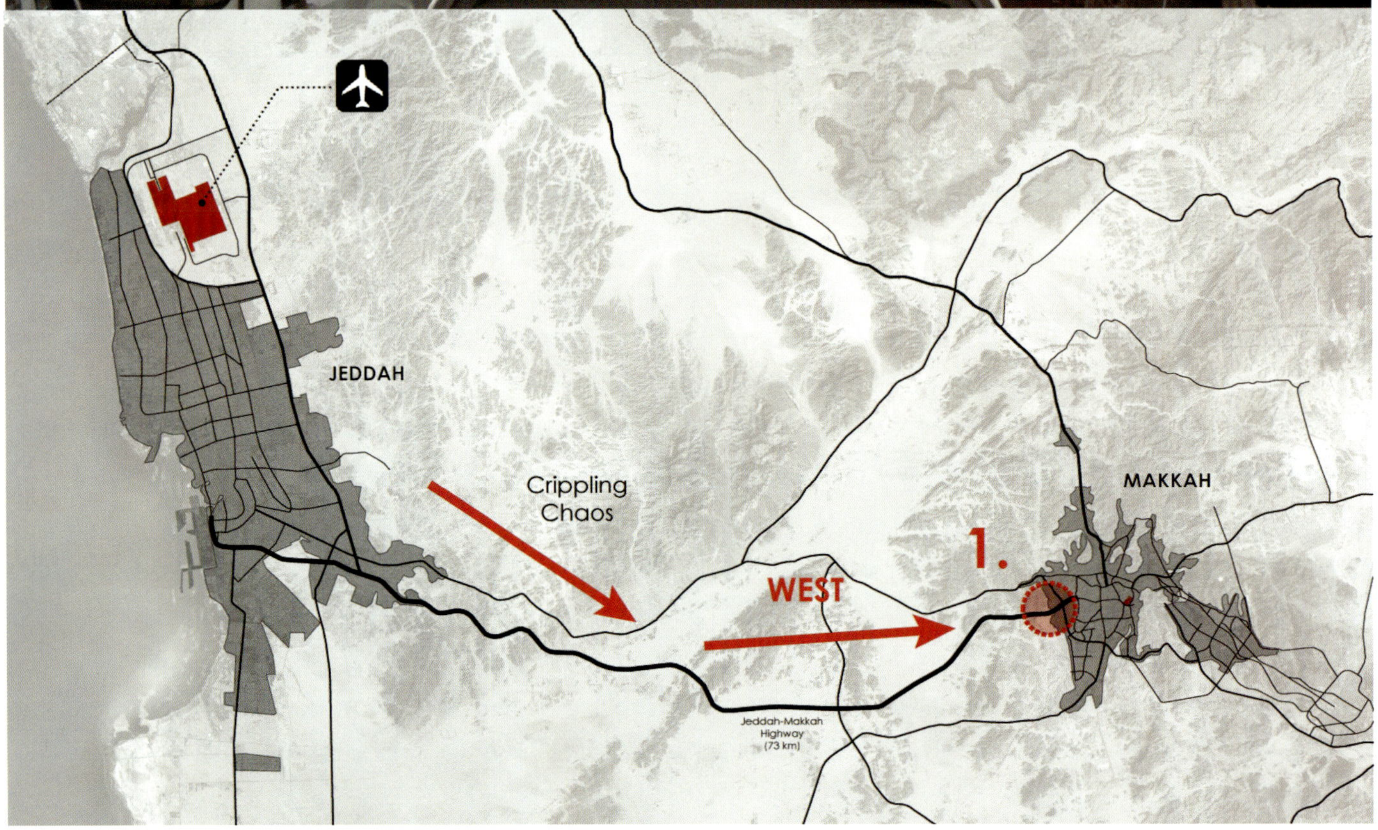

Current Hajj Terminal:
Failure to Address Surge Distribution

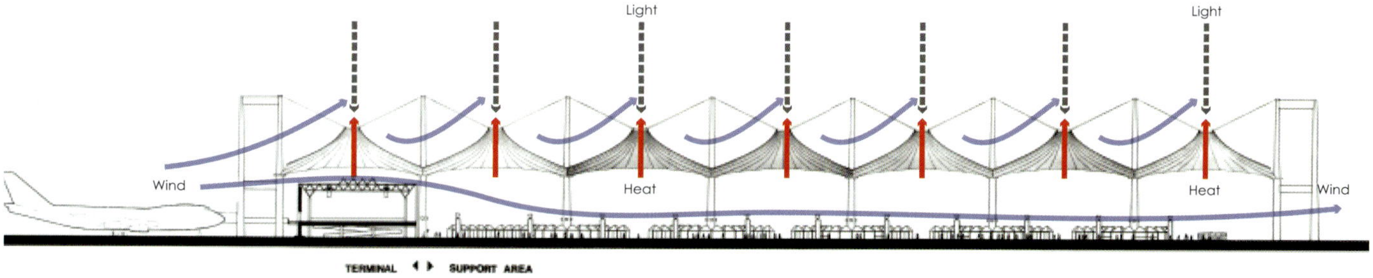

Section: Natural Ventilation and Solar Radiation

Historic: Multidirectional Entries + Multifunctional Stations

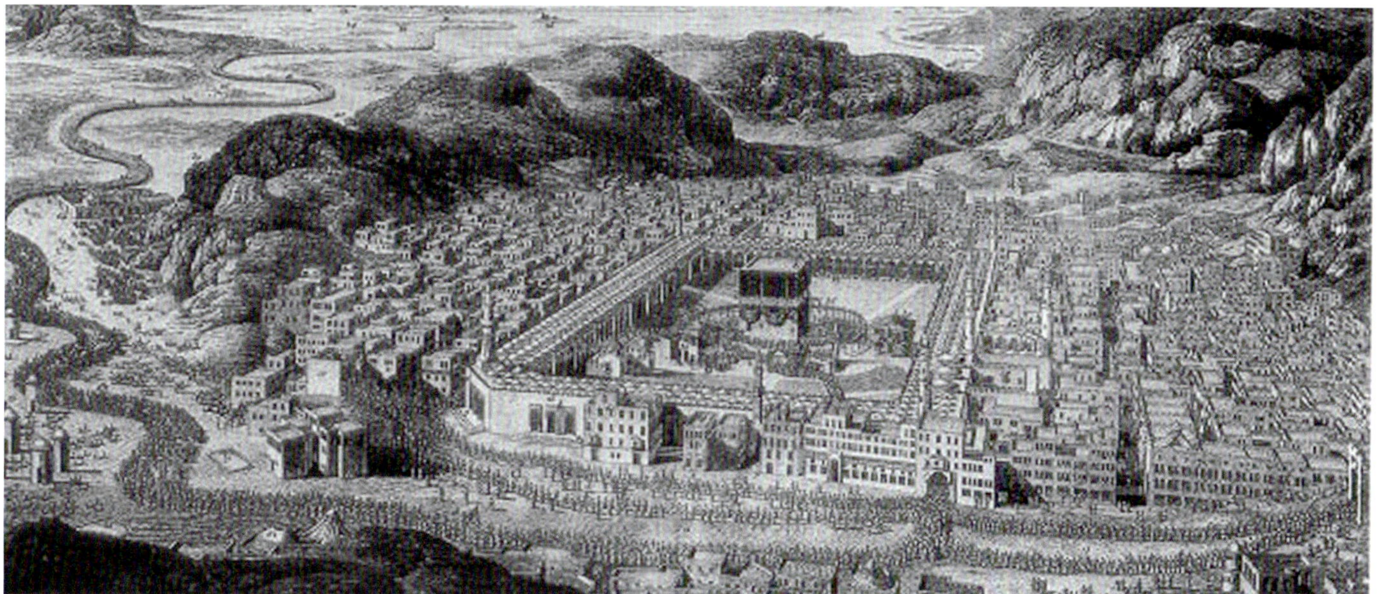

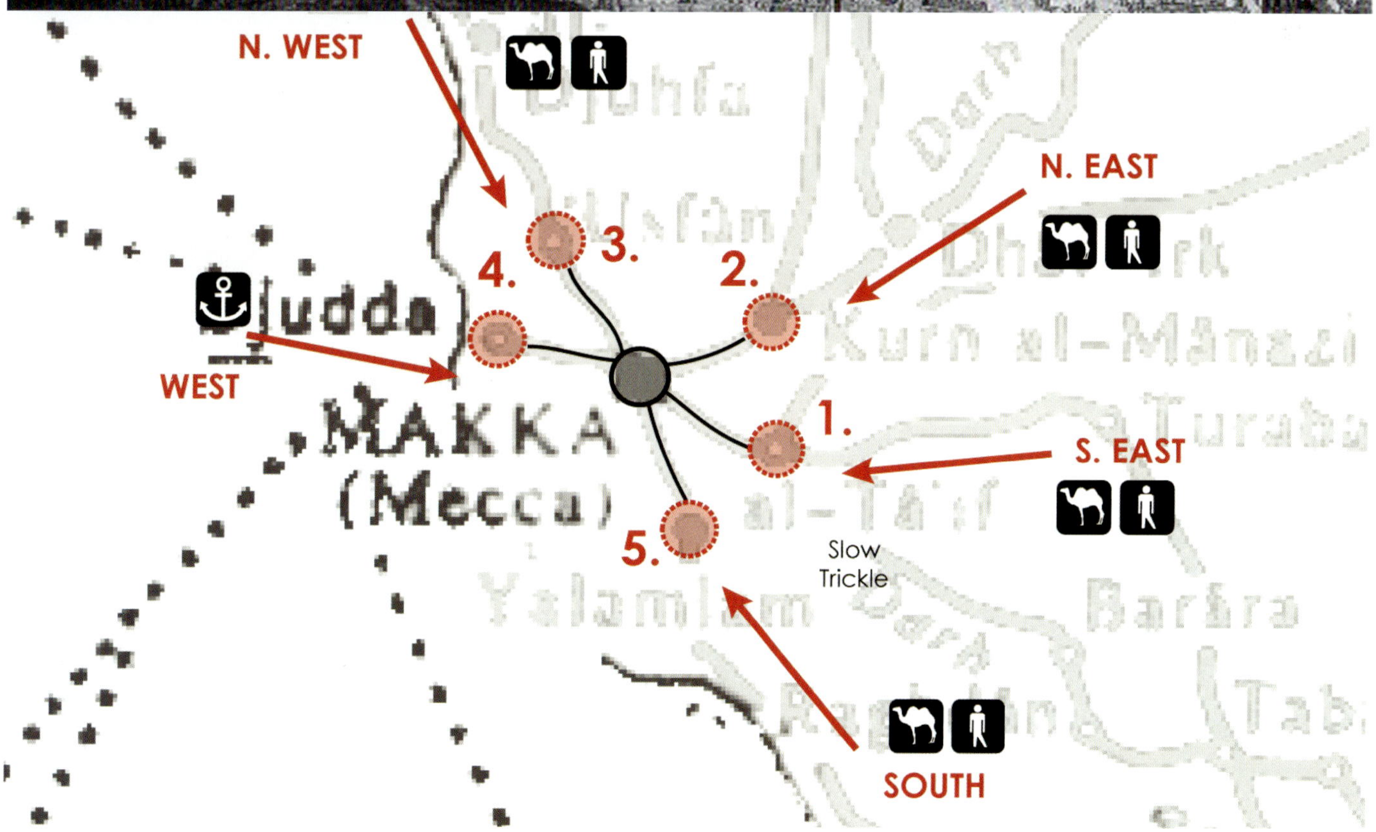

2.2 The Pilgrims Are Our Crops

Extreme Surges: A Financial Disaster

Despite the logistical chaos of the Hajj, ironically the only condition worse than overactivity is underactivity. The seasonal pilgrimage means that infrastructure oscillates between total desolation and crushing congestion, a financial disaster to maintain. As the current subsizing oil revenues are becoming unsustainable, a new knowledge economy is forcefully being sewn through the rapid proliferation of Western-style Education Cities. Four of the multi-billion dollar, tabula rasa projects are already underway within Makkah's vicinity.

"We Sow No Wheat, the Pilgrims Are Our Crops"
- Makkan Proverb

Survey Reveals Alarming Facts of Life in Makkah
Maha Akeel, Arab News

JEDDAH, 22 March 2005 — A new study on the social and economic characteristics of Makkah residents reveals some surprising and alarming facts of life in the holy city.

30% Unemployed
92% in Service Sector

40% High School Graduates
30% Homes No Sewage
25% Homes No Municipal Water

↓

"Towards a Knowledge Economy"
- Jeddah, 2008

A NEW ERA OF VISIONARY EXPANSION

Oil revenues are facilitating several mega projects with a combined value of one trillion riyals

1st Annual Conference of the Arabian Knowledge Economy Association
In collaboration with Al-Aghar Strategic Think Tank Group and Jeddah Chamber of Commerce and Industry
"Towards a Knowledge Economy"
12-13 January 2008, Jeddah Hilton, Saudi Arabia

Saudi Arabia: Knowledge Economic Cities

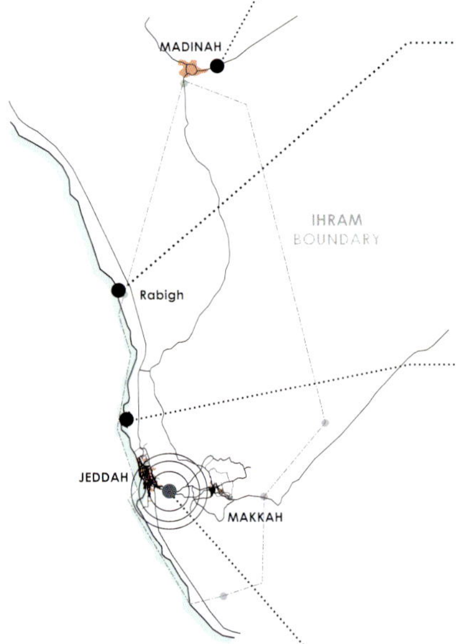

1. MKEC: Madinah Knowledge Economic City

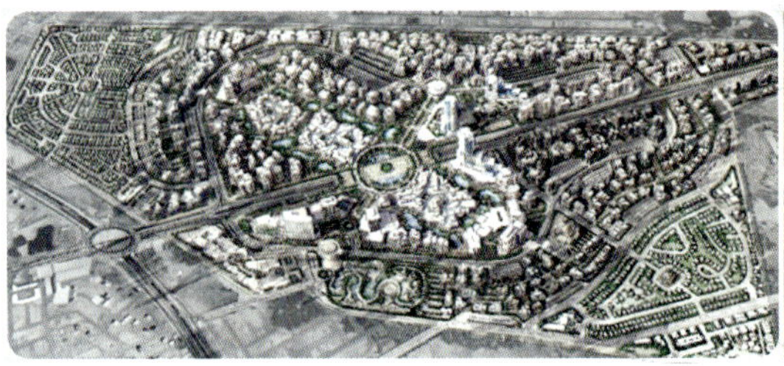

2. KAEC: King Abdulla Economic City

3. KAUST: King Abdulla University of Science and Technology

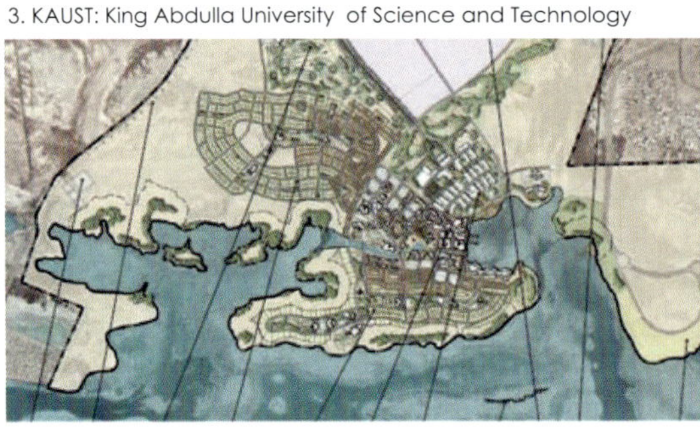

4. **Plans Under Way for SR10 Billion Makkah Knowledge Economic City**

Be the first to comment on this article NEW!

21 August 2006
JEDDAH -- Plans are under way to establish a new economic city in Makkah at a cost of SR10 billion. The Um Al-Qura Economic City project, which has been presented to Custodian of the Two Holy Mosques King Abdullah for his approval, is expected to create 15,000 new jobs, informed sources said.

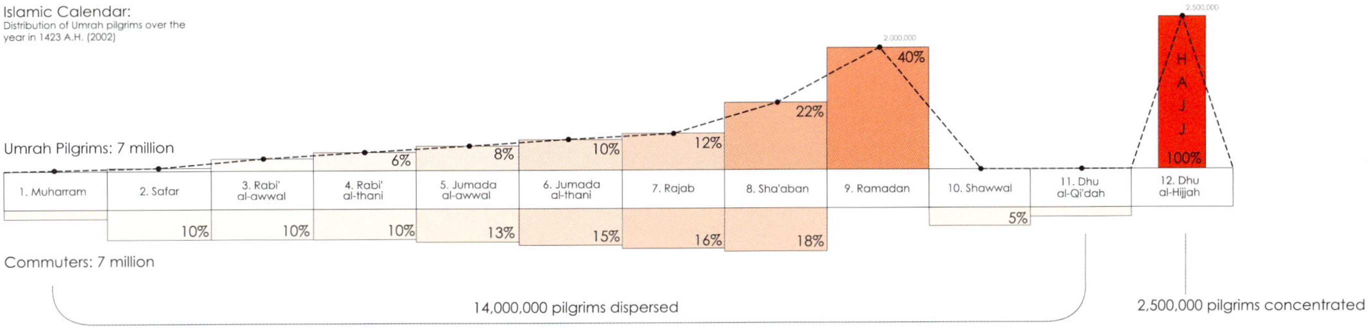
Islamic Calendar:
Distribution of Umrah pilgrims over the year in 1423 A.H. (2002)
Umrah Pilgrims: 7 million
Commuters: 7 million
1. Muharram
2. Safar
3. Rabi' al-awwal
4. Rabi' al-thani
5. Jumada al-awwal
6. Jumada al-thani
7. Rajab
8. Sha'aban
9. Ramadan
10. Shawwal
11. Dhu al-Qi'dah
12. Dhu al-Hijjah
6%
8%
10%
12%
22%
40%
2,000,000
HAJJ
100%
2,500,000
10%
10%
10%
13%
15%
16%
18%
5%
14,000,000 pilgrims dispersed
2,500,000 pilgrims concentrated

OFFPEAK:
Total Desolation
HAJJ PEAK:
Crushing Congestion

The Mosque as a Resilient Type:

The imported "Education City" type is based on a tabula rasa campus masterplan composed of a patchwork of monoprogrammatic zones or districts, such as a commercial core, science and technology parks and residential neighbourhoods.

The Islamic University however, always grows directly around a central mosque and its courtyard, which generally form city centers and are the primary public spaces of an Islamic Community. A resilient and lasting type, the uniquely flexible and magnetic mosque has persisted as an effective social and economic generator.

Not only is adopting an Islamic mosque-based university for Islam's Holiest City more symbolically appropriate, but it also creates a favourable and competitive regional difference to the growing number of Education Cities in the Middle East.

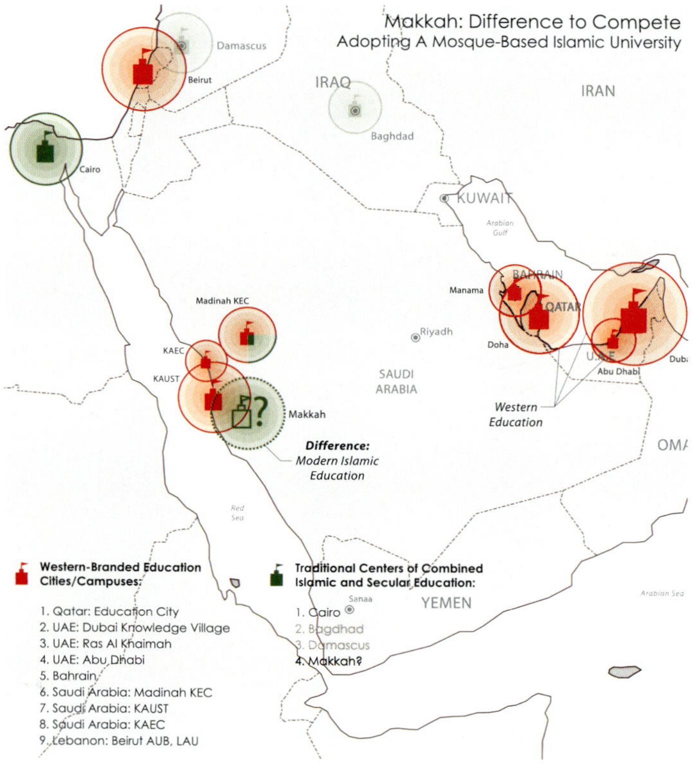

Knowledge City vs Traditional Islamic University

Madinah Knowledge Economic City: 2006 AD

Al Azhar University (Relgious and Secular), Cairo: 988 AD

Masterplanned Campus

Monoprogrammatic, Zoned Districts
Educational Cores require supplementary programs

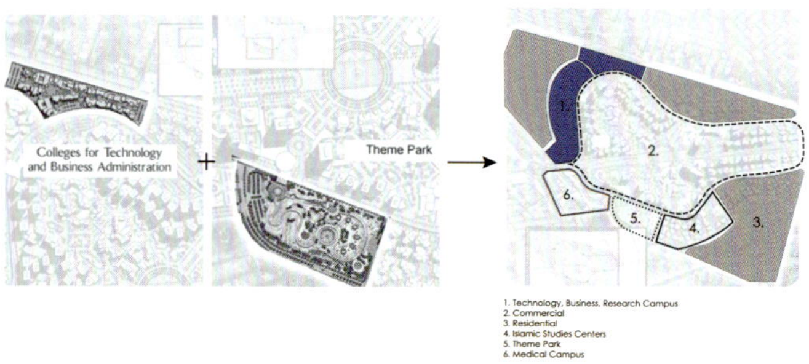

Organic Growth From Mosque Core

Mosque = Adaptable Dominant Type
Organization + Structure Allow Absorption of Various Programs (reflected in schools and housing)

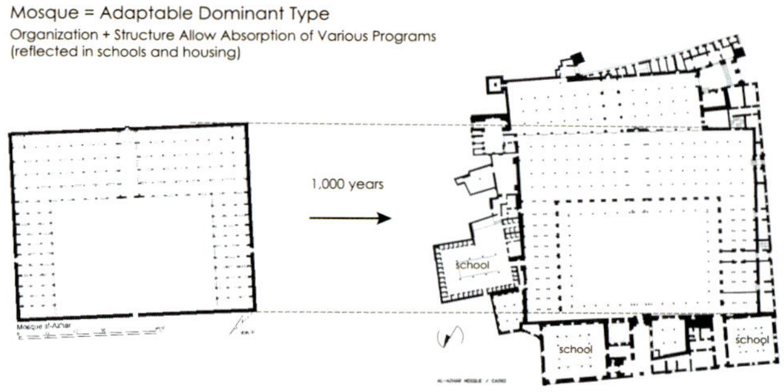

a. Original Mosque
(970)

b. Expansion around Mosque core for schools and libraries
(988-1900s)

A Resilient Type: Mosque

1. Flexible Space:
Absorb Surges and Varying Programs

2. Attractor:
Magnet for Critical Mass

3. Strategy: Aviation Outposts

3.1 Combined Types: Airport-Mosques
3.2 Type Analysis
3.3 Typological Change: Airport as Walled City

Aviation Outposts: The Airport-Mosque

In response to the twin crisis facing Makkah, the project proposes twin strategies. The first disperses the airport. The second integrates airport with knowledge industry in the form of a mosque-based Islamic University.

The currently singular bottleneck airport and its haemorrhaging infrastructural corridor will be abandoned – to be replaced by a series of dispersed, multidirectional airports. Distributing the extreme pilgrim surges, they restore the historic gateways and pilgrim outposts of Makkah. Lying outside of Makkah's holy boundary, they occupy the three main entries to the city, the western (towards Jeddah), northern (towards Madinah) and eastern (towards Taif). As smaller transport nodes, they break down the mammoth airport scale required to process the Hajj pilgrims, reducing redundant infrastructure into more manageable and even reprogrammable chunks. While the current Hajj Terminal lies inactive for the majority of the year, the proposed dispersed airport nodes are instead opportunistically combined with new integrated education city in the form of Islamic Mosque-Universities: Airport-Mosques. The airports are then activated throughout the year by a programmatic switch that accommodates both the fleeting pilgrim surges and the more permanent student populations - eliminating the airport's off-peak burden while generating new forms of growth to rehabilitate a stagnating Holy City in the era of aviation. Finally, the Airport-mosques disperse Makkah's centralised growth within new limited nodes that crucially resist over-congestion and over-development.

Type Analysis:

So how to combine two seemingly opposite types: the dense, courtyard based mosque with the linear and sprawling airport? Taking the 1970's circular airport of Berlin Tegel, which unlike today's monolithic examples was deliberately growth limiting, a suitable non-sprawling airport model needed by Makkah is found. However, the juxtaposition of the two types raises a number of design and performance issues. The Islamic university is based on a gradual aggregation of programs around a magnetic mosque core that is engrained with a clear linear orientation to Makkah. Following a rigorous structural and geometric grid, the mosque establishes a clear spatial hierarchy. The central public courtyard and mosque are the most accessible while the peripheries, containing student cells, are the most private.

In contrast, the non-directional circular airport is purely about a democratic efficiency of flows between two boundaries – the landside and airside, which are separated by a security buffer. Originating from a time when security was less of a problem, Berlin Tegel's optimally compressed and over-efficient security line (with a threshold of less than 3 meters separating secured and unsecured areas) has made it vulnerable today and lead to its proposed replacement.

Airport as Islamic Walled City:

Yet, the security buffer is the organisational key to marrying mosque to airport. The growth-limiting but over-compressed airport is no longer viable in today's high security airport culture. But, by inserting a mosque core and stretching this buffer, the Tegel diagram is reversed and an excessively thickened security zone emerges where educational programmes can aggregate within a new internal pocket. What used to be a problematic line, is now a volumetric and programmatic ring: a campus, mediating between traditional land and airside functions. Airport then, becomes analogous to Islamic walled city, which established a central mosque and a bounding wall – allowing the city to eventually grow within these two critical elements. The mosque, as the public and political heart of the community, acts as the attractor for growth, while the bounding wall is its regulator.

Unlike today's massive and ever-expanding airport types, Berlin Tegel was based on a highly controlled system of limited growth nodes – where each airport had a fixed capacity that deliberately opposed expansion to prevent bottlenecks and the logistical mayhem seen in ubiquitous sprawling airports such as London Heathrow. Adopting this strategy for Makkah, the aerotropolis' cancerous sprawl is eradicated, replaced by a controlled, polynodal growth. With the fixed cap on Hajj pilgrim numbers, the three airport nodes proposed cater to the current demand. However, if the Hajj grows above its 3 million limit, then to accommodate that growth new airport nodes can be created at different gateway sites into Makkah to further decentralize the surges (historically there were 6 main entries surrounding the Holy City). Hence, Makkah's currently escalating centralized congestion is resisted. And instead, a long-term functional, dispersed infrastructure and diversified growth established.

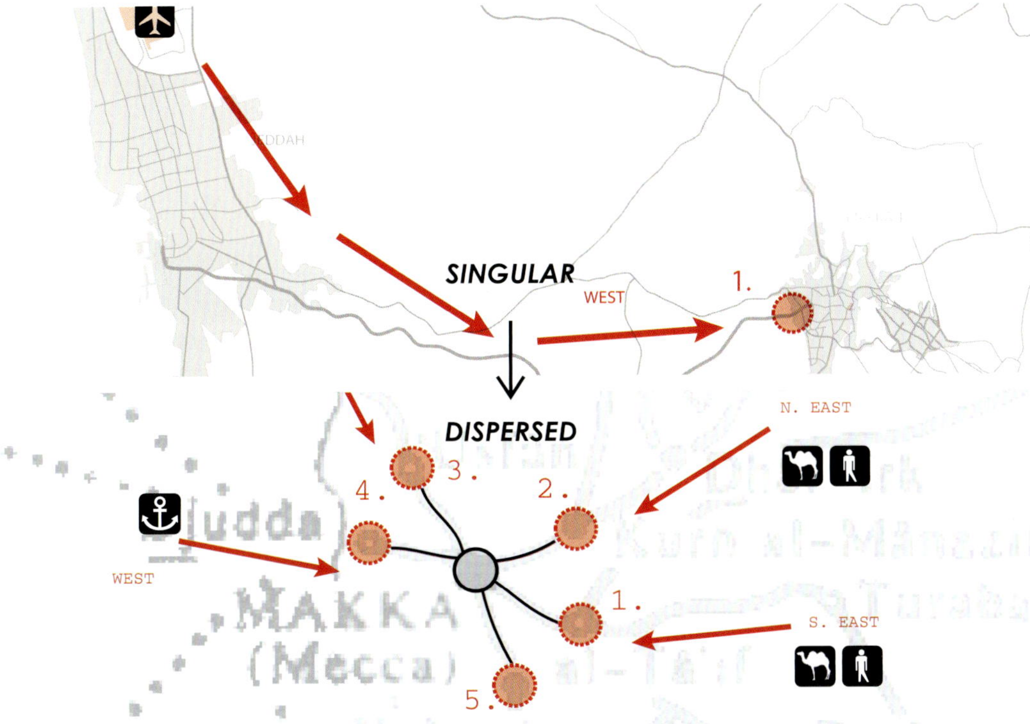

Deena Fakhro

Regional Strategy: Polynodal Gateway Airports

Type Analysis: Mosque vs Airport

3.1 Combined Types: Airport-Mosques

Strategy:

The project investigates the opportunity rather than the burden of redundancy through integrating the airport with a new knowledge industry centred around a flexible mosque.

The singular haemorrhaging airport becomes multi-directional, restoring the gateways, and dispersing congestion. Furthermore, the currently redundant airport is opportunistically combined with a new Islamic Mosque-University: an airport-mosque, switching between pilgrim surges and student populations.

Off-Peak: Programmatic Switch
From Airport to Mosque and University

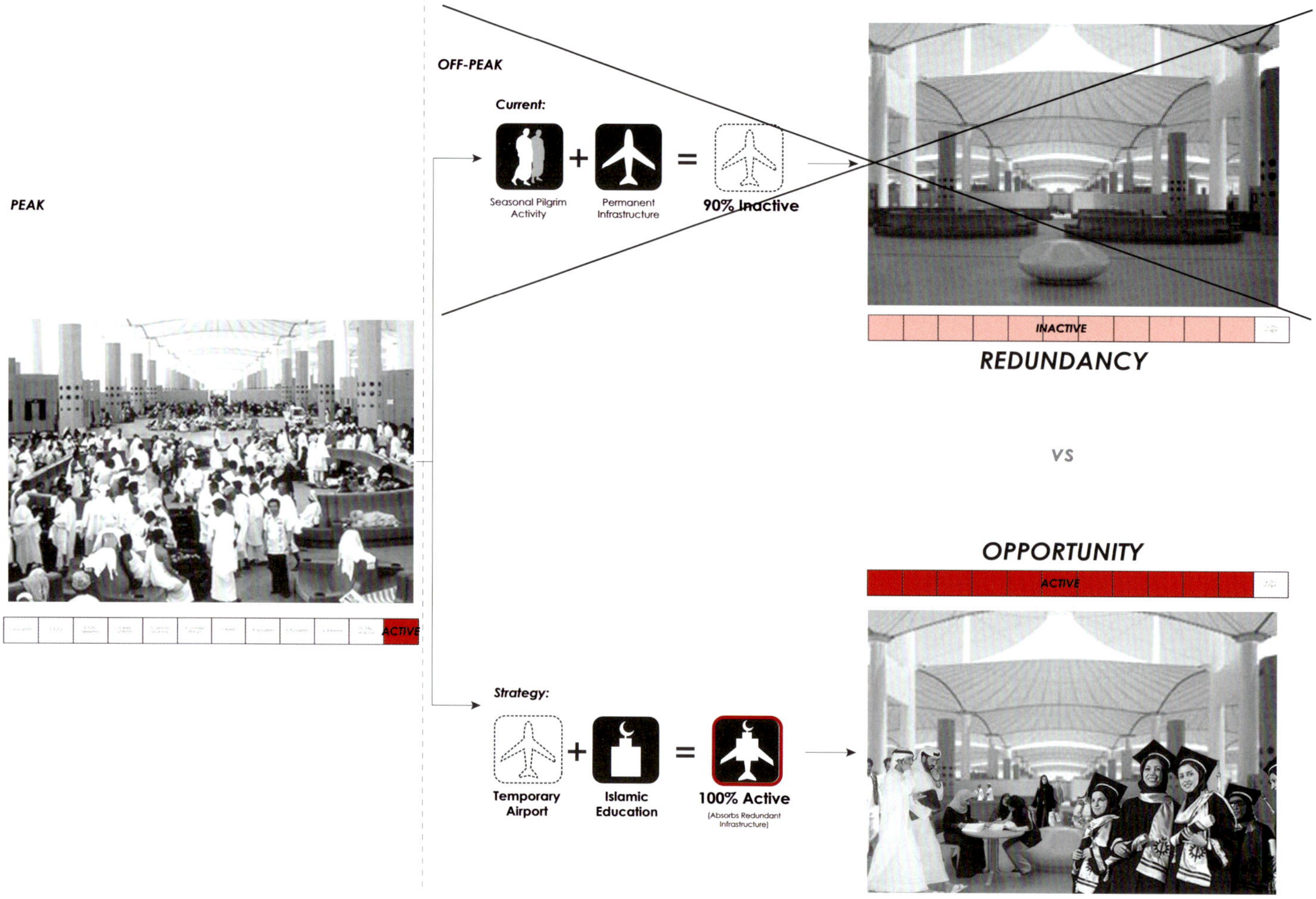

Balanced Peaks of Activity: Pilgrims vs Students
Absorb off-peak redundancy through educational programs

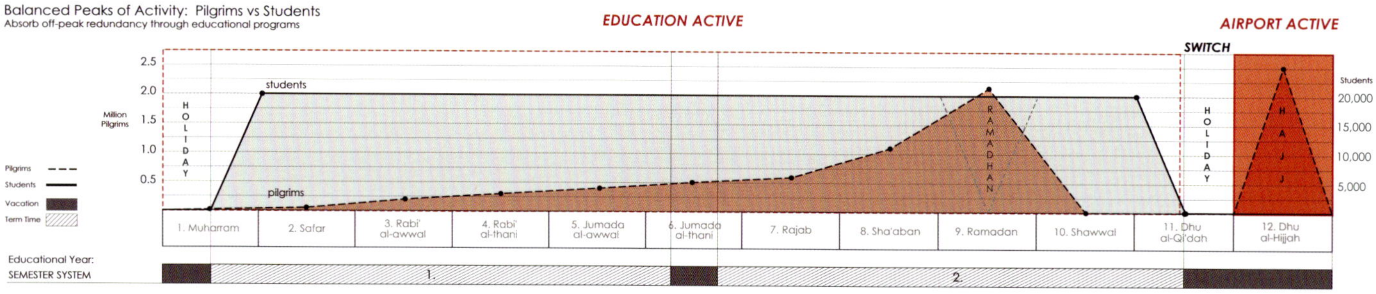

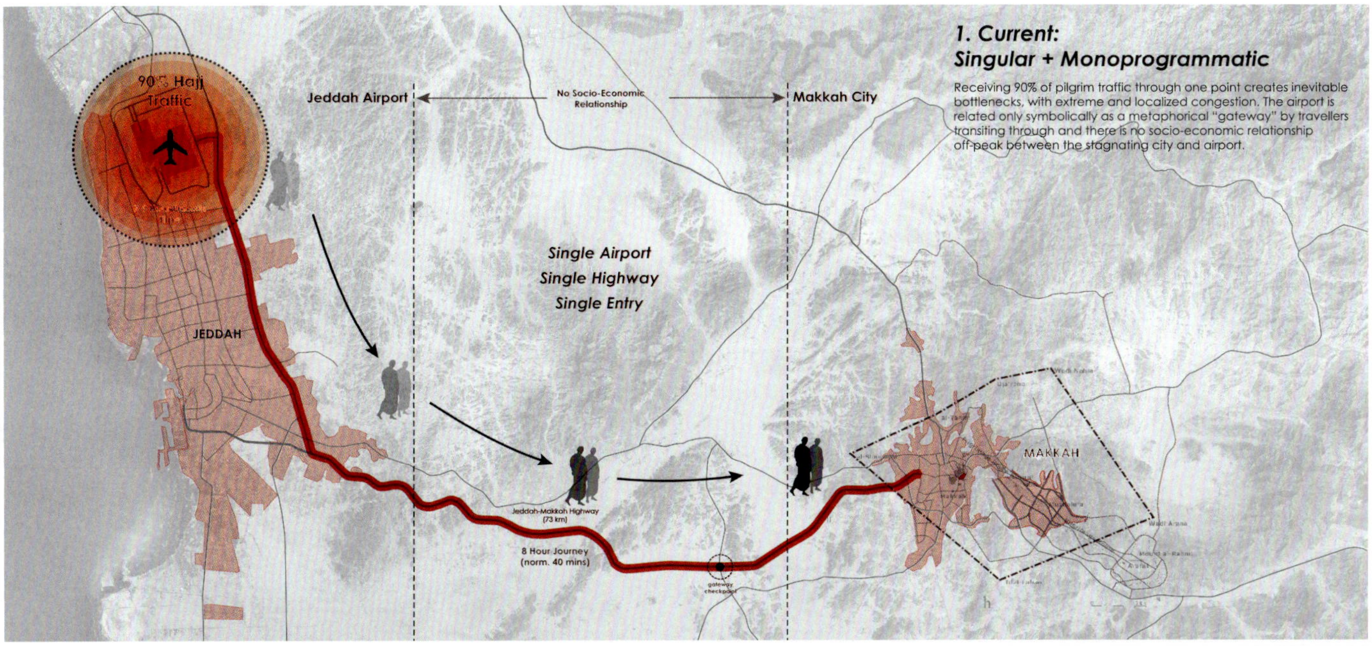
1. Current:
Singular + Monoprogrammatic
Receiving 90% of pilgrim traffic through one point creates inevitable bottlenecks, with extreme and localized congestion. The airport is related only symbolically as a metaphorical "gateway" by travellers transiting through and there is no socio-economic relationship off-peak between the stagnating city and airport.
90% Hajj Traffic
Jeddah Airport
No Socio-Economic Relationship
Makkah City
Single Airport
Single Highway
Single Entry
JEDDAH
MAKKAH
Jeddah-Makkah Highway (73 km)
8 Hour Journey (norm. 40 mins)
gateway checkpoint

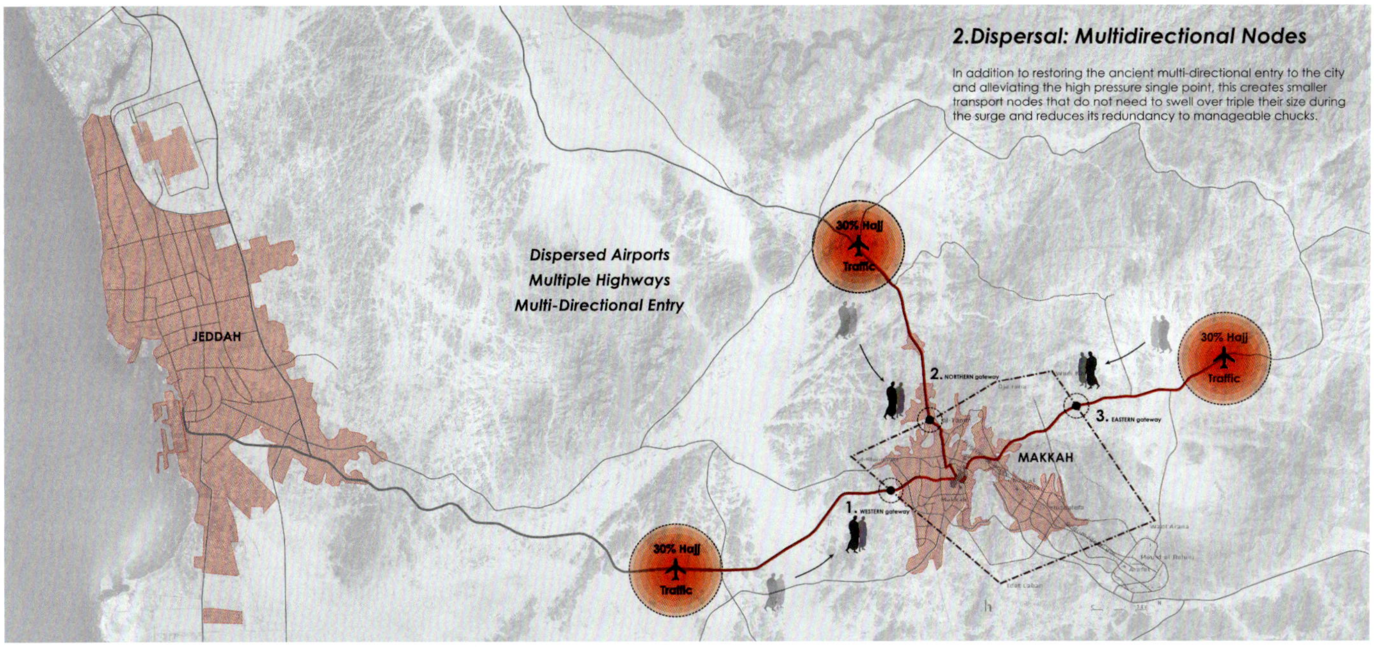
2.Dispersal: Multidirectional Nodes
In addition to restoring the ancient multi-directional entry to the city and alleviating the high pressure single point, this creates smaller transport nodes that do not need to swell over triple their size during the surge and reduces its redundancy to manageable chucks.
Dispersed Airports
Multiple Highways
Multi-Directional Entry
JEDDAH
30% Hajj Traffic
30% Hajj Traffic
30% Hajj Traffic
2. NORTHERN gateway
3. EASTERN gateway
1. WESTERN gateway
MAKKAH

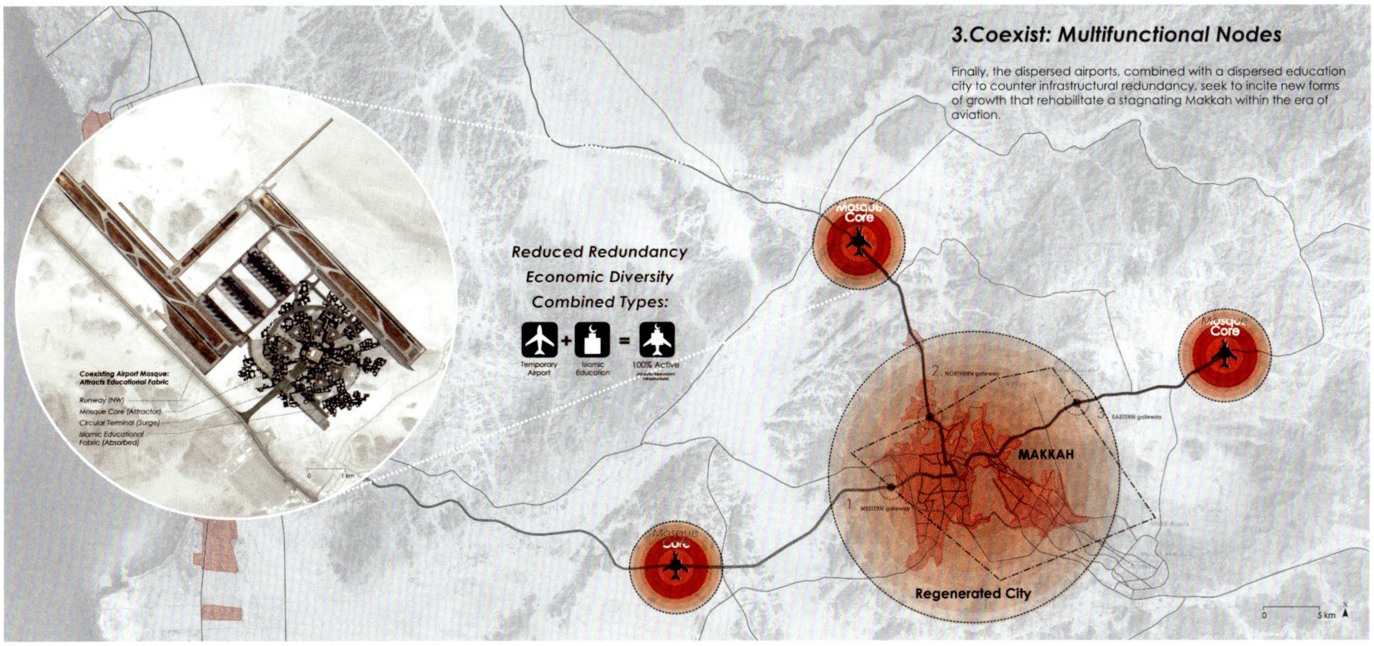
3.Coexist: Multifunctional Nodes
Finally, the dispersed airports, combined with a dispersed education city to counter infrastructural redundancy, seek to incite new forms of growth that rehabilitate a stagnating Makkah within the era of aviation.
Reduced Redundancy
Economic Diversity
Combined Types:
Temporary Airport
Islamic Education
100% Active
Coexisting Airport Mosque: Attracts Educational Fabric
Runway (NW)
Mosque Core (Attractor)
Circular Terminal (Surge)
Islamic Educational Fabric (Absorbed)
Mosque Core
Mosque Core
Mosque Core
MAKKAH
Regenerated City
0
5 km

3.2 Type Analysis: Islamic University vs Circular Airport

Brief History of Type: Mosque

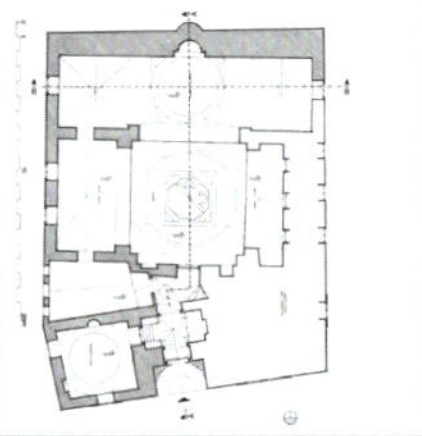

Origins: Local Mosque
Informal Madrasas

950: Grand Mosques Containing Schools
Adopt Formal University Status

1400+: Mosque Complexes
Purpose-built Schools Around Mosque

Mosque Type Analysis

Plan:

Programs:

Circulation:

Al Azhar Mosque, Cairo 988

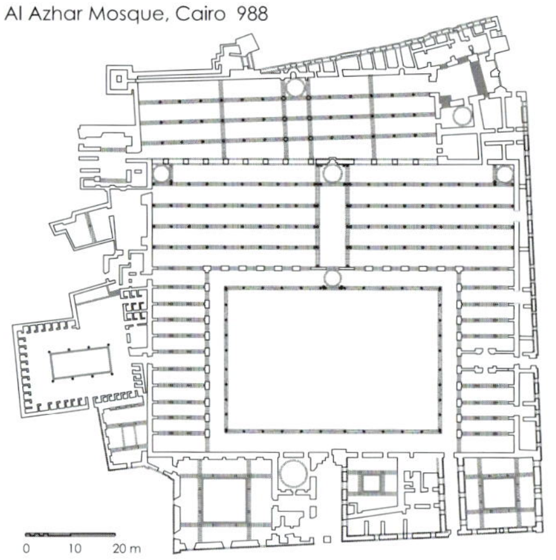

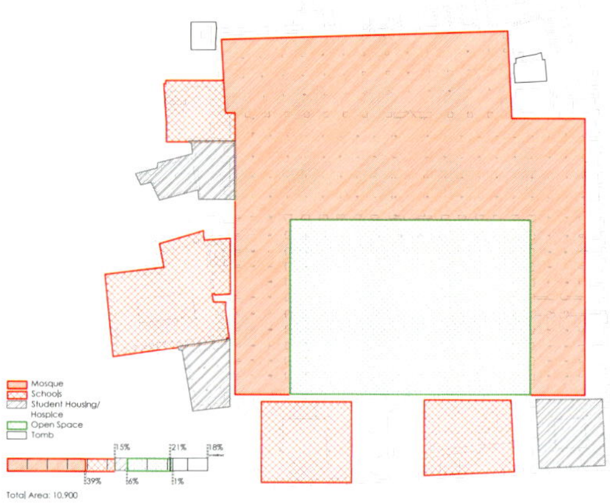

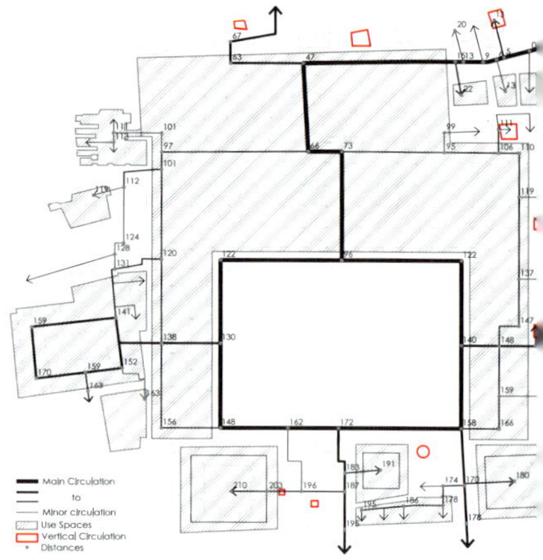

Corban Mustafa Pasa Complex, Turkey 1523

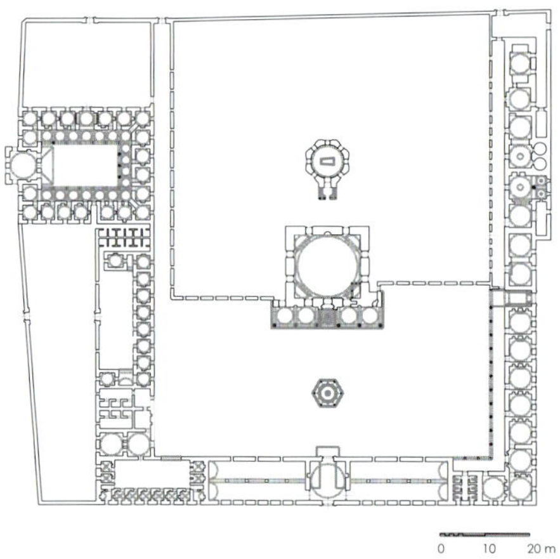

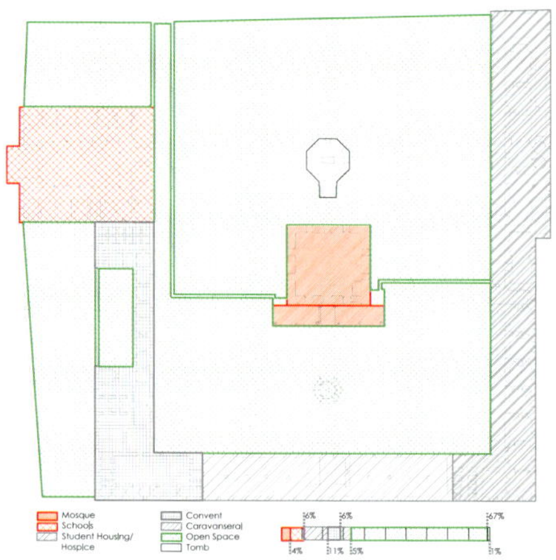

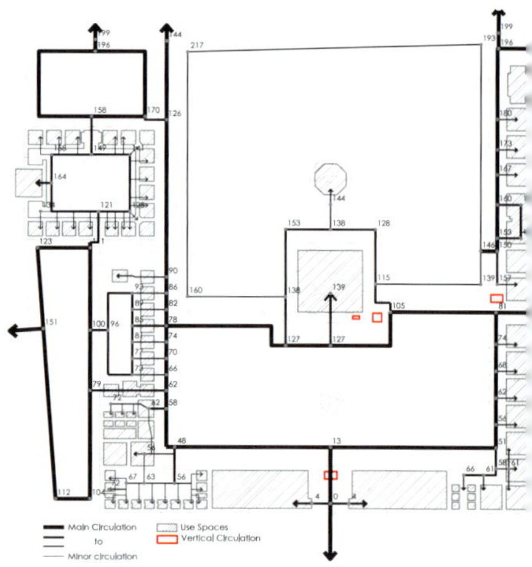

Sokollu Mehmet Pasa Complex, Turkey 1571

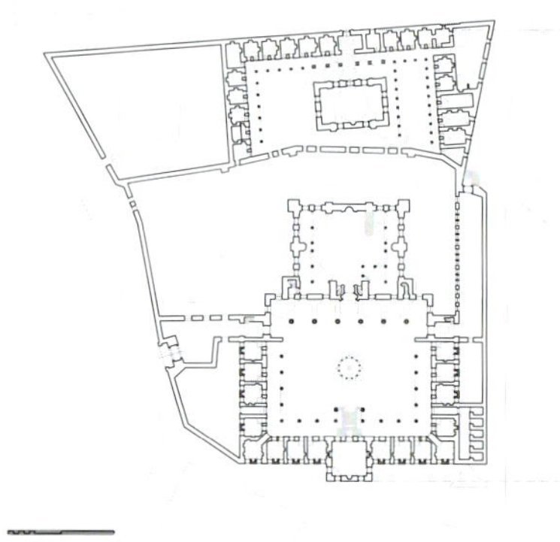

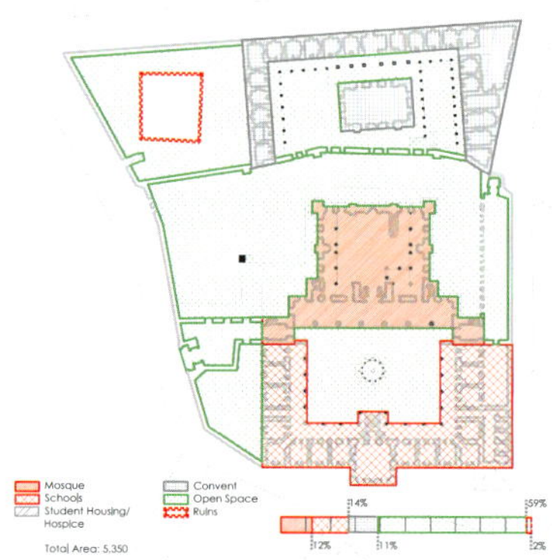

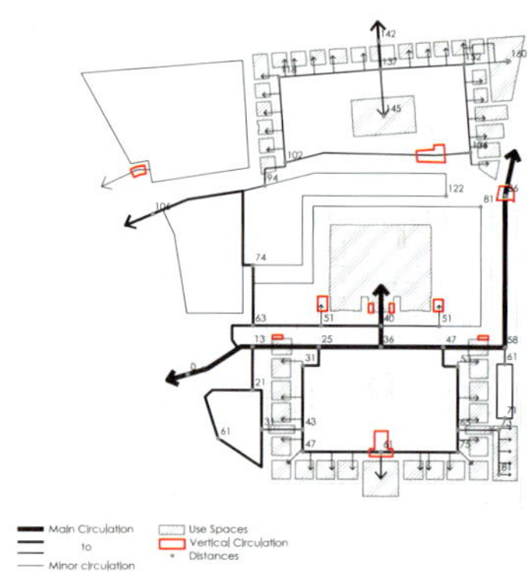

Brief History of Type: Circular Airports

Origins: 1900s Oval Racetracks
Airshows
Efficient Gathering Places

1930s: Gatwick "Beehive"
Hoar, Marlow and Lovett

1970s: Circular Airport Movement
Berlin Tegel (GMP), Cologne-Bonn Airport, Charles De Gaulle T1 (Andreu)

Accessibility (Privacy):

Geometry (Grids):

Symmetry:

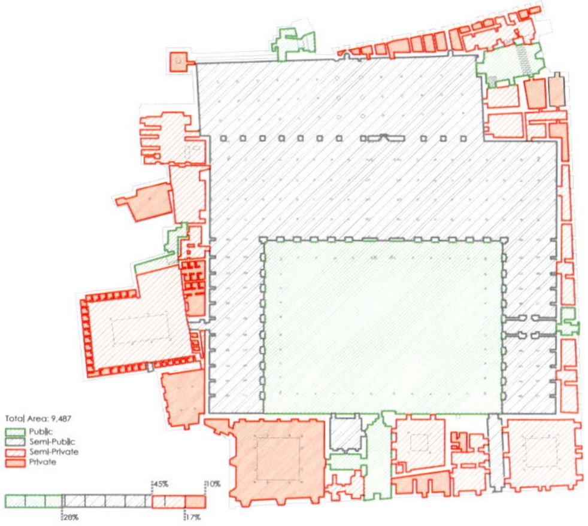

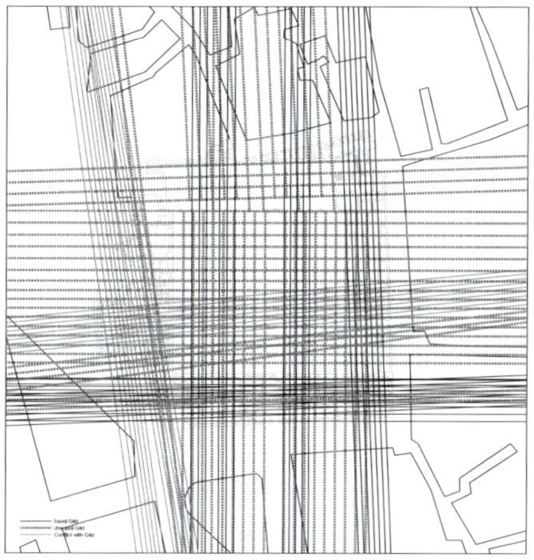

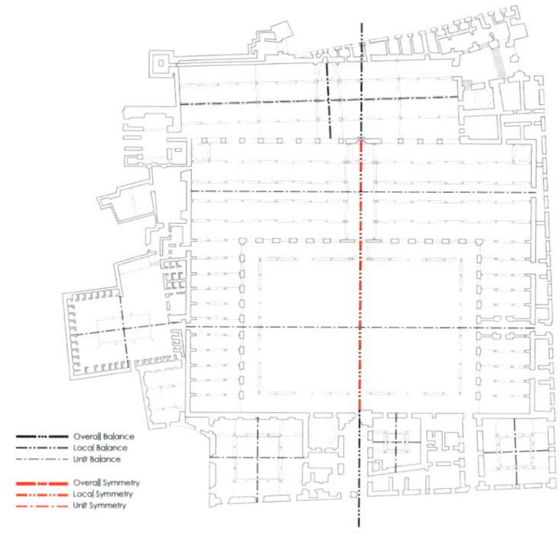

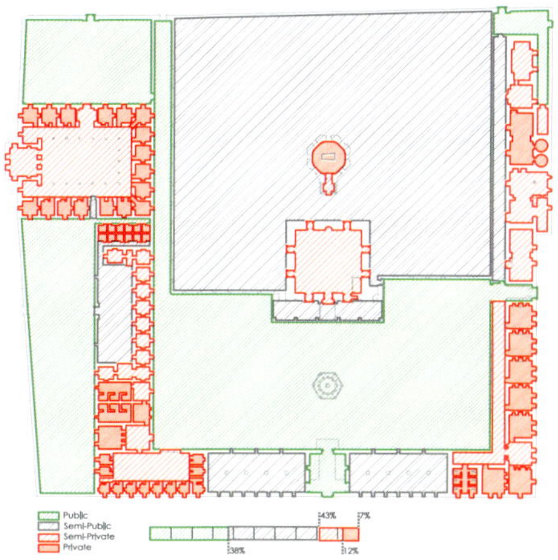

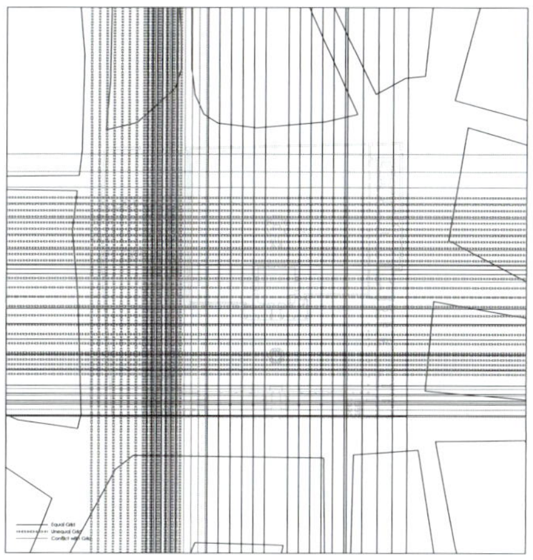

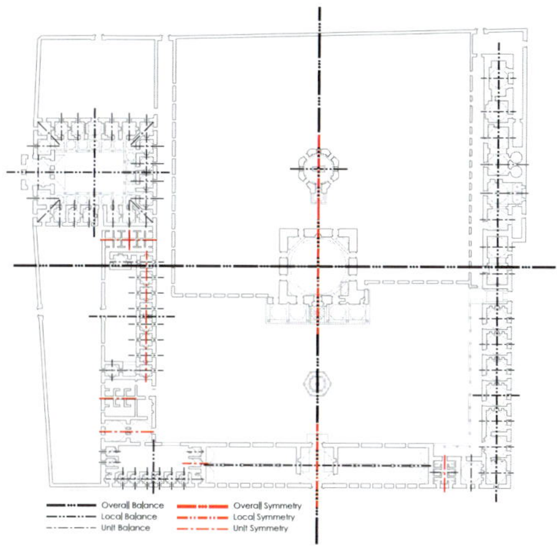

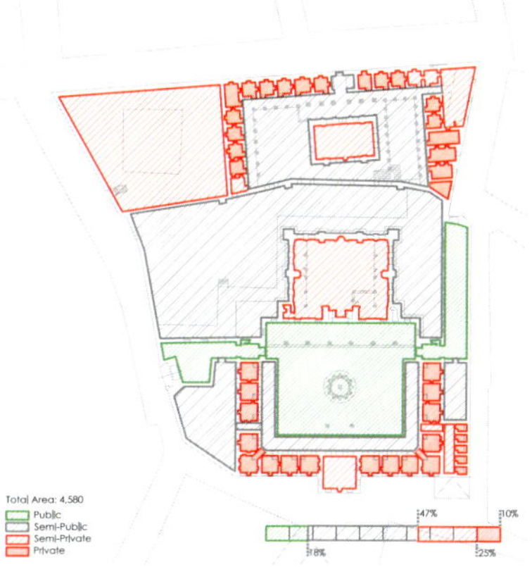

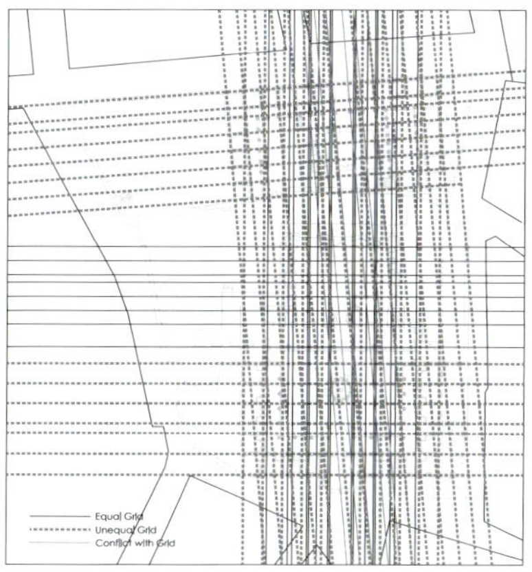

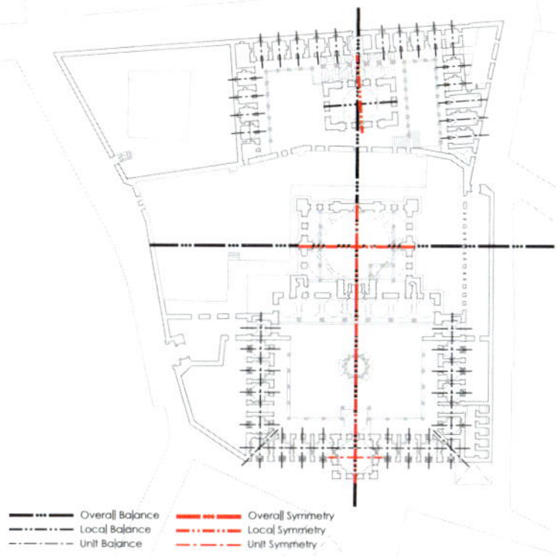

Type Analysis: Circular Airport

Charles De Gaule T1

Plans: Programs: Circulation:

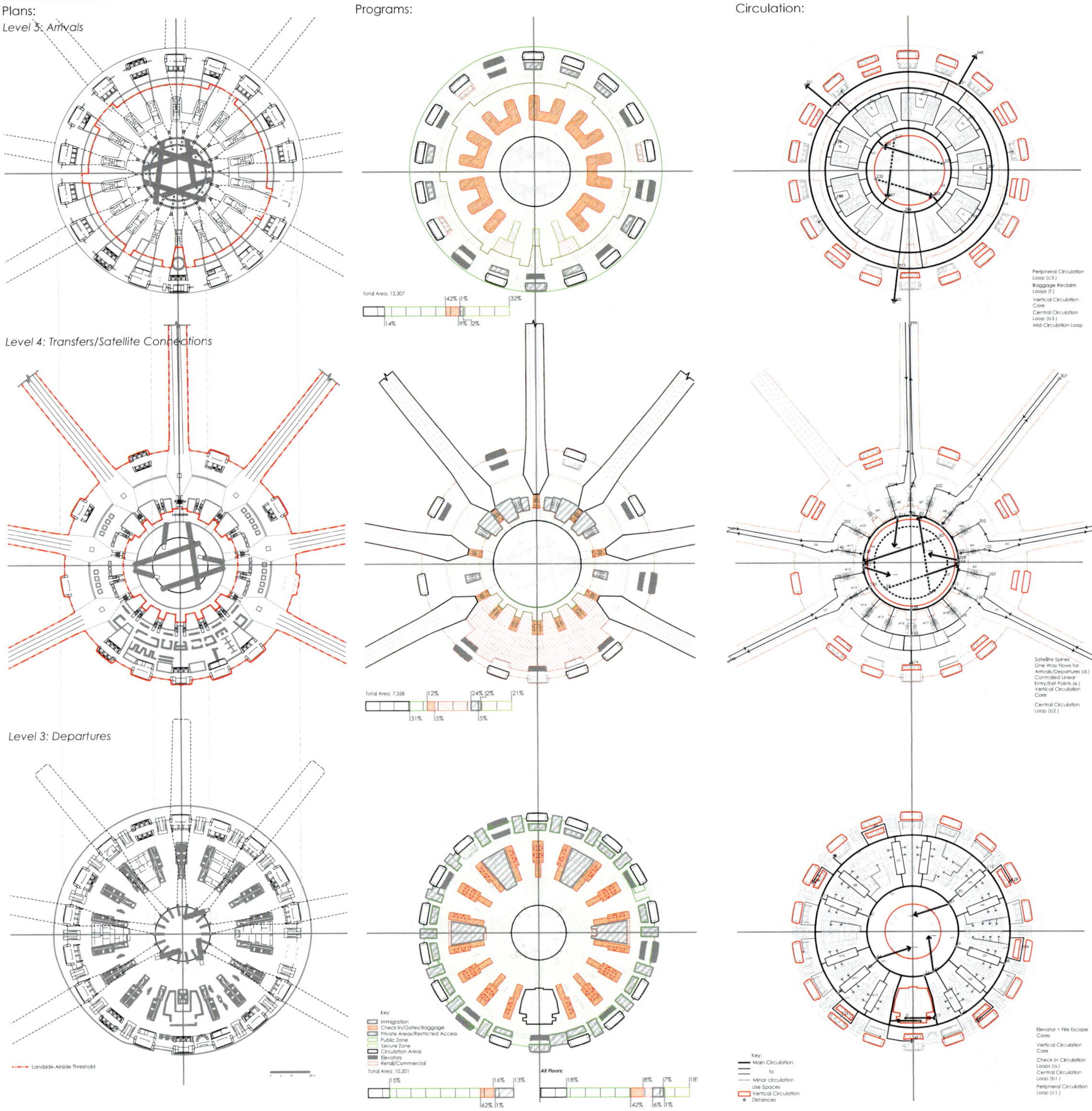

Berlin Tegel

Plan: Programs: Circulation:

Level 1: Departure + Arrivals

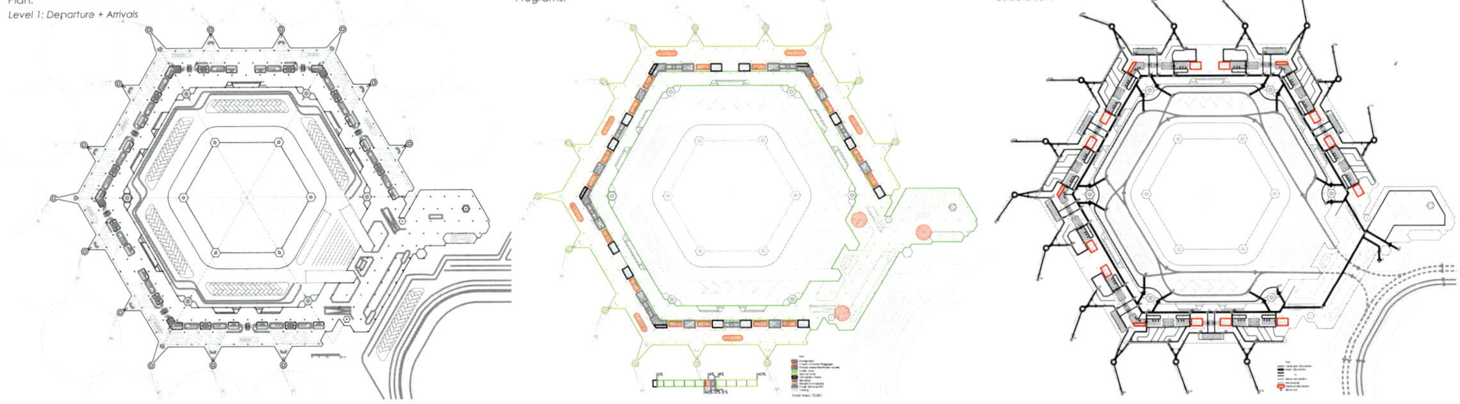

Geometry:

Accessibility (Privacy):

Symmetry and Balance:

Level 5: Arrivals

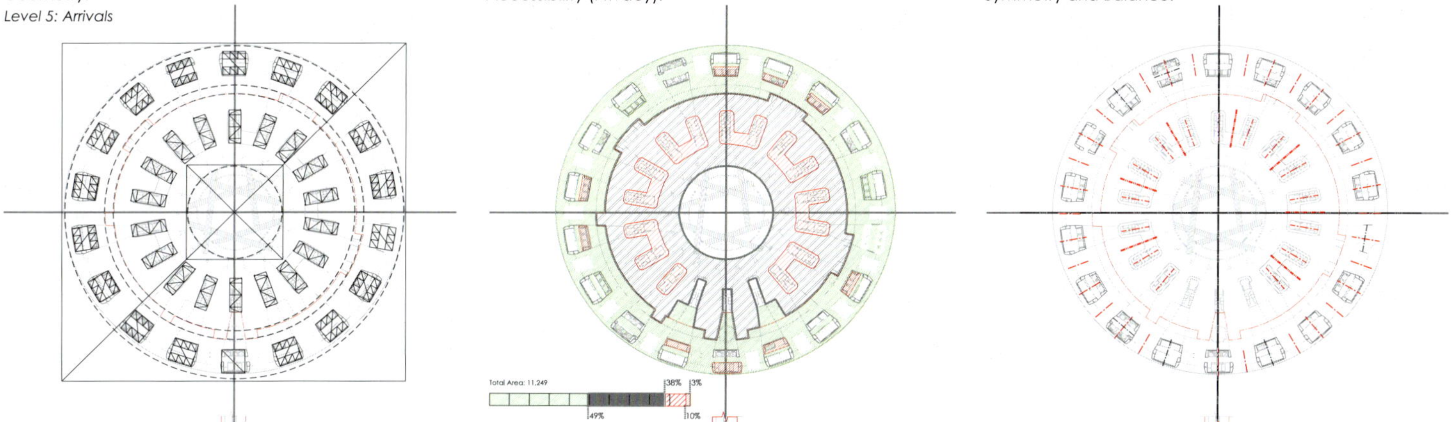

Level 4: Transfers/Satellite Connections

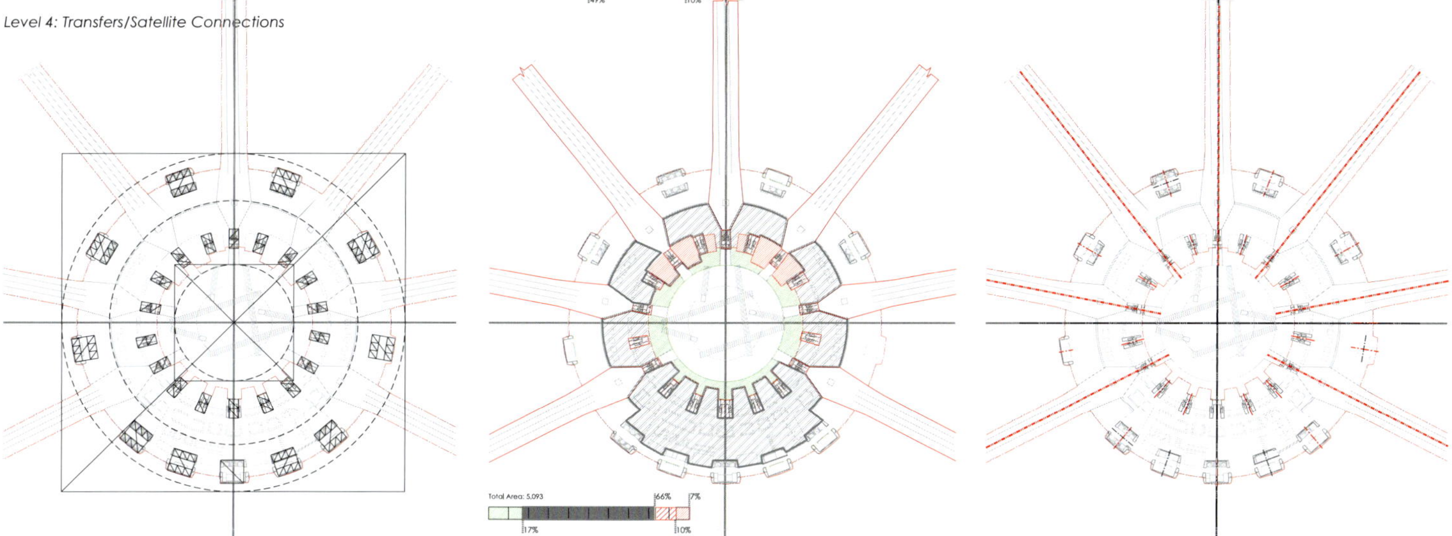

Level 3: Departures

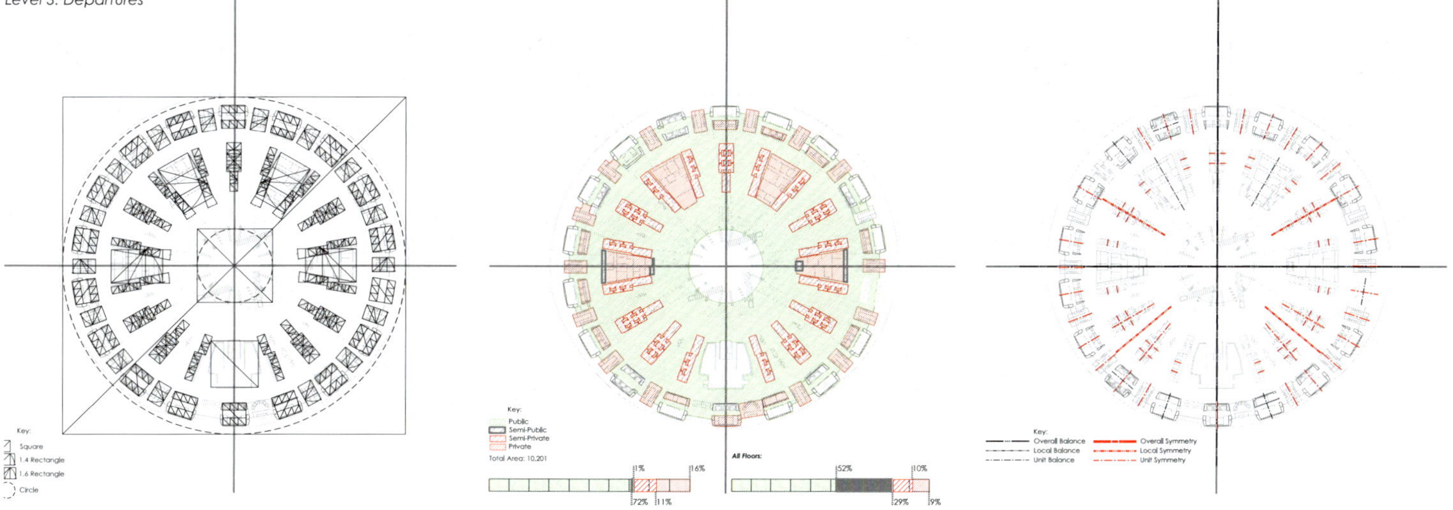

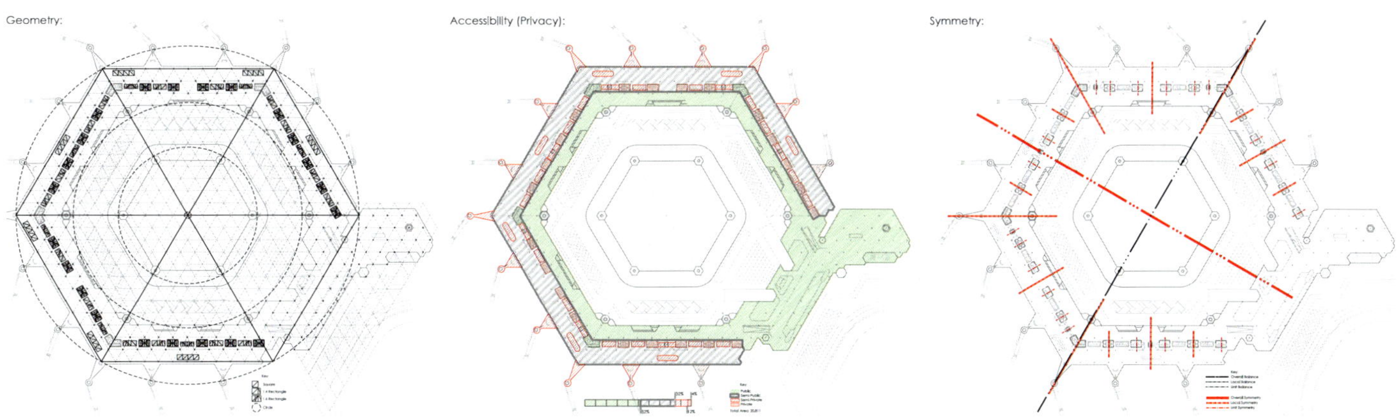

Deena Fakhro

3.3 Airport as Islamic "Walled City"

By inserting a mosque core and stretching the security buffer, an excessively thickened security zone emerges within the circular airport type, where educational programmes can aggregate within a new internal pocket. And what used to be a vulnerable security line, becomes a volumetric, programmatic ring - a campus.

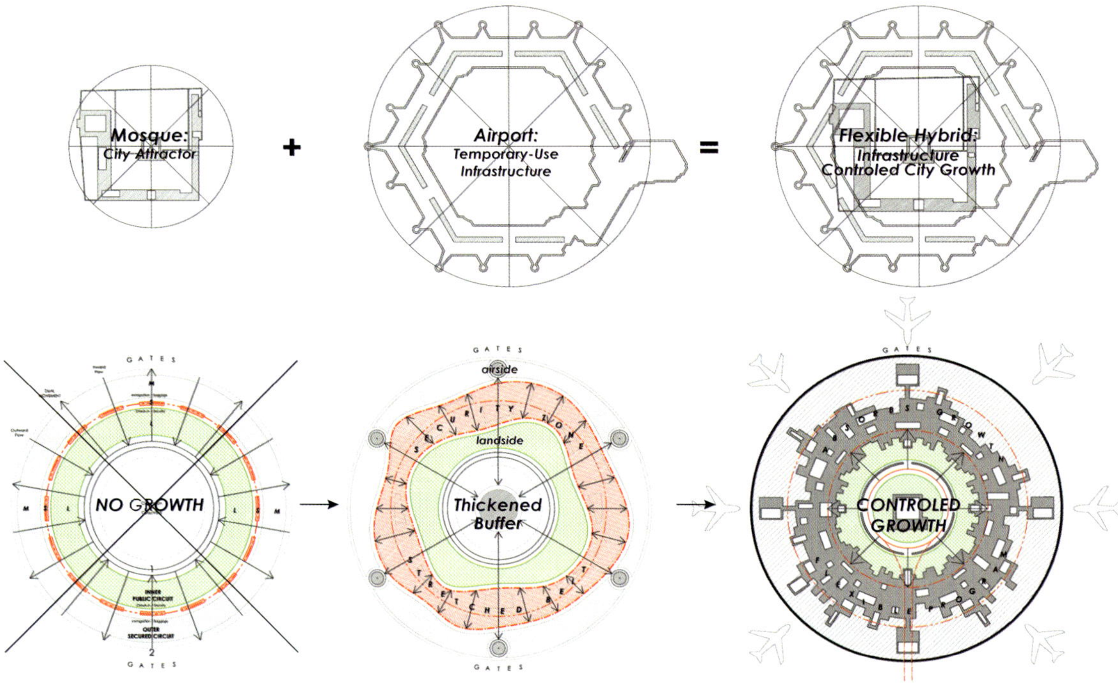

Typological Change: Thickened Security Buffer = Internal Growth Pocket

Interiorized Aerotropolis as Walled City

a. Typological Change Diagram:

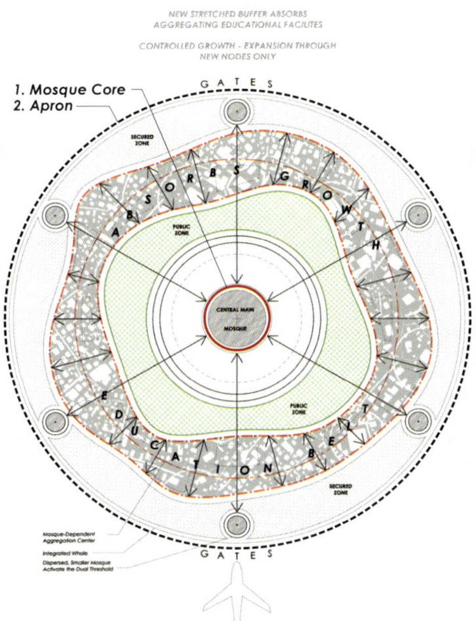

b. Damascus: Historic Walled City

Airport as Wall: Fixed Growth Avoids Bottleneck

With the airport, as a temporary-use, permanent infrastructure, it replaces the wall as a controller and predefiner of growth. The capped growth of the airport nodes, means that authorities are forced to build autonomous nodes elsewhere around the city, avoiding the same bottleneck situation from occuring again.

Islamic Walled Cities: Mosque Core, Walled Dense Growth

Historic planned Islamic cities, such as the plan for Baghdad were often based on the circle as a pure geometric representation of divine, unified cosmic order. The circle was viewed as a just and democratic form for its centrally equidistant nature. At the heart of all typical cities lay the grand mosque, around which concentric rings of plots were parcelled out to various communities to develop. An external defensive wall contained growth increasing maximization and utilization of space.

Growth: Sprawling Growth vs Autonomous Nodes

London Heathrow: Sprawling Aerotropolis

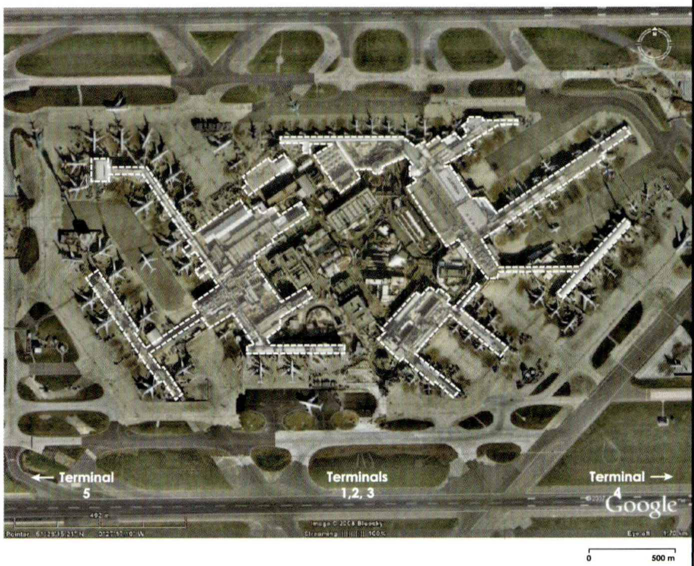

Berlin Tegel: Controlled, Nodal Growth

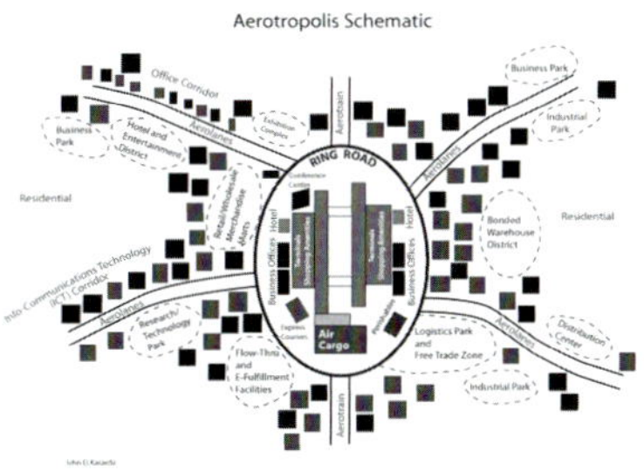

1. Aerotropolis Scheme: Cancerous Sprawl

Continuous growth and expansion of terminals and aviation dependent community on site

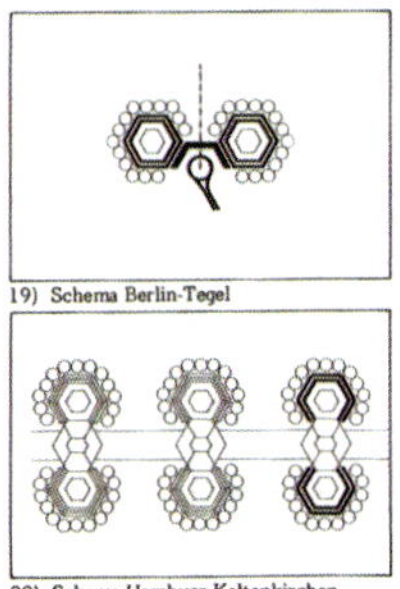

19) Schema Berlin-Tegel

20) Schema Hamburg-Kaltenkirchen

1. Limitation: No-Expansion Terminal

Actual Terminal Cannot Grow, Growth Only Accommodated By New Node

2. Possible Expansion Through New Node

Mirrored, Autonomous Growth Node Along Infrastructural Corridor

Airport becomes analogous to Islamic walled city, establishing a central mosque and a bounding wall, within which the city eventually grew. The aerotropolis' cancerous sprawl is eradicated and replaced by a controlled, polynodal growth.

4.0 The Hajj-Airport Anomoly

4.1 A Decentralized Spiral Framework
4.2 Inscribing Orientation
4.3 Interiorized Aerotropolis: A Suspended City

An Anomaly:

The Hajj Airport is an extreme anomaly: it's traffic is specific to the pilgrimage; it's location is a sacred gateway; and it's activity is seasonal. Unlike any other in the world, it's use, flows, and demands are highly predetermined, controlled and fixed. Revolving completely around the six day pilgrimage, beginning on the eighth day of the 12th Islamic month Dhul-Hijjah, the airport's activity is momentary but more intense than any other in the world. These unique circumstances of the Hajj airport then, force a series of spatial transformations within the generic circular airport type.

A Decentralized Spiral Framework:

Berlin Tegel's efficient and decentralized circulation framework needs to be adapted to the specific Hajj traffic. Unlike a typical airport that processes both arriving and departing flights, flows within the Hajj airport are largely one way – with the majority of pilgrims either arriving to or departing from the Hajj. In Tegel, linear and segregated arrivals and departure routes on the ground level alternate with each other. First, through twisting the arrivals and departure routes, and overlapping them, the compressed floor area is maximized, catering entirely to switching one-way flow systems. Pre-Hajj, the arriving pilgrims move in parallel across one system of spiral ramps, while post-Hajj the now departing pilgrims circulate across an inverse spiral system. Furthermore, through a sectional layering of spiral ramps more than one access route to each gate is opened up. Critical in easing pressure in extreme bottlenecks, this avoids the dangerous stampedes that occasionally occur at the current Hajj Terminal's gates. Thus, the unstoppable pilgrim waves, peaking at 9,000 passengers an hour, are optimally dispersed within a new decentralized spiralling framework.

Inscribing Orientation:

Based on a democratic efficiency of flows, the completely symmetrical and homogenous circular airport type is non-hierarchical and often disorientating to its users. On the other hand, to the mosque orientation is the most crucial element; mosques allude to a greater relationship, that of a person physically to Makkah, and spiritually to God. On its threshold, Makkah's Gateway Airports are a symbolic extension to the Holy City, marking the sacred beginning of the Hajj. Through offsetting the central core of the circular airport and varying sectional heights accordingly, a linear axis is inscribed that stresses a critical orientation to Makkah. Reflecting this axis are the circulation patterns. As arriving pilgrims move from the back of the airport, they spiral in unison towards the mosque core, and then out to the city – echoing the very rituals of the pilgrimage (circumambulation of the Ka'aba) upon arrival. And thus the powerful sense of arrival towards divine destination is embedded within the airport.

Interiorized Aerotropolis:

Once the short pilgrimage is over, the mosque and its volumes begin to emerge from the framework. Undulating, arching layers weaving between and structurally supporting the segregated airport levels, seed pockets for an off-peak academic community to inhabit. An arching, open hall extends throughout the public lower ground level of the airport, reminiscent of the typical mosque's expansive pillared halls. Above the arching hall sit the archetypal Islamic courtyards, which absorb the airport functions, and all enclosed within a climate-sheltering roof – similar to the entire networks of covered streets and open spaces in the desert-climate. The airport then is no longer simply a massive transport node, but interiorizes and encapsulates within it several urban elements of a traditional Islamic city.

CENTRALIZED

DECENTRALIZED

DISORIENTATING

ORIENTATING

STRUCTURE

SPACES

Hajj Terminal — Berlin Tegel

CDG Airport — Mosque, Madinah

Mosque, Cordoba — Interiorized Aerotropolis

4.1 A Decentralized Spiral Framework

The hajj anomaly forces a series of spatial transformations within the airport: to cater to one way flows, arrivals and departures are overlapped as a system of inverse spirals; to further decentralize flows, ramps are layered sectionally; and to orient users to Makkah, the central core is offset. Varying the rotation and slope of the spiral ramps creates a different distribution of the 3 critical airport zones (landside, security buffer, airside). While the airport's fixed capacity means that each model must provide the same amount of floor space to process the pilgrims. Therefore varying the number of distribution routes changes the grain of the courtyard fabric from a larger, more collective model to a smaller and more dispersed model.

Spiralling Circulation: Rotations and Slopes

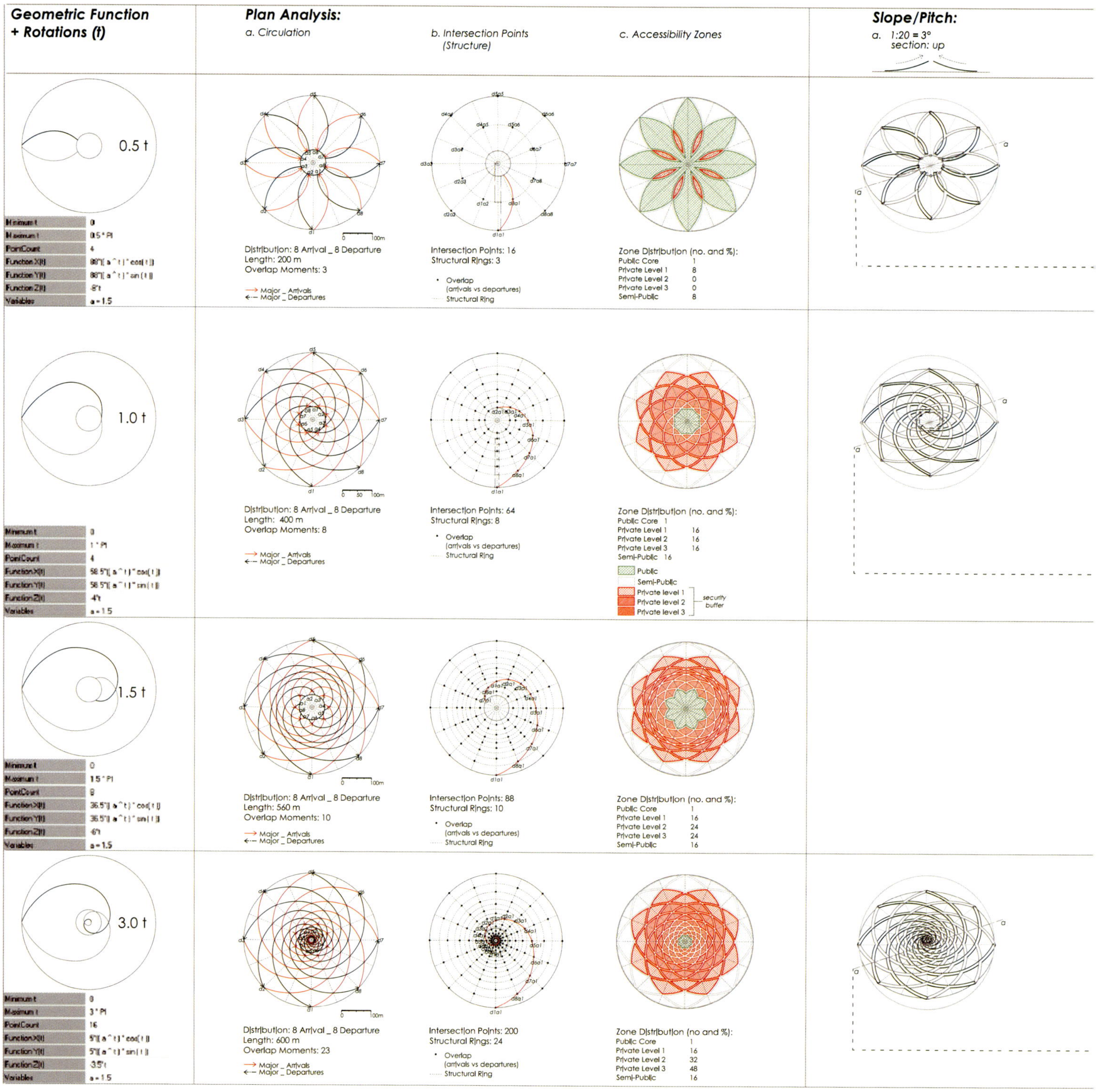

Circular Airport Circulation Transformation:

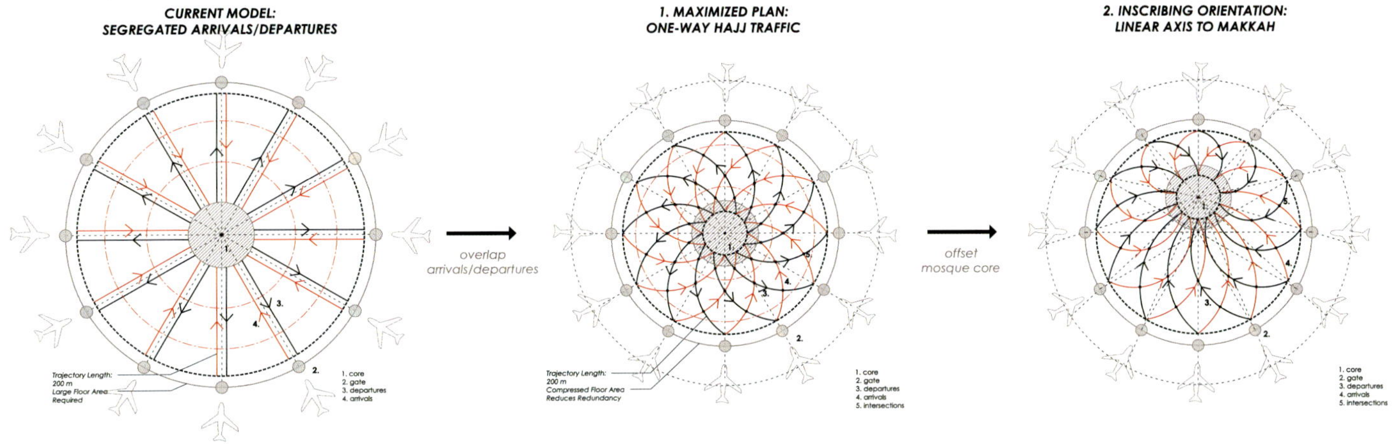

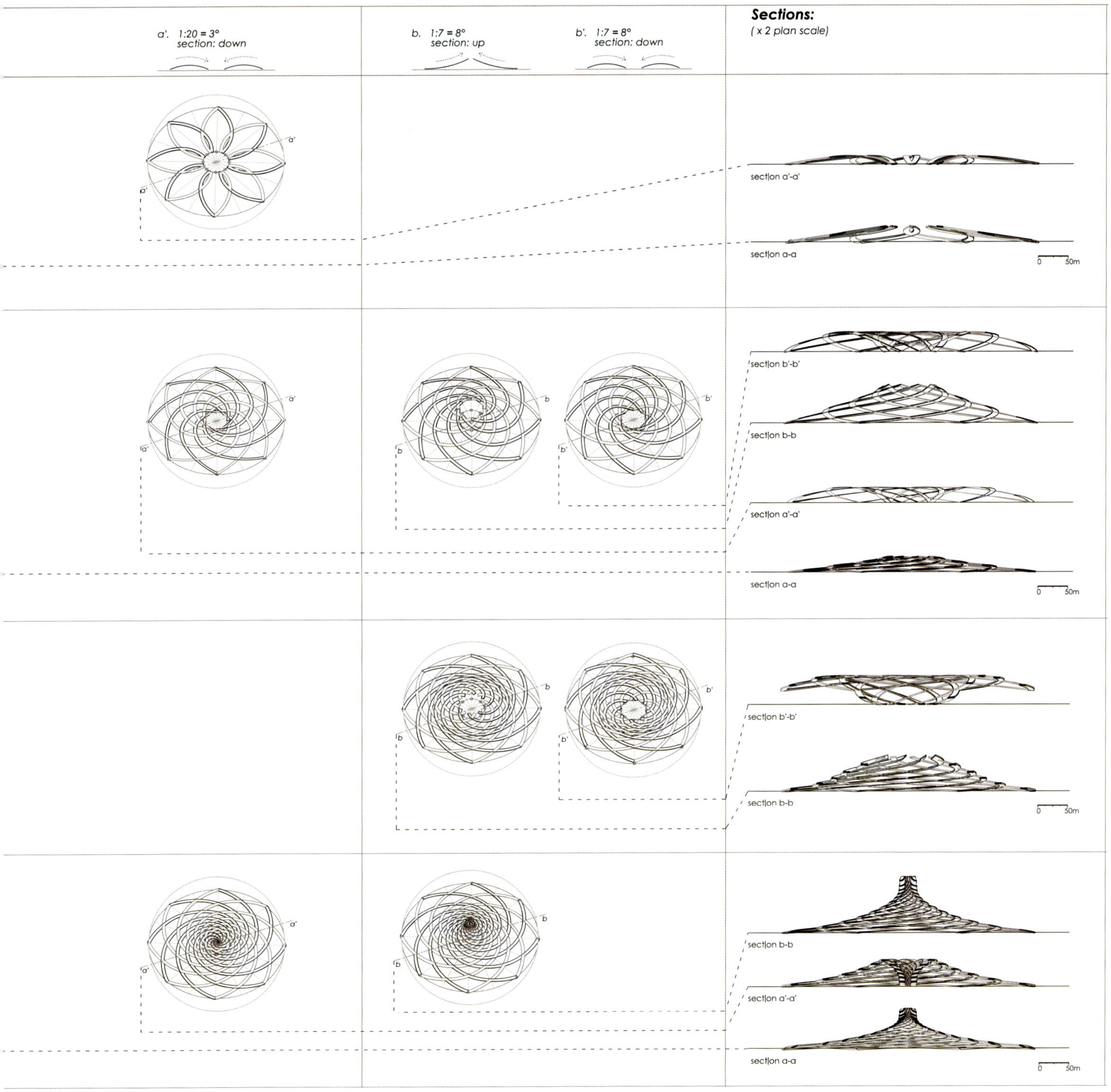

Pilgrim Surge Dispersal: Model Comparison

Constant Parameters
Peak Capacity: 3,000 passengers/hour
Built Floor Area: 125,000 m²
Areas: Circulation = 1.5m²/person
Bag Claim = 1.4m²/person
Check-In = 1.2m²/person

Model 1: 8 Dispersal Routes > 375 passengers each

Least number of parcels, most parcel size variation, results in extreme height differences as each parcel accommodates the same open space and built volumes

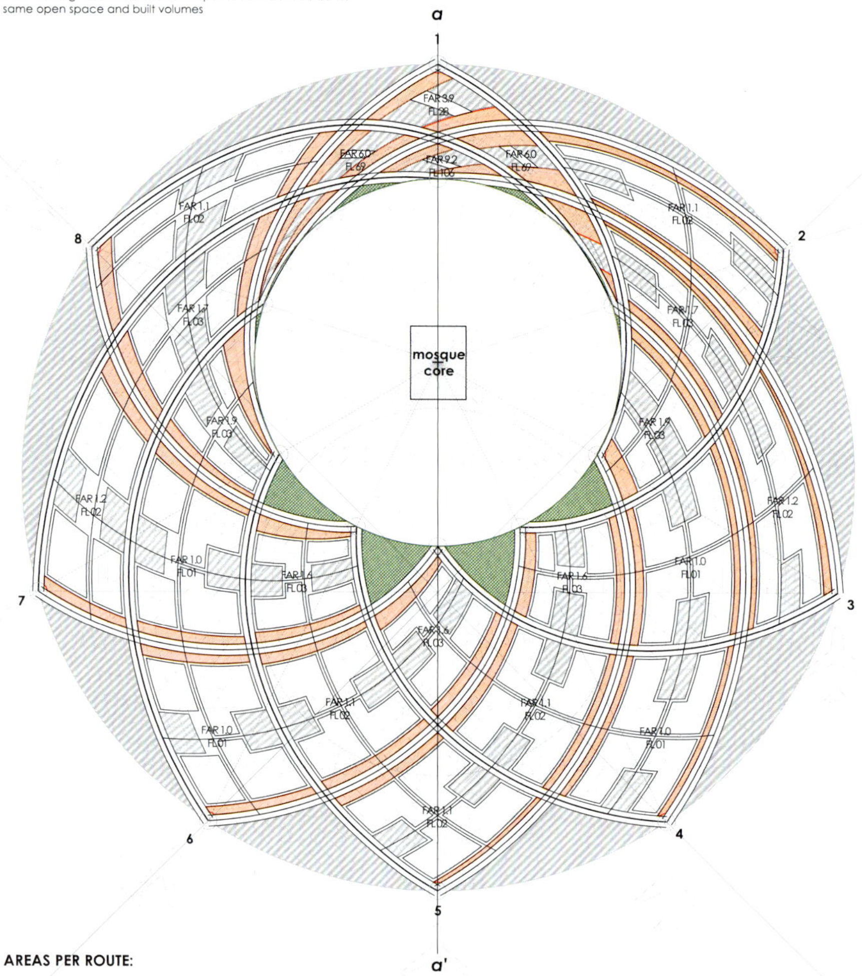

AREAS PER ROUTE:

Built Floor Area = 124,800 / 24 plots = **5,200 m²**
Departure Area = **1,350 m²**
Arrival Area = **1,575 m²**

Model 2: 12 Dispersal Routes > 250 passengers each

Significant parcel size variation, results in extreme height differenc[e] as each parcel accommodates the same open space and built volumes

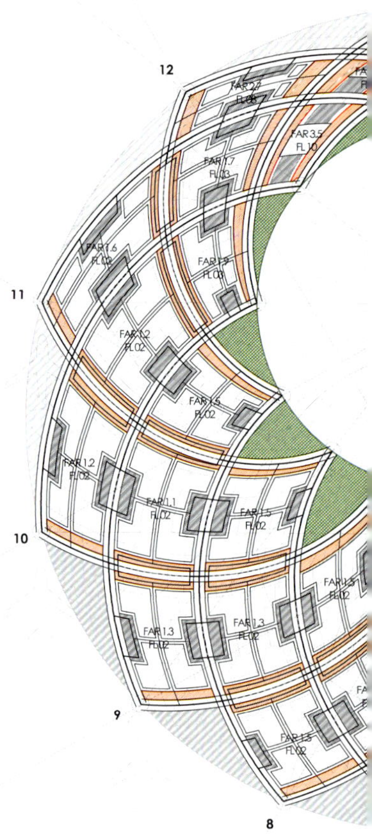

AREAS PER ROUTE:

Built Floor Area = 124,800 / 36 = **3,500 m²**
Departure Area = **900 m²**
Arrival Area = **1,050 m²**

Illustrative Sections: Massing Heights

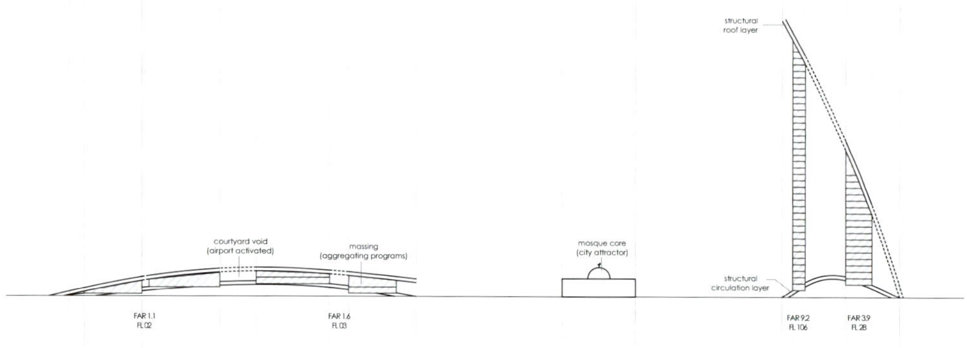

Section a-a'

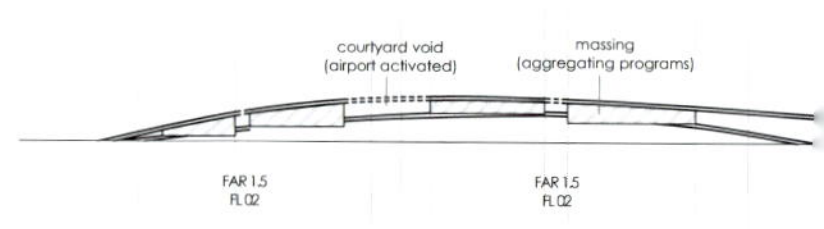

Section b-b'

Model 3: 16 Dispersal Routes > 188 passengers each

Most number of parcels, least parcel size variation, results in moderate height differences as each parcel accommodates the same open space and built volumes

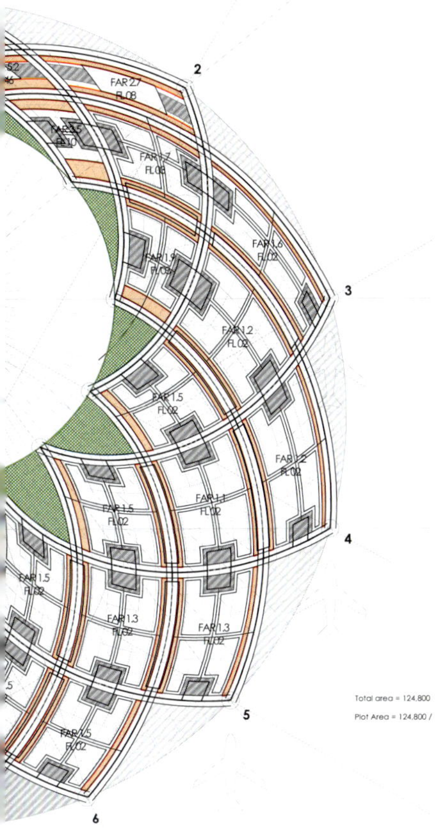

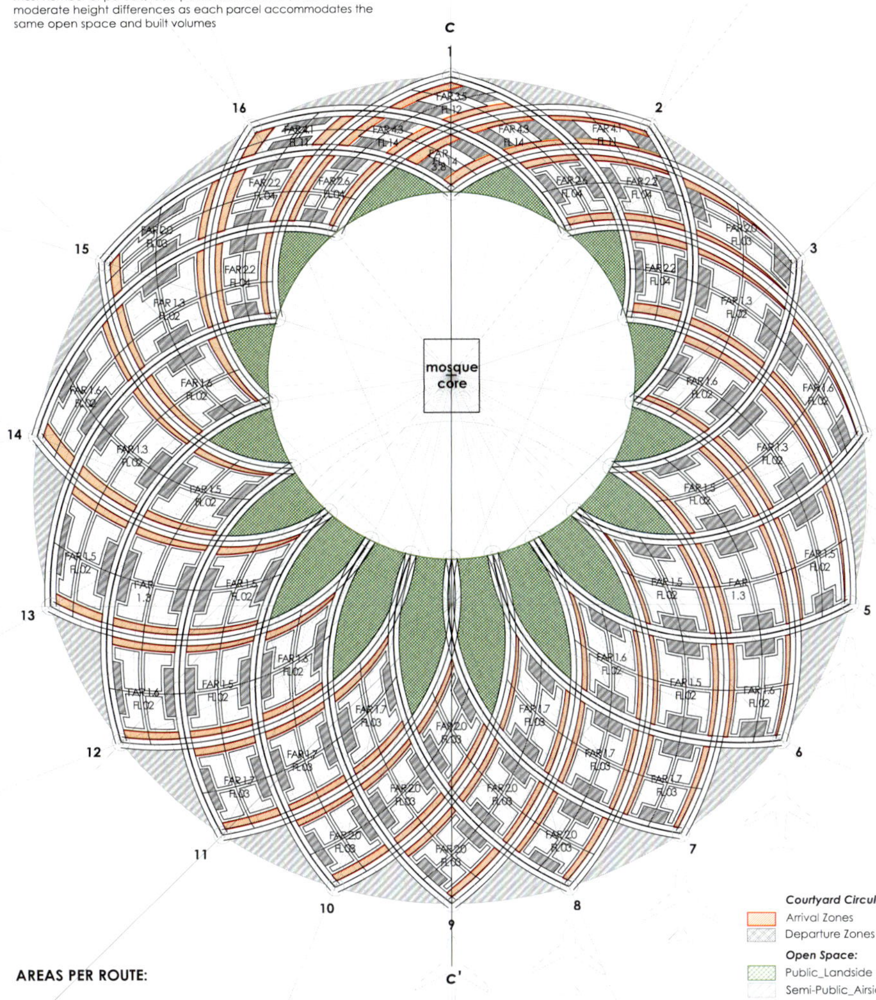

AREAS PER ROUTE:

Built Floor Area = 124,800 / 48 = **2,600 m²**
Departure Area = **675 m²**
Arrival Area = **789 m²**

Courtyard Circulation System:
- Arrival Zones
- Departure Zones

Open Space:
- Public_Landside
- Semi-Public_Airside

Circulation:
- Airport_Departures
- Airport_Arrivals
- Mosque_Secondary
- Building Footprints

0 — 50 m

Makkah

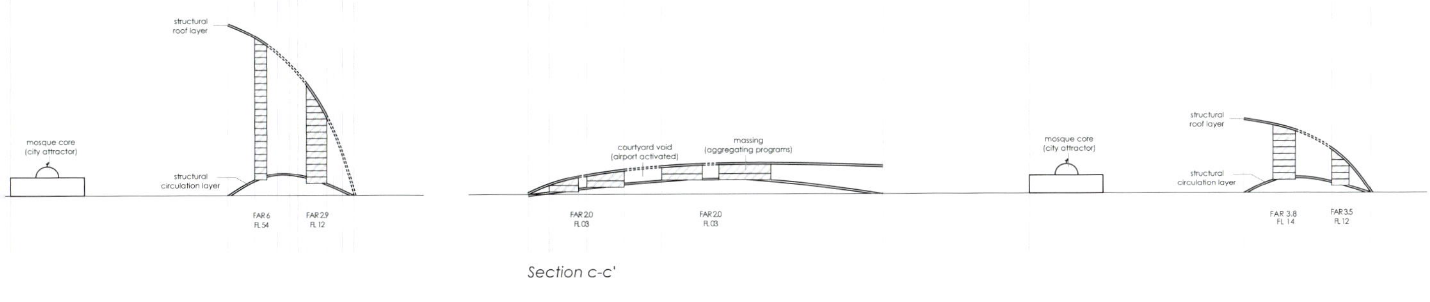

Section c-c'

Deena Fakhro

Model 3: 16 Dispersal Routes
Sectional Dispersal of Circulation

To further decentralize circulation, the spiralling circulation bifurcates sectionally, opening up two access routes to every gate. During the extreme peak of the Hajj surge, the secondary circulation route to and from each gate would be vital in preventing the stampedes that currently occur at Jeddah's Hajj Terminal.

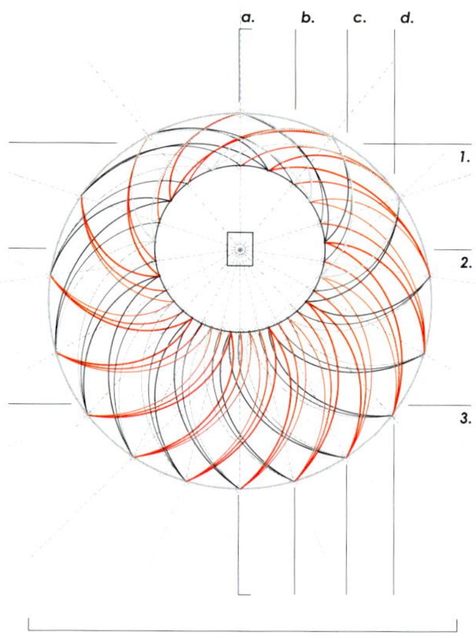

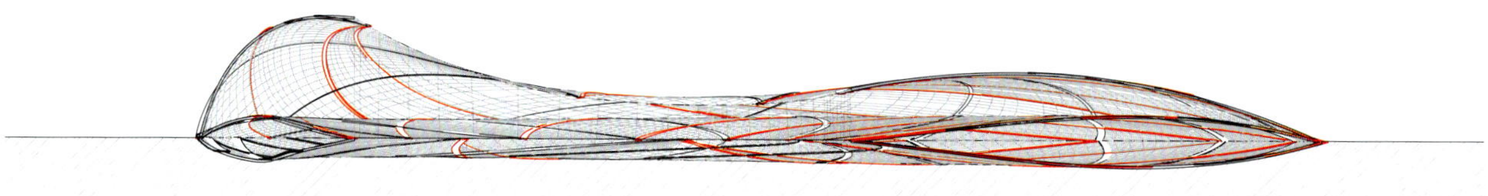

section a.

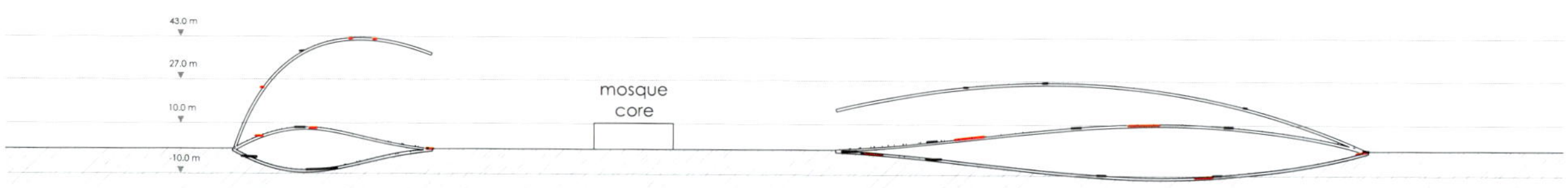

section b.

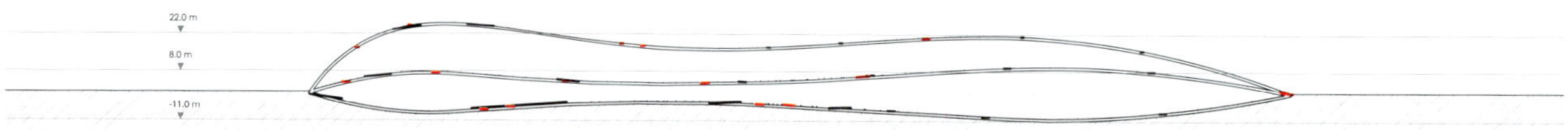

section c.

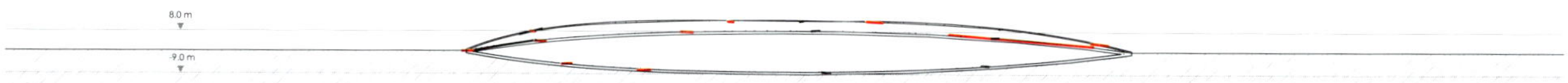

section d.

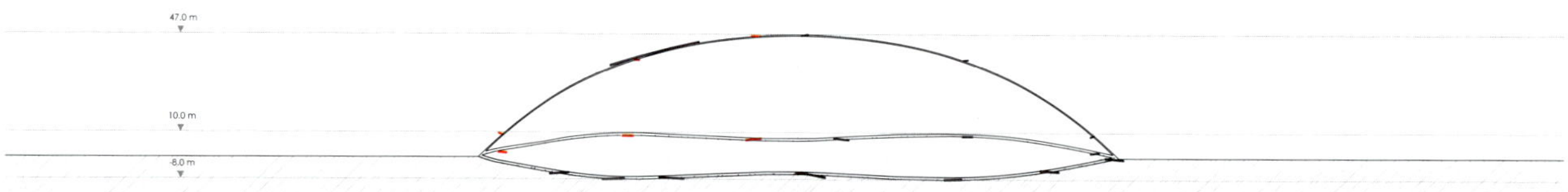

section 1.

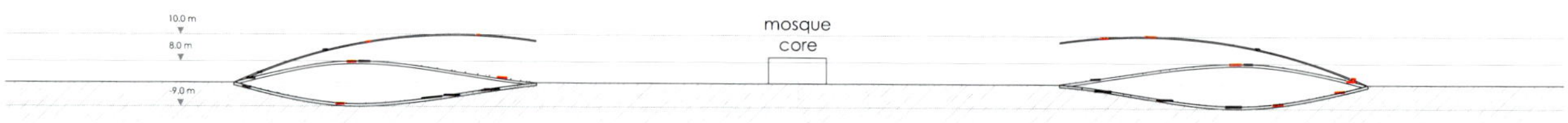

section 2.

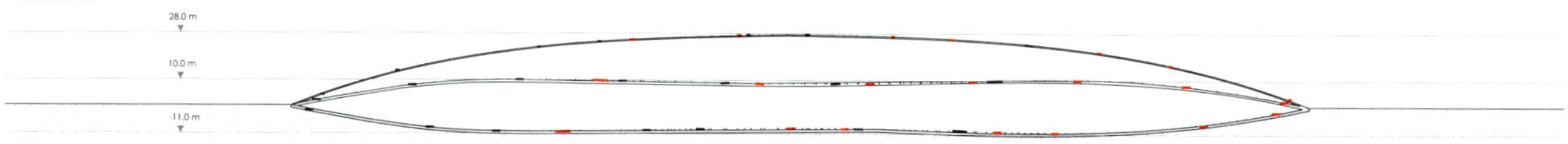

section 3.

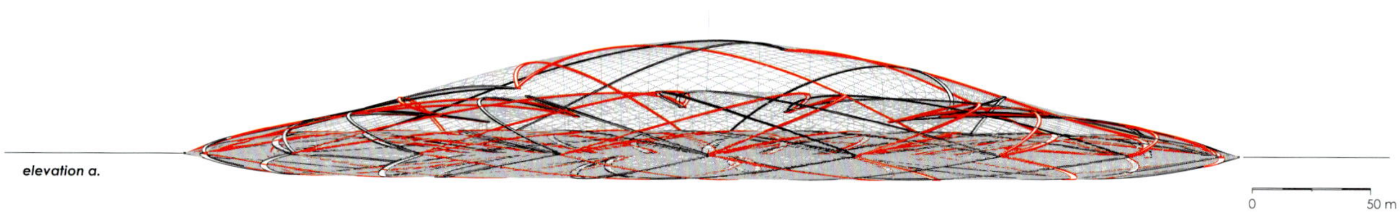

elevation a.

c. Roof
(Structure/Massing)

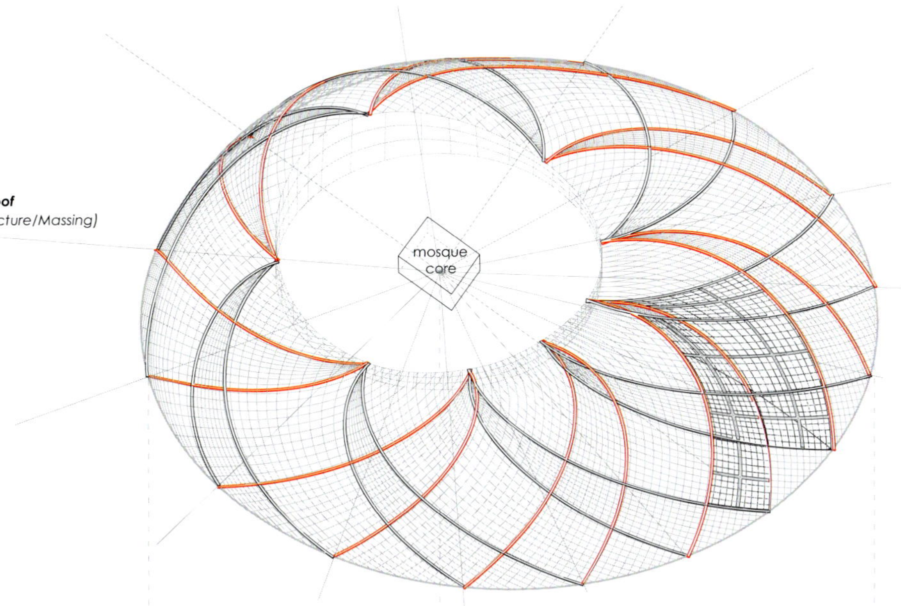

b. Level 0
(Upper Ground)

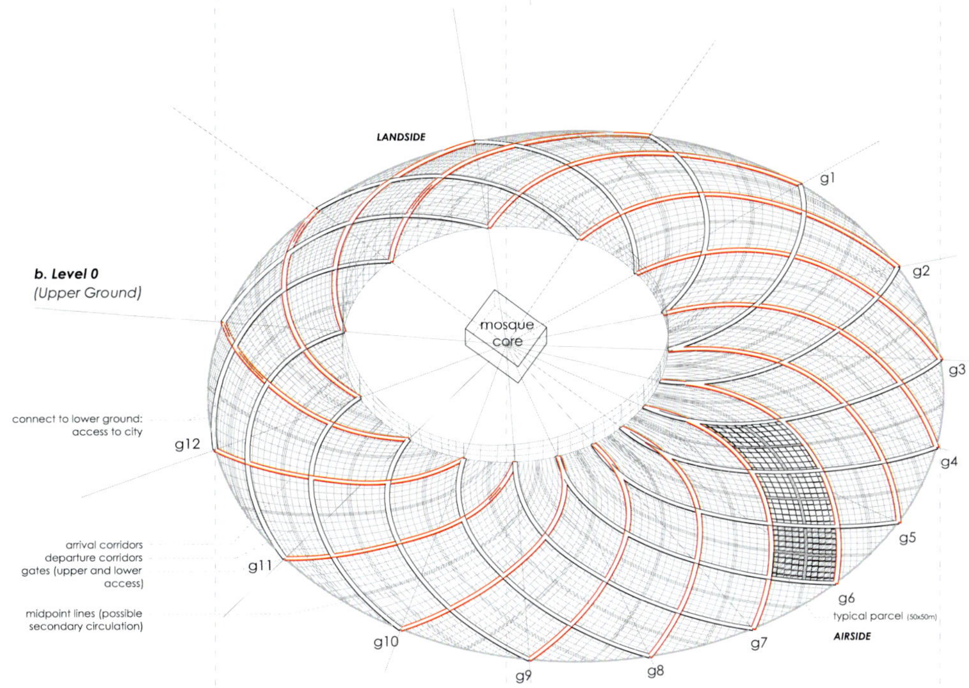

a. Level -1
(Lower Ground)

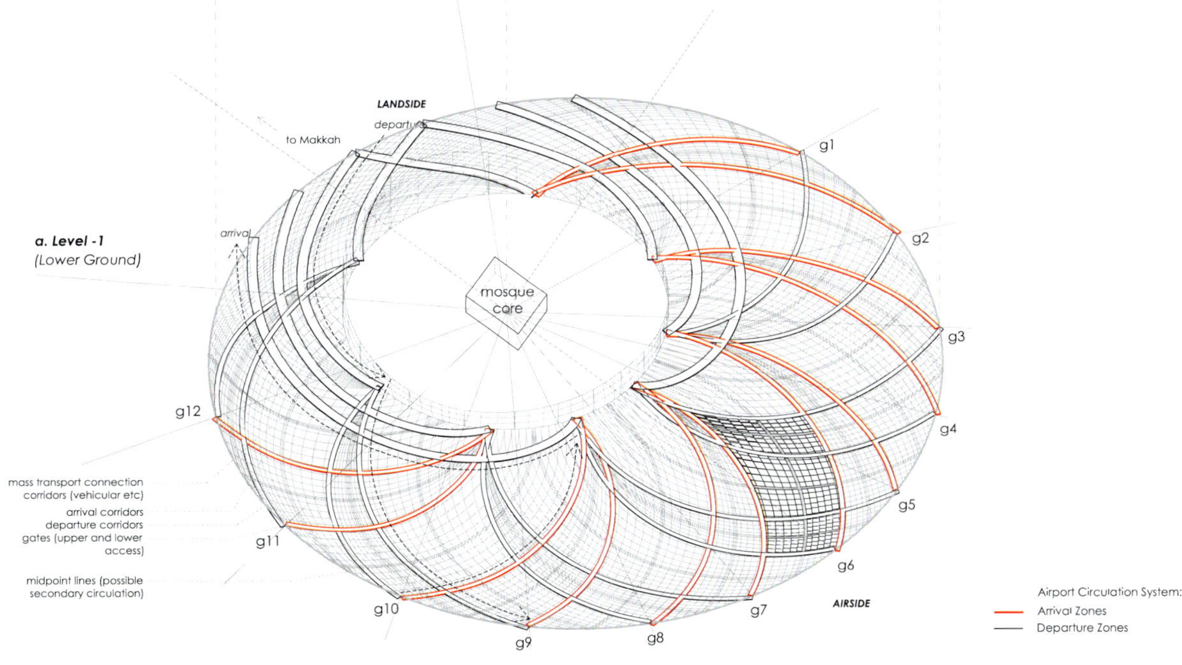

4.2 Inscribing Orientation

Extension to Holy City:

Where Circular airports are efficiently non-heirarchical, Makkah's Gateway Airports should be a symbolic extension to the Holy City, marking its threshold. Offsetting the central core inscribes a linear axis in the airport, stressing a critical orientation to Makkah that is reflected in the circulation patterns. As pilgrims move from the back of the airport, facing Makkah, they spiral in unison towards the mosque core, and out to the city – echoing the rituals of the pilgrimage upon arrival. The powerful sense of arrival towards divine destination is embedded within the airport type.

Regional Strategy: Restoring Valley Gateways

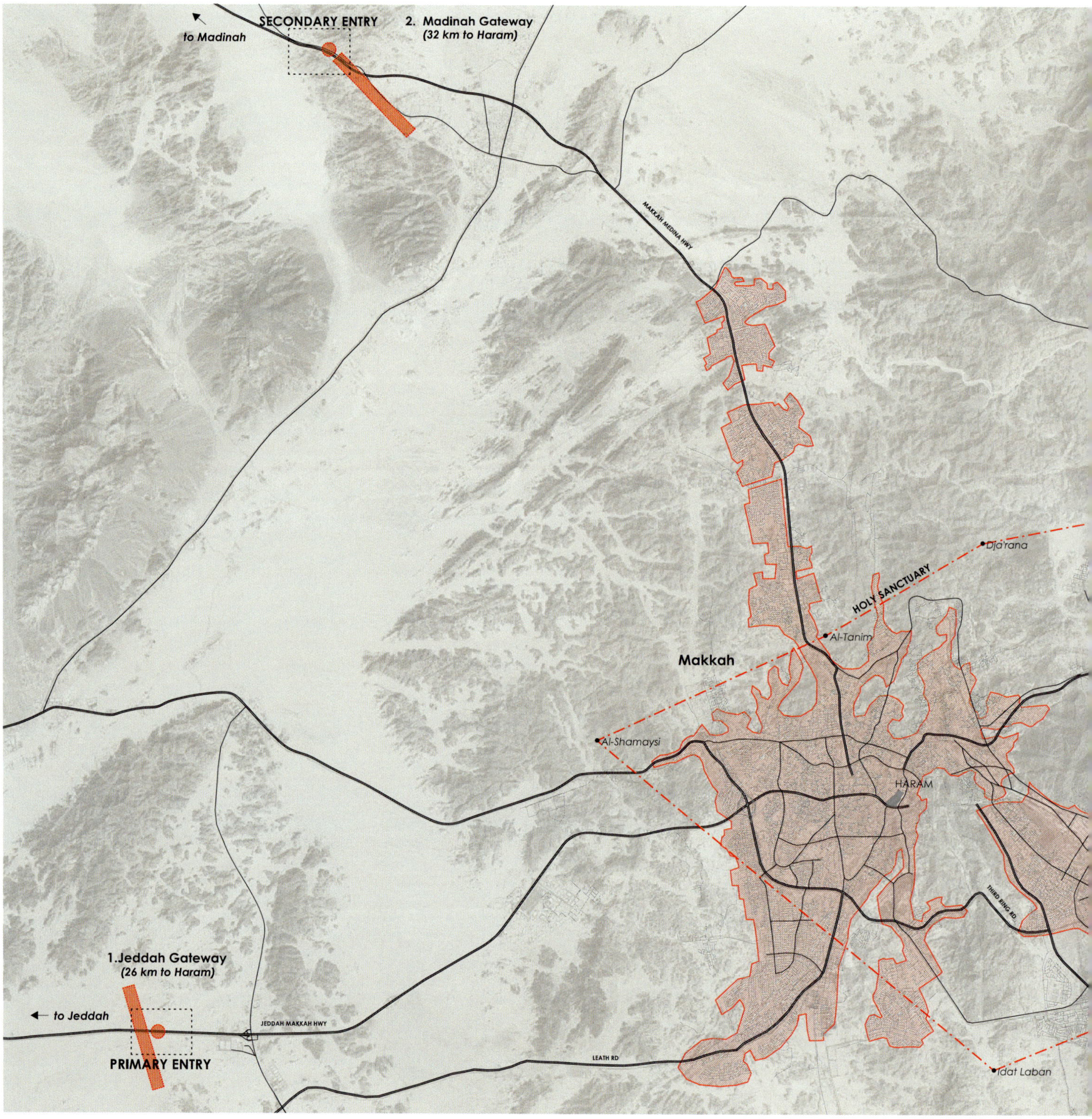

Orientation: Core and Massing Frame the Gateway
Conceptual Plans and Sections

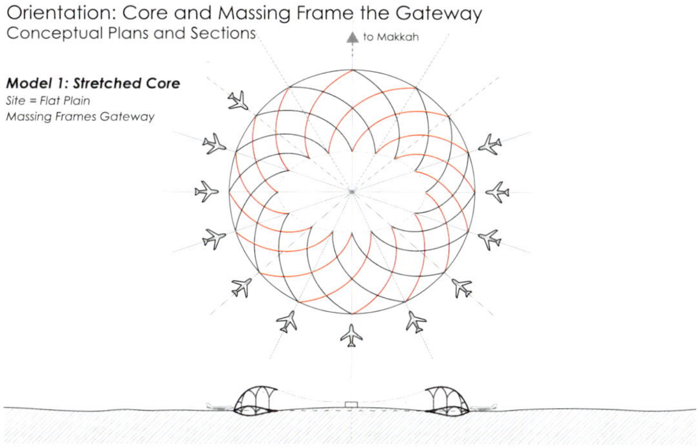

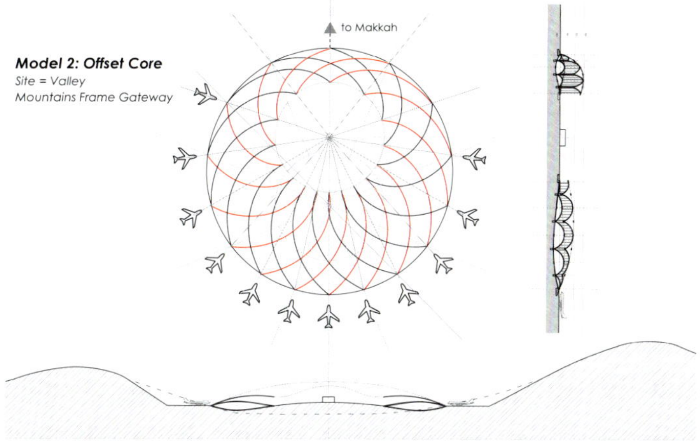

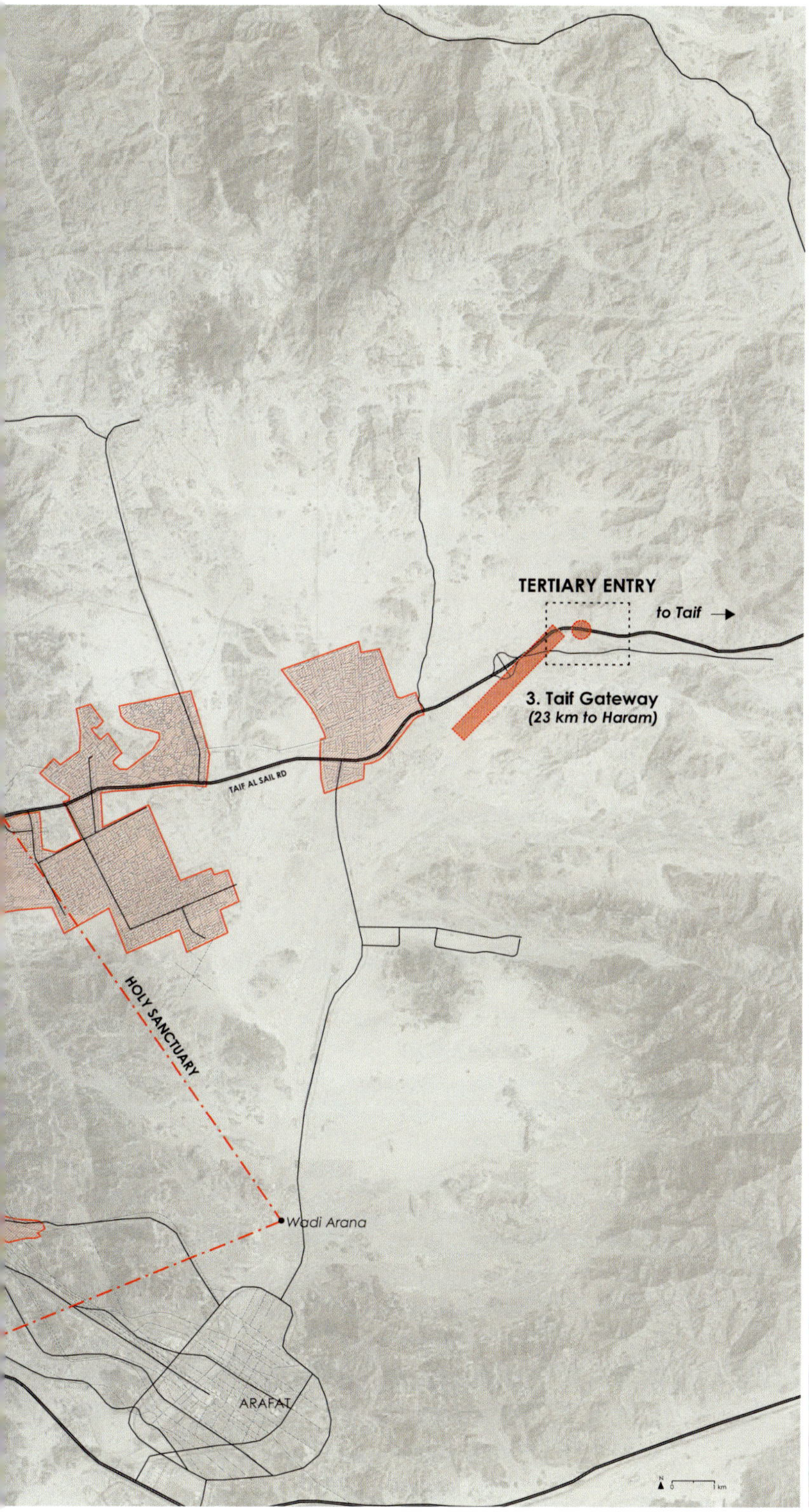

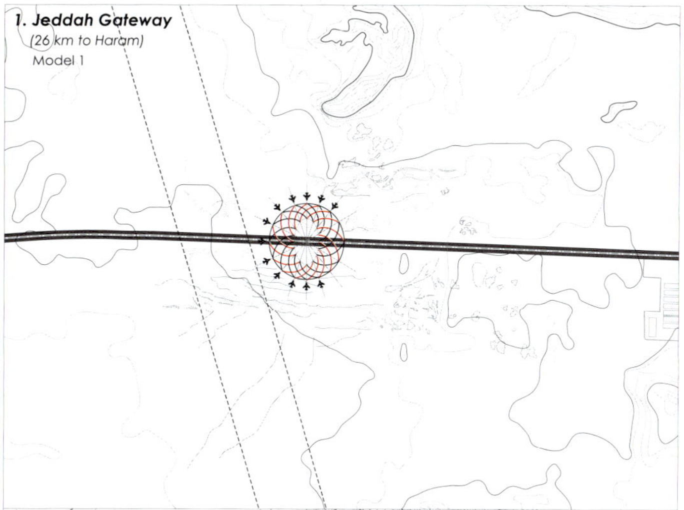

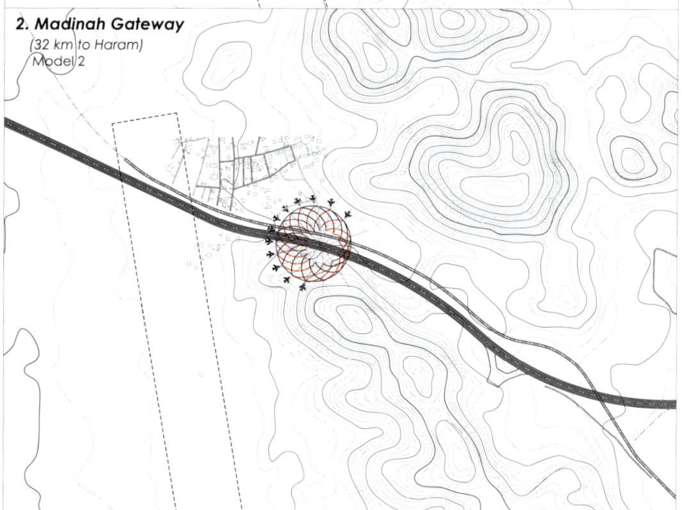

Circulating Around the Mosque, Echoing the Pilgrimage

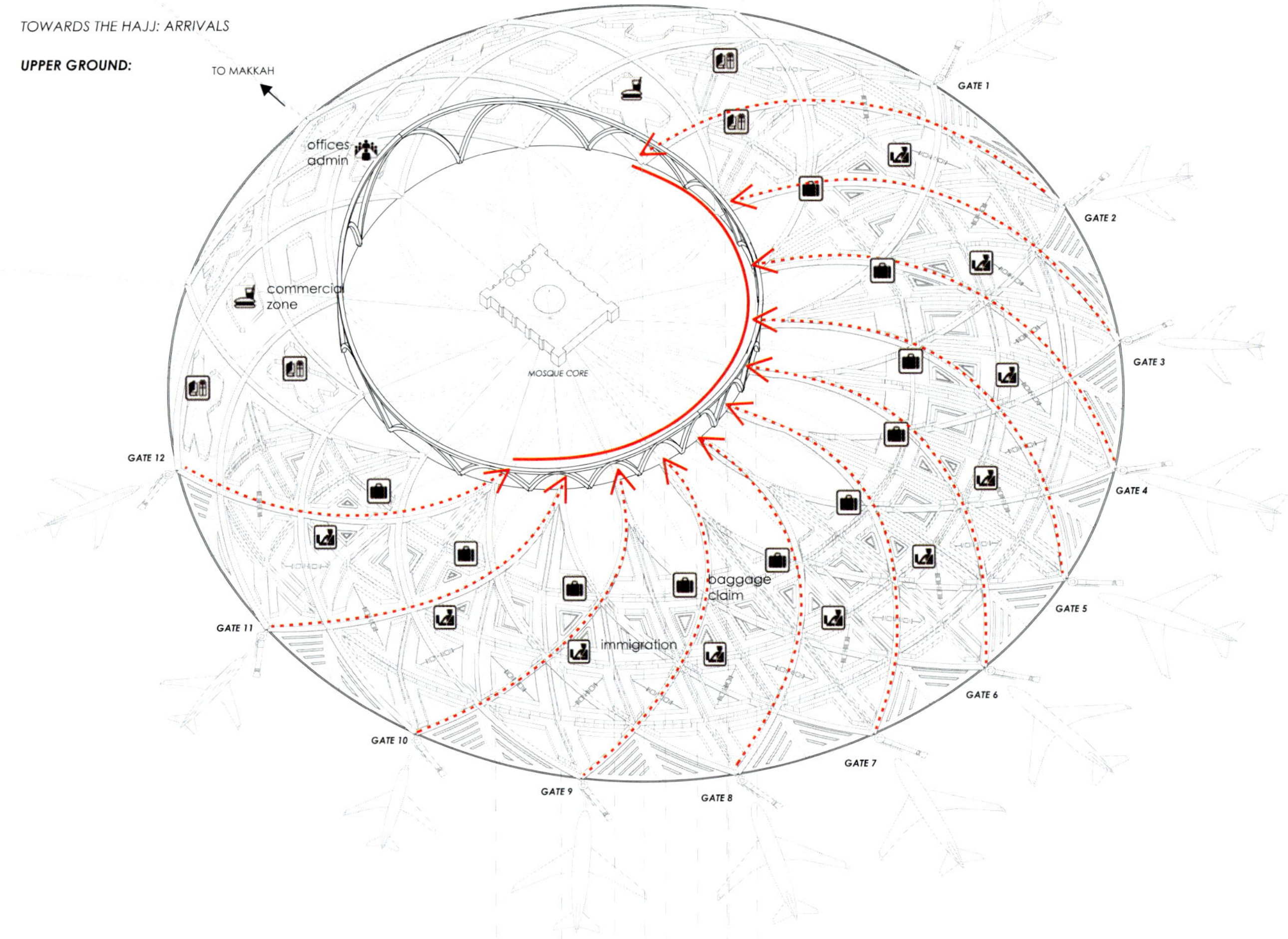

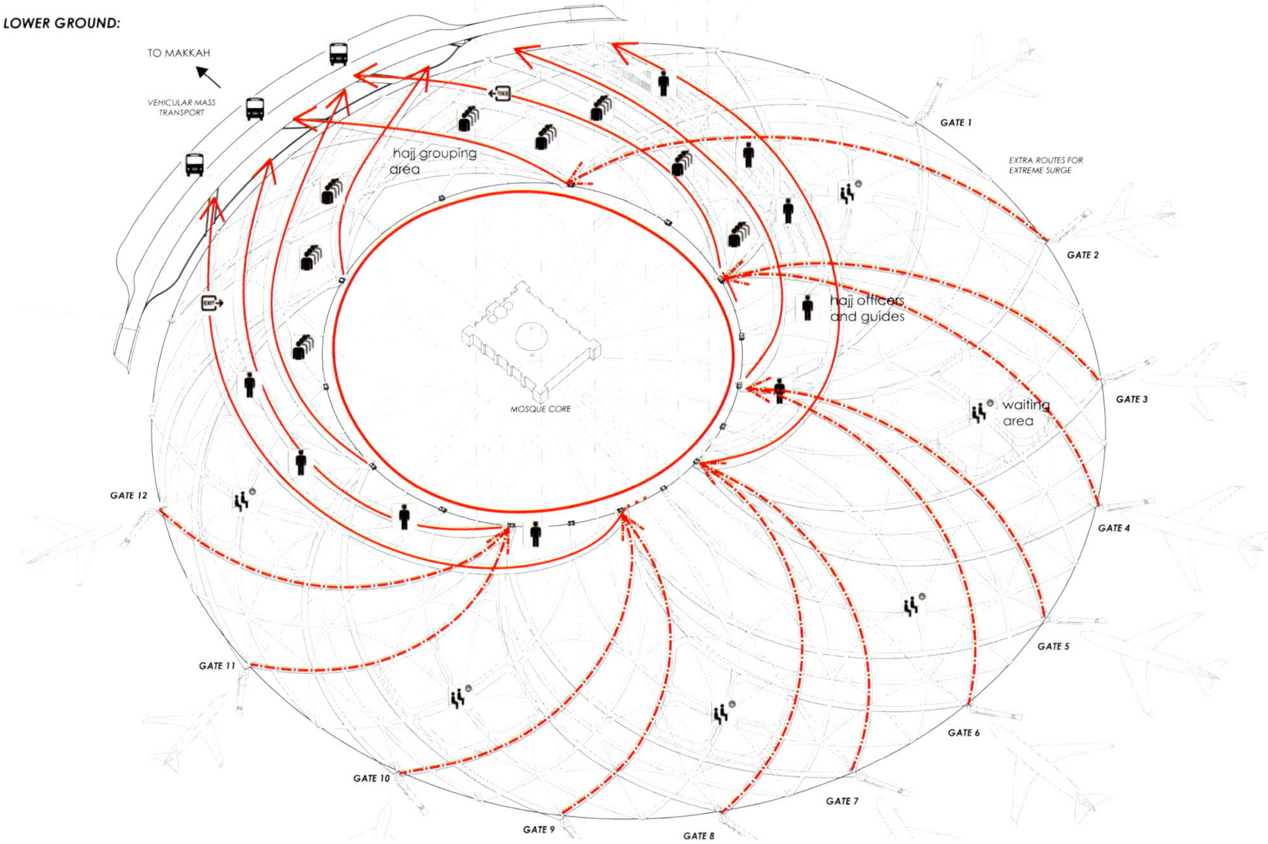

FROM THE HAJJ: DEPARTURES

UPPER GROUND:

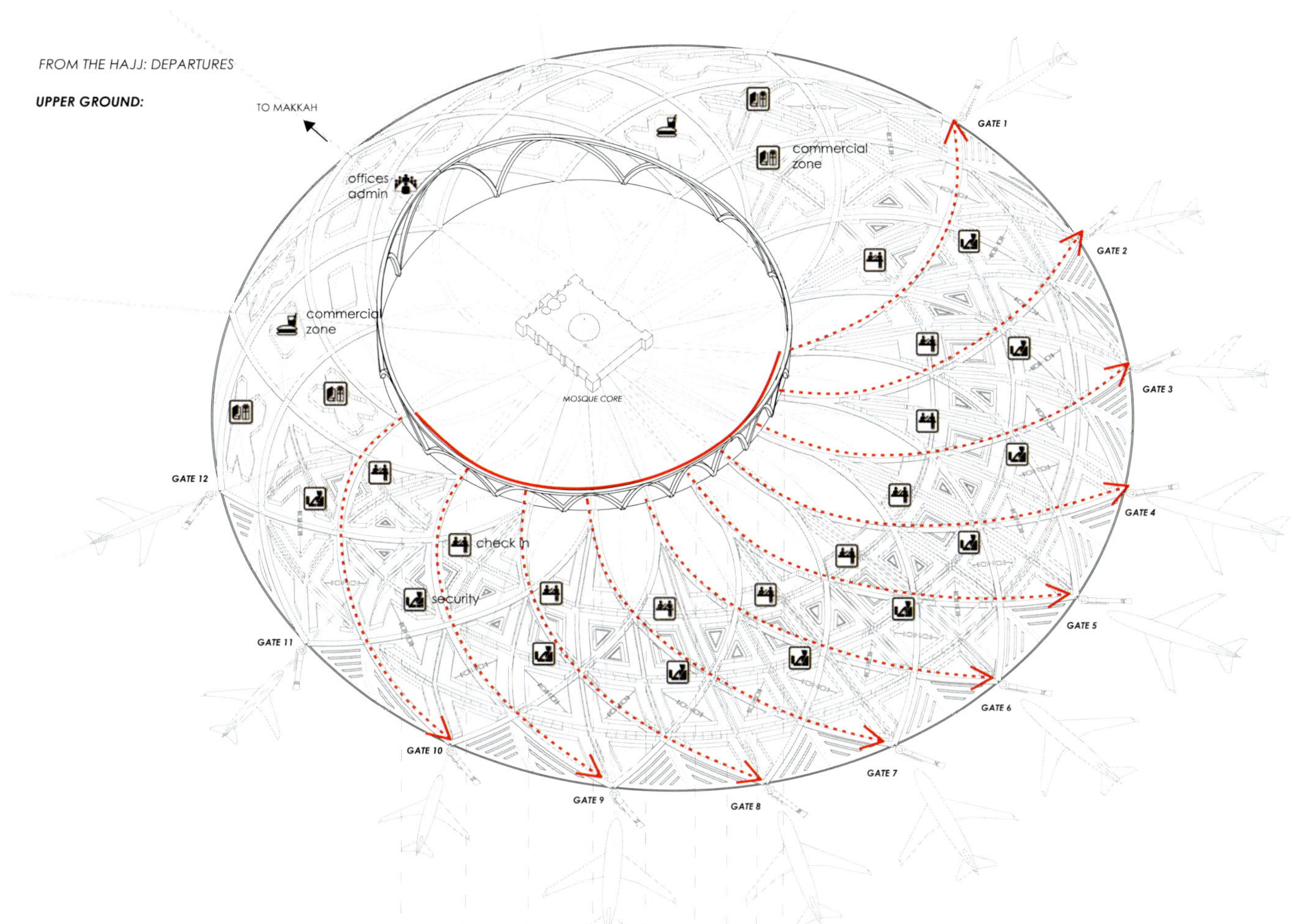

LOWER GROUND:

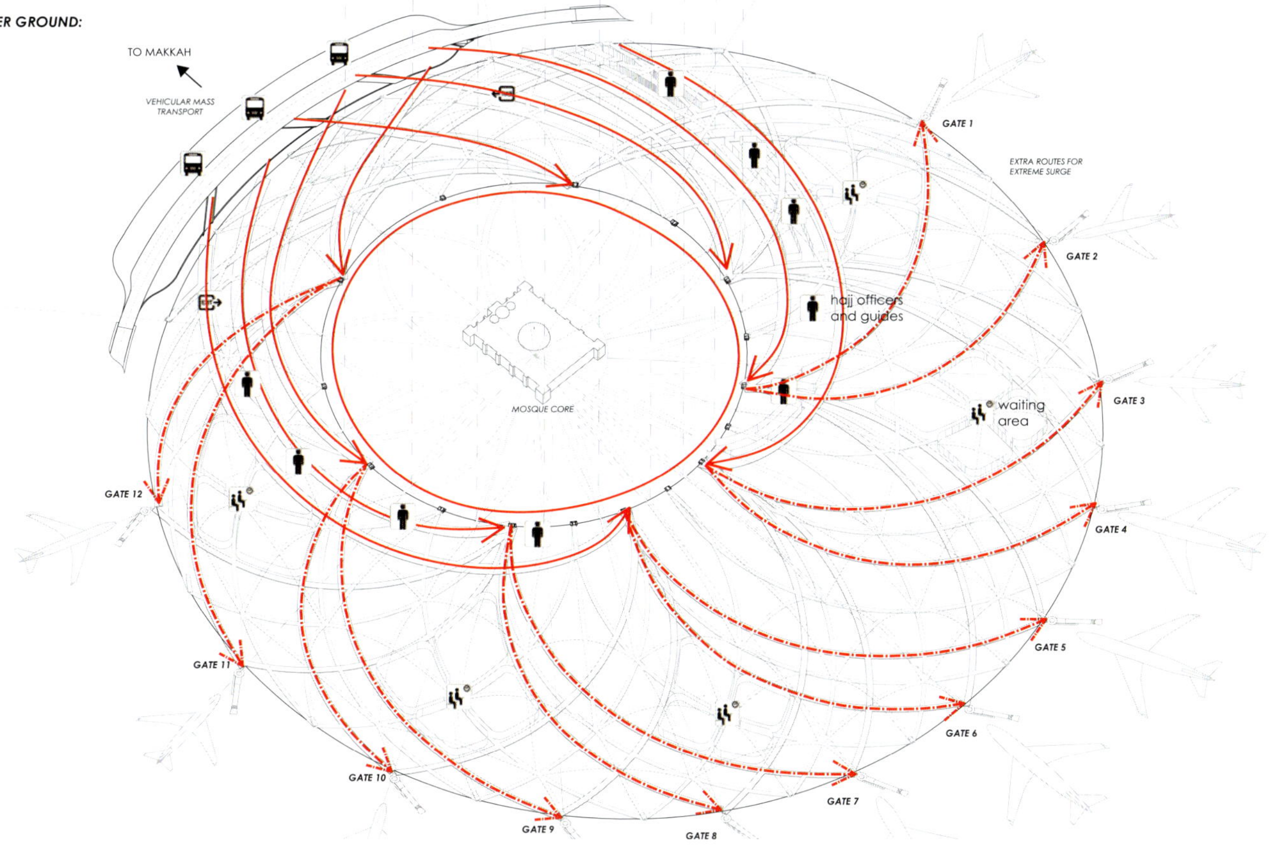

Deena Fakhro

4.3 Interiorized Aerotropolis: A Suspended City

Once the short pilgrimage is over, the mosque campus emerges from the framework. Arching layers, weaving between the segregated airport upper and lower ground levels, seed inhabitable pockets for an off-peak academic community. The marrying of two structurally ornamental types, the airport and the mosque, is articulated through a layered and woven reading of both. The airport's decentralized circulation patterns form the primary structural system, a spiral diagrid, which is overlaid on an undulating secondary arching structure - reminiscent of the mosque's expansive pillared halls.

STL Model (Scale 1:1000)

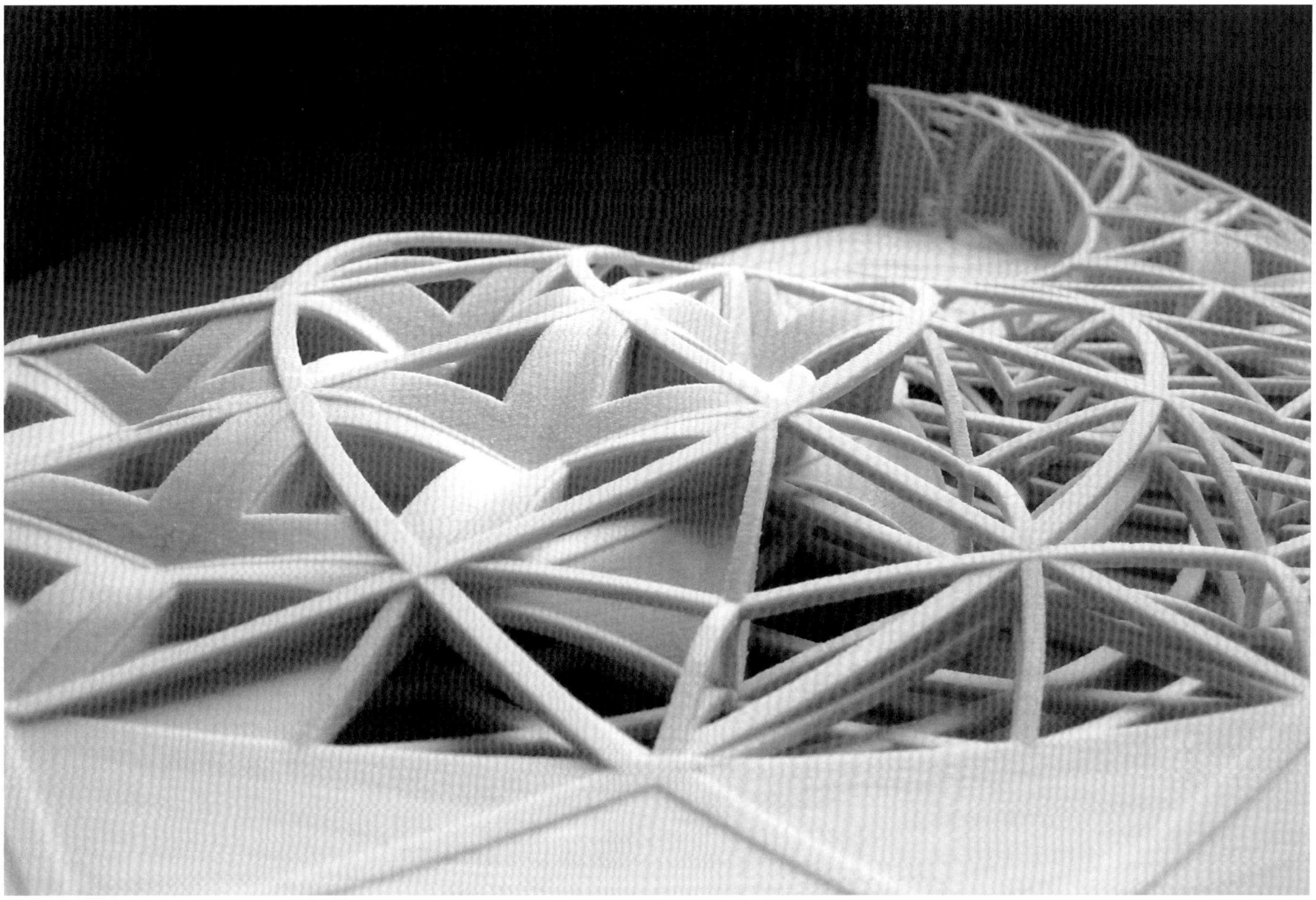

Arching Structure: Reminiscent of the Mosque's Pillared Halls

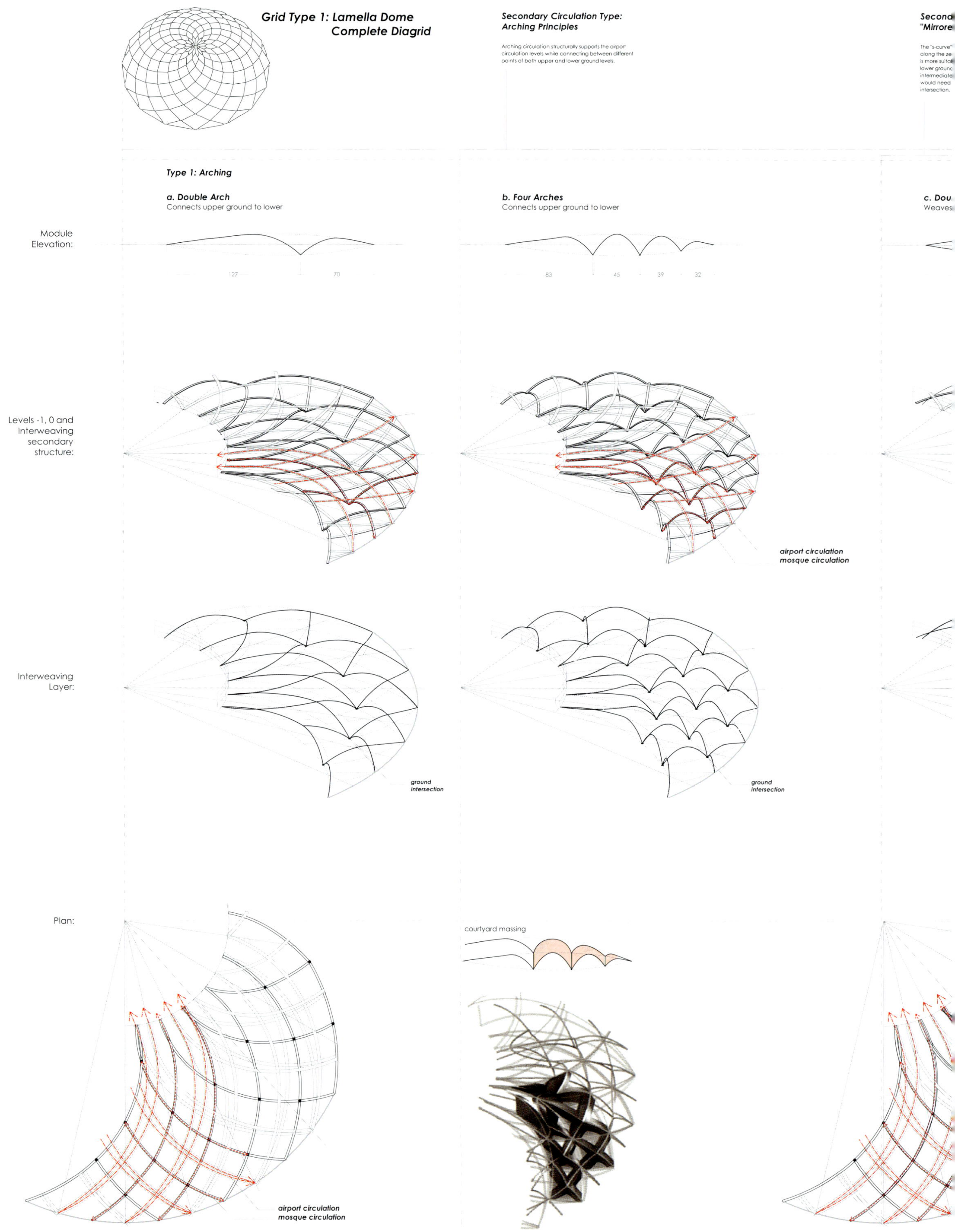

that

etry

Grid Type 2: Schwedler Dome
Ribbed Diagrid

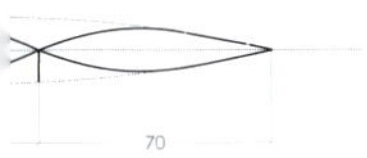

d. Four Arches Mirrored_
Weaves both levels

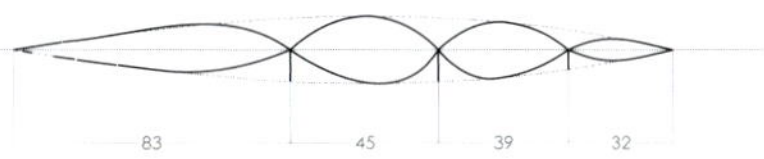

Type 2: Arching

a. Double Arch
Connects upper ground to lower

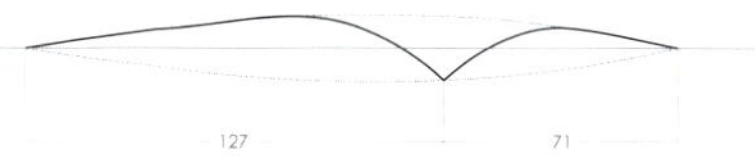

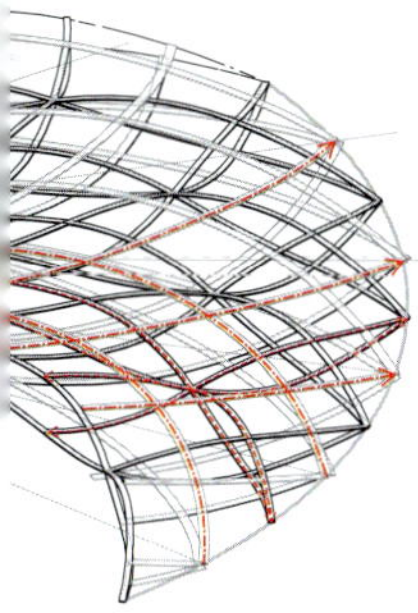

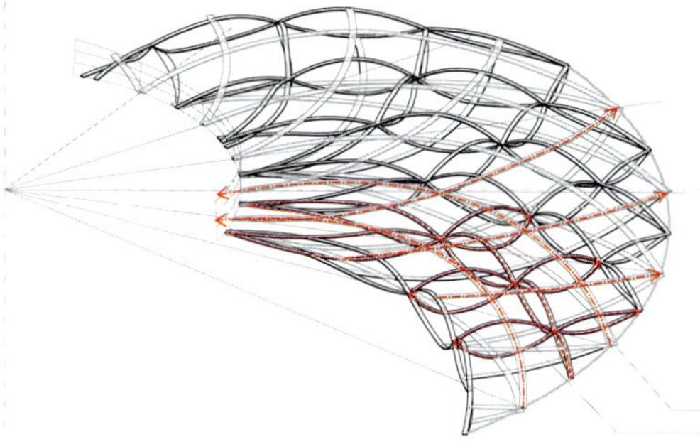

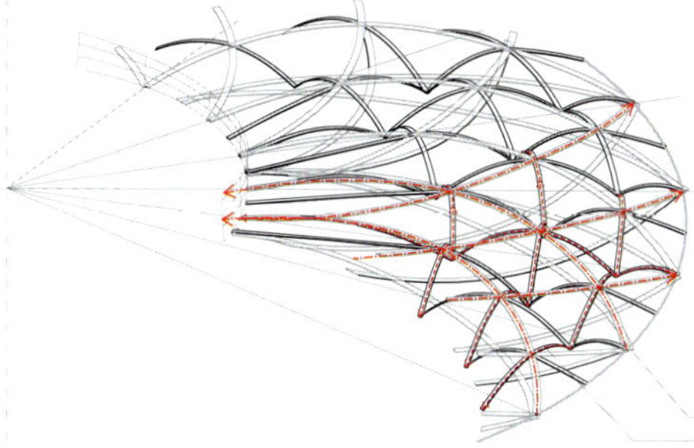

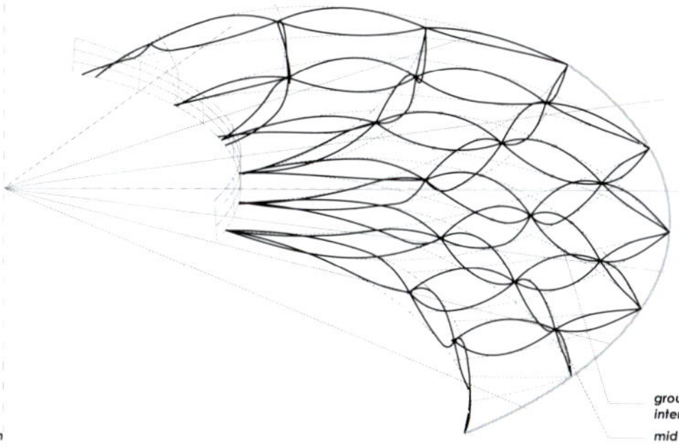

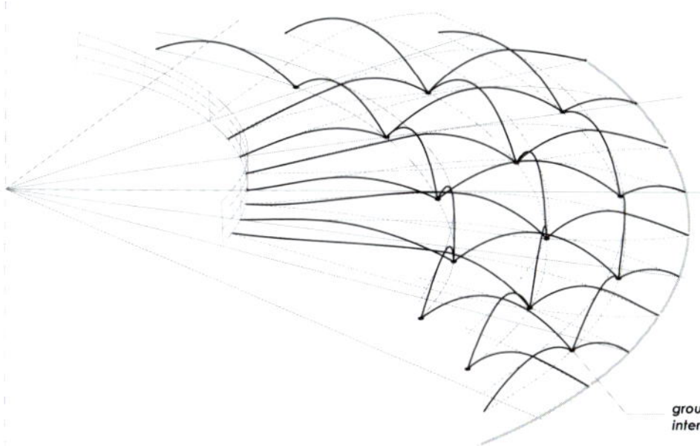

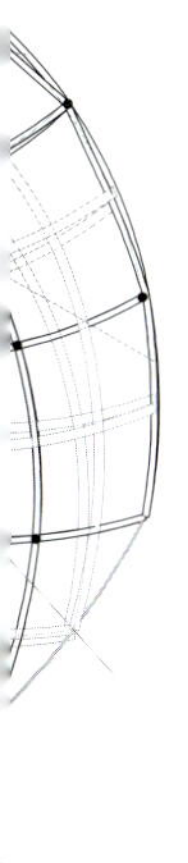

courtyard massing

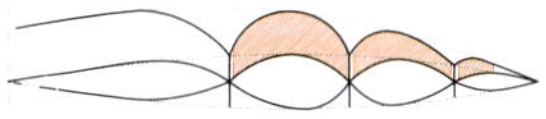

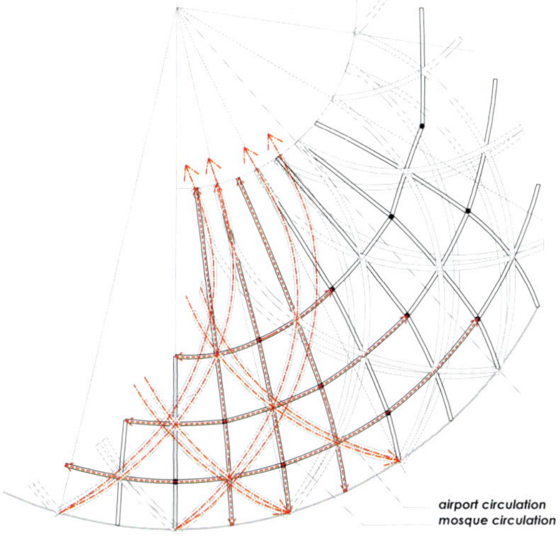

The mosque's undulating pillared hall is extended below a raised ground level containing the archetypal courtyards, all enclosed within the climate-sheltering airport skin. The airport, no longer simply a massive transport node, encapsulates several elements of an Islamic city.

Overlapping Plans:

a b c

b. Level 0
(Upper Ground)

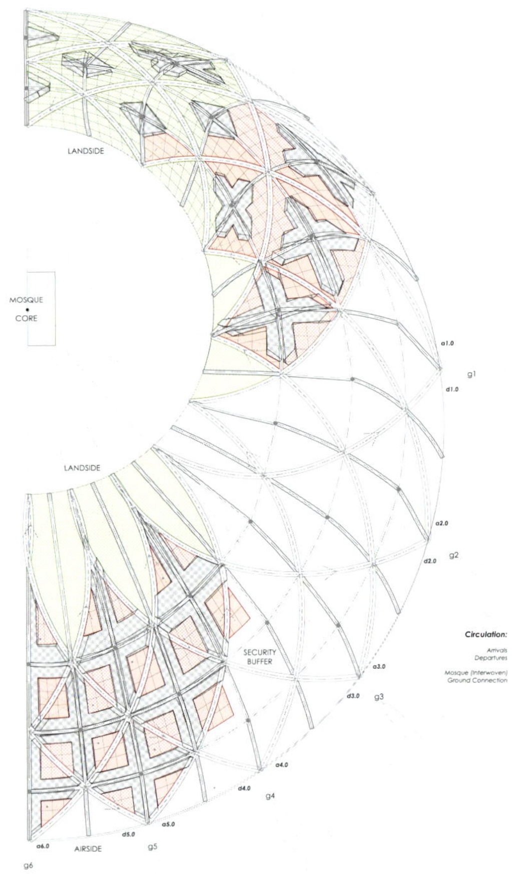

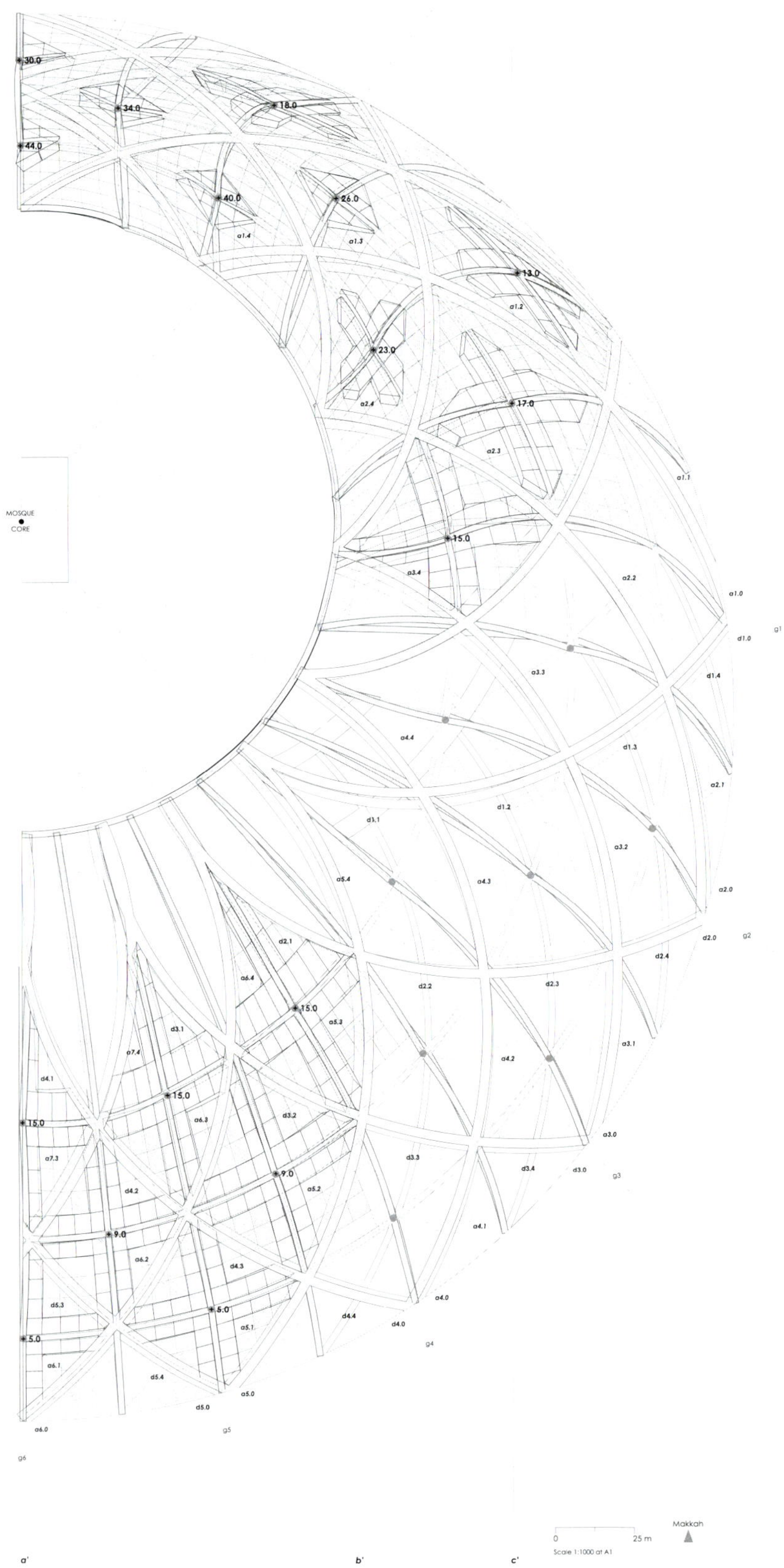

a' b' c'

Upper Ground Plan:

Switch Diagrams: Seasonal Hajj Airport vs Islamic University

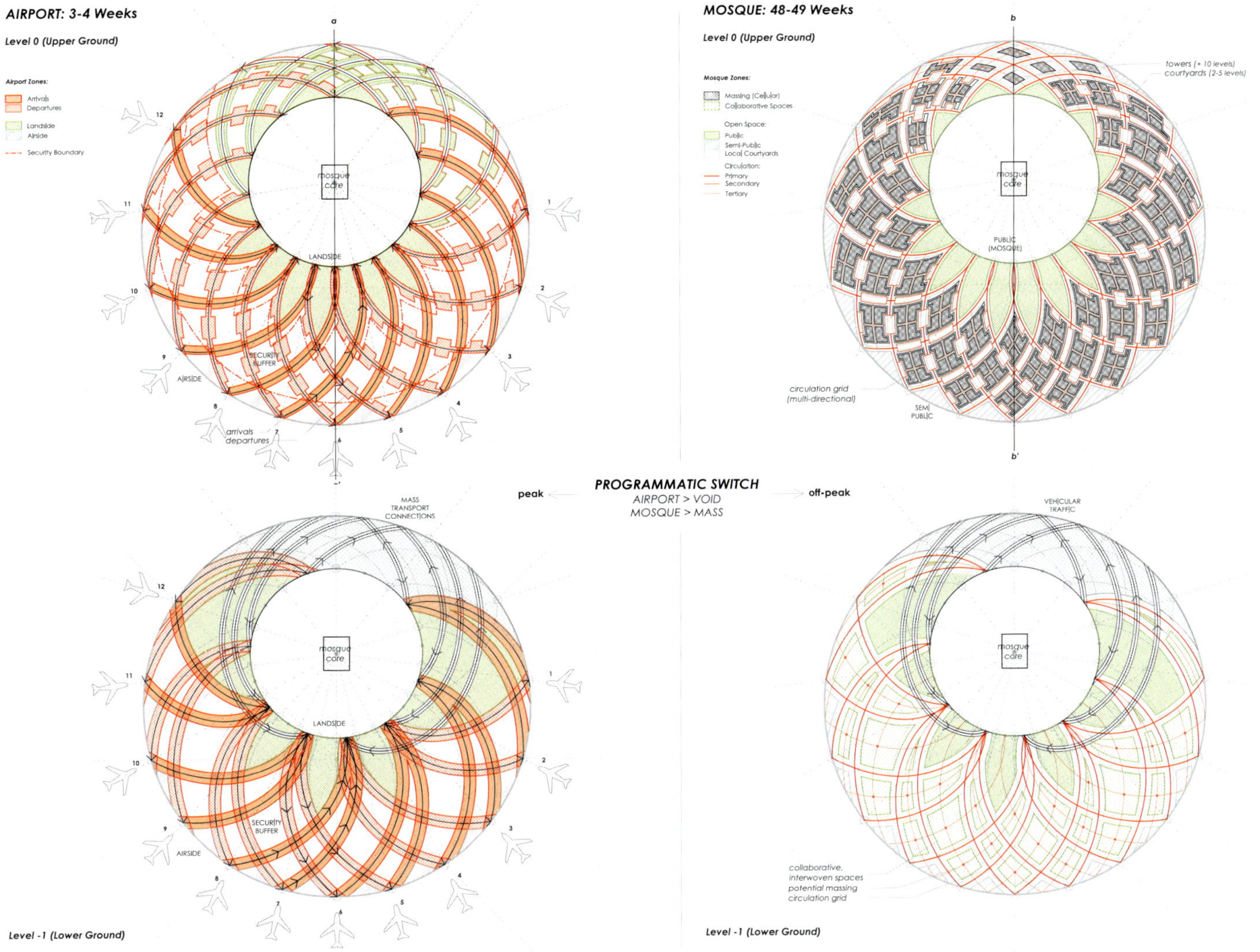

Conceptual Sections:

Section a-a'

Section b-b'

Section c-c'

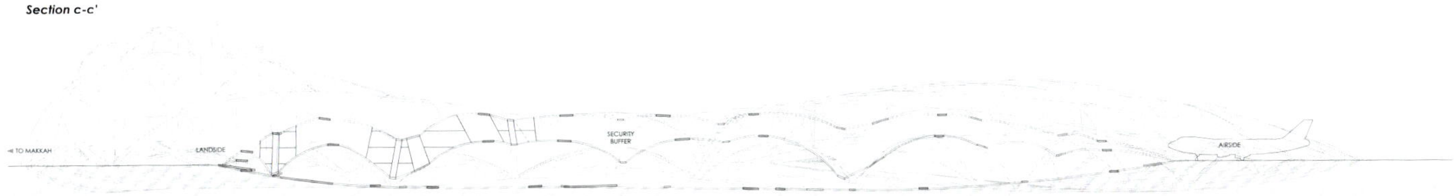

Deena Fakhro

Encapsulating Elements of a Traditional Islamic City withing the Airport Type

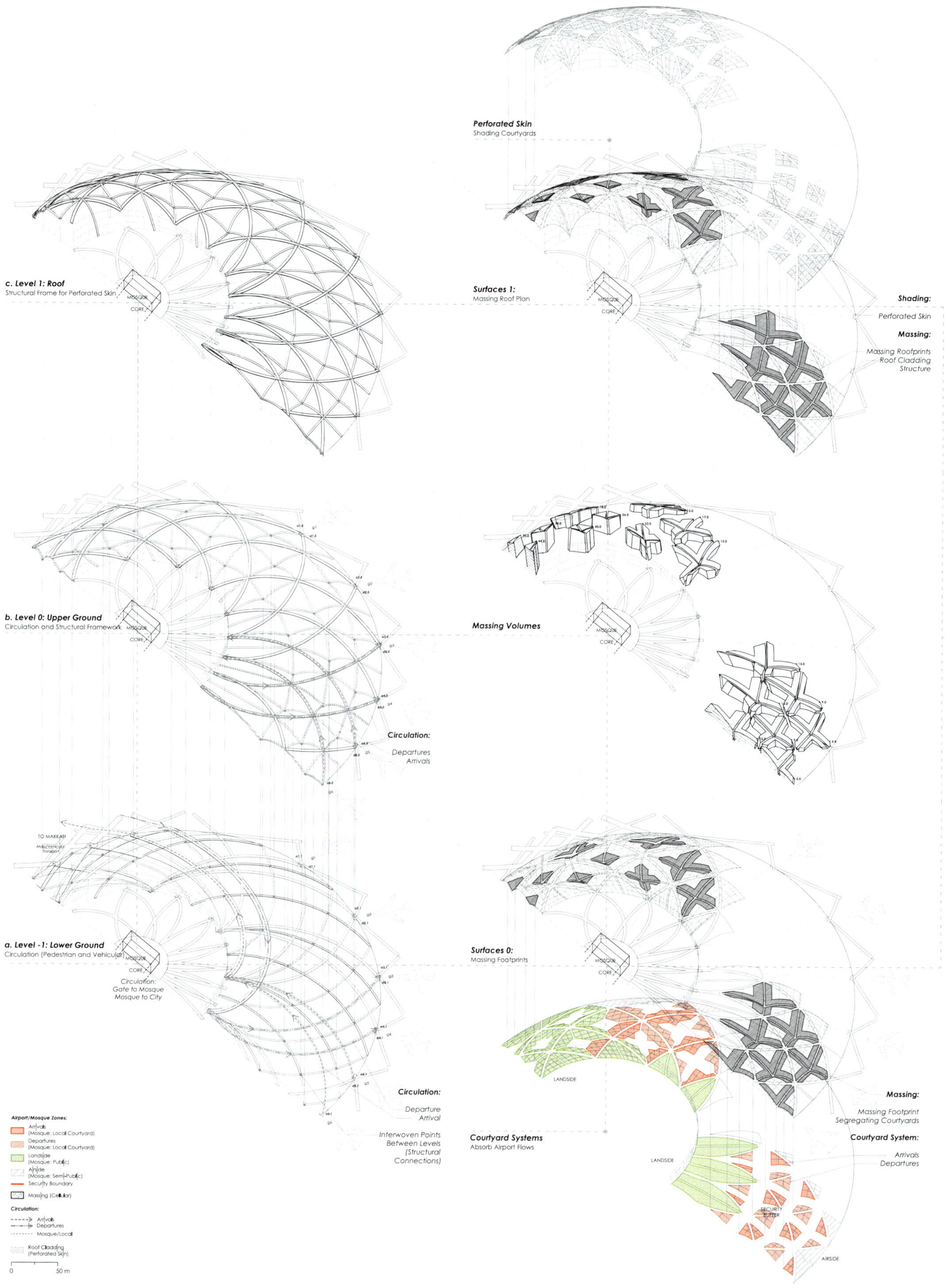

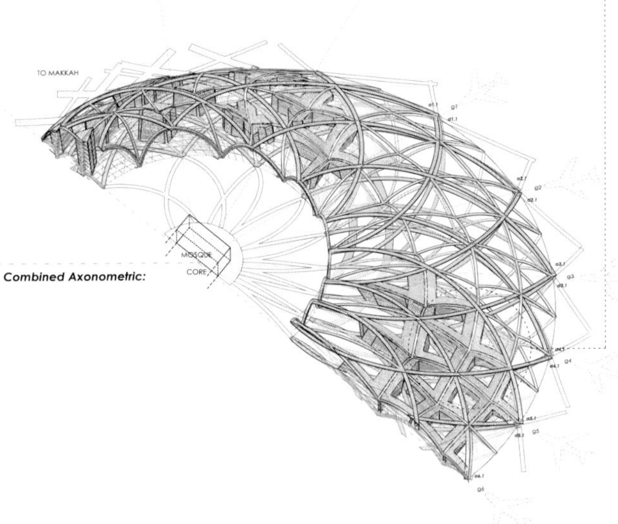

Combined Axonometric:

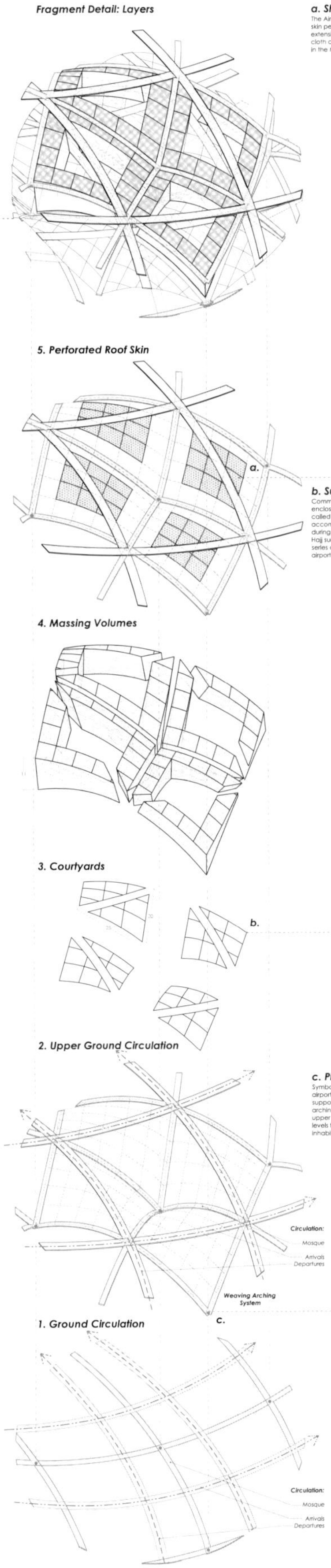

a. Shaded Streets:
The Airport's perforated roof skin performs similarly to the extensive arched, woven or cloth covered souks common in the Middle East.

b. Surge Courtyards:
Common in most mosques, an enclosed large open space, called the Sahn, is used to accommodate excess capacity during Friday prayers. Similarly, the Hajj surge is absorbed through a series of linked courtyards in the airport.

c. Pillared Hall:
Symbolic of the mosque, the airport layers will be supported by a weaving arching system, connecting upper and lower ground levels for off-peak inhabitation.

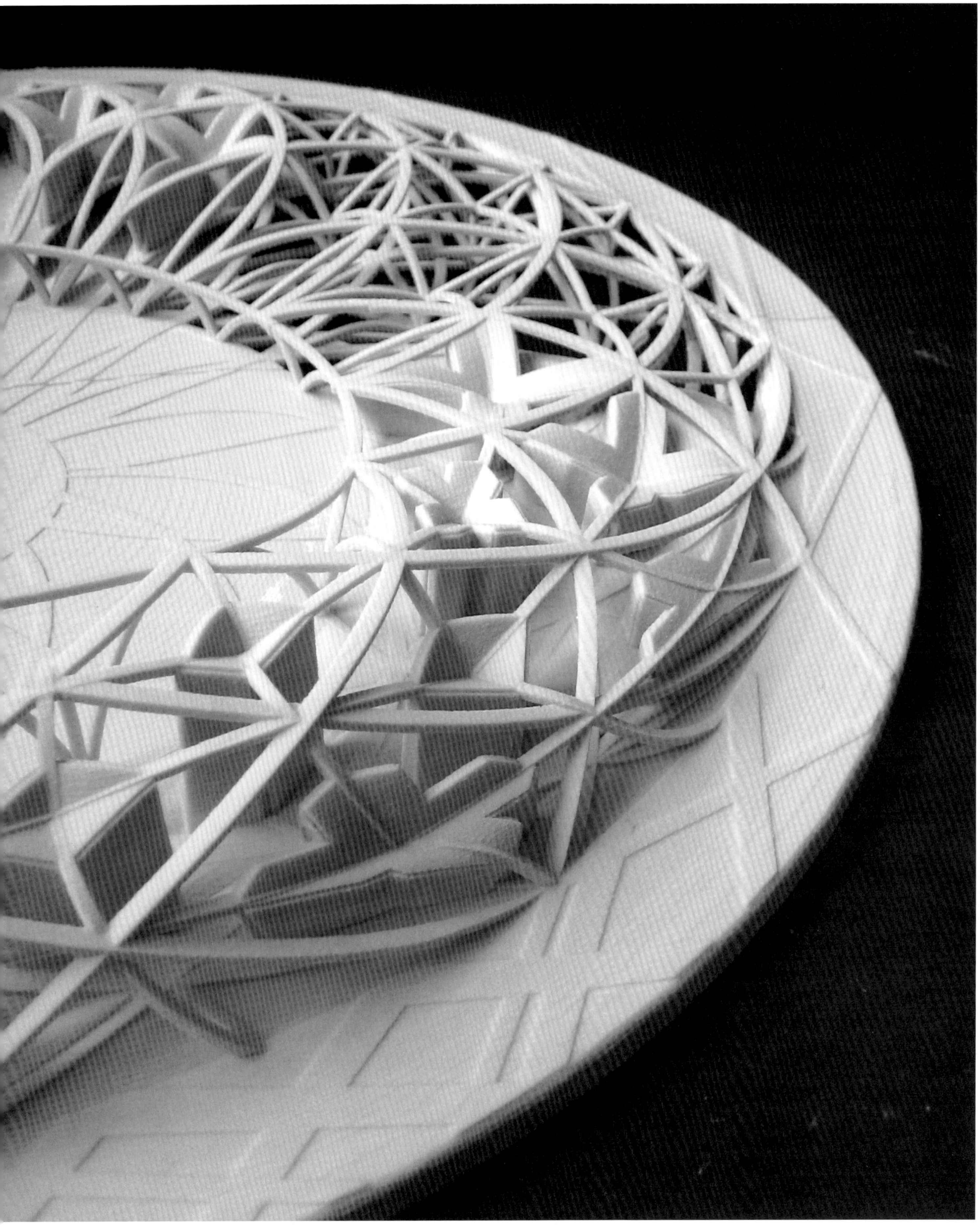

5. Switch

5.1 Structural Layering
5.2 Airport to Islamic University

Structural Layering:

Working on a detailed fragment of the Airport-Mosque, the layering and assembly of different structural and circulation components is developed further. The fragment is composed of four superimposed layers that encapsulate both airport and campus spaces: the ground, the primary arching layer, the courtyard volumes, and the secondary arching layer. Beginning with the ground, a sequencing illustrates the fabrication, materiality and gradual build up of the double arching tiers, concluding with a 1:200 laser cut fragment model illustrating the process. The airport's decentralized spiral circulation patterns translate into a structural diagrid dome, where a series of inverse and intersecting spirals form a lamella grid. Therefore, there is a direct relationship between the airport's functional organization and its structural one. However, the circulation requirements dictate that the spiral ramps remain at a soft, walkable slope, resulting in a shallow dome that is prone to local or global buckling failure. To compensate for the dome's shallow curvature, far from ideal catenary or parabolic geometries, a series of superimposed and perpendicular arches is used to break its large span at several points and connect to pile foundations in the ground. In terms of materiality, the ground and primary tier of arches are formed from both precast concrete members and insitu cast concrete to allude to the traditional heavy masonry pillars of mosques. While the upper tiers are formed from lighter and stiffer steel members.

Airport to Islamic University:

Yet the switch moment itself, when the structure shifts from Islamic university to Hajj airport, must be minimally disruptive as it occurs bi-annually. Therefore, the layered architecture remains exactly the same, while different parts of the plan and circulation grid are activated for the two different programs.

The airport is read as a series of linked courtyards, forming a connected network of open spaces able to temporarily accommodate the airport functions, from check-in desks to immigration booths. Borrowing from the logic of the typical mosque courtyard or "sahn" where during Friday prayer surges the excess worshipers filter out into open spaces, the Airport-Mosque absorbs the momentary Hajj surges through a larger system of courtyards. Here, circulation is blocked off to form a corridor like movement within the open spaces – processing pilgrims linearly through airport technicalities.

While off-peak, the educational programs emphasize the massing, or cellular courtyard volumes, and the circulation grid is fully opened up to allow for maximum freedom and interaction. Maintaining the organizational logic of its historic precedent (the Islamic university) the aerotropolis' campus volumes progress in a gradual linear hierarchy, moving from most public towards the central mosque core to most private at the outer periphery. The levels of privacy ensure that public concentration is maintained within the central core area while filtering out towards a non-growing and unchangeable edge.

Thus, by activating different parts of the plan and circulation grid the two programs coexist symbiotically, and opportunity surfaces where redundancy lay.

Airport (1 Month)	Campus (11 Months)
Closed Circulation	Open Circulation
Waiting Halls	Lecture Halls

5.1 Structural Layering

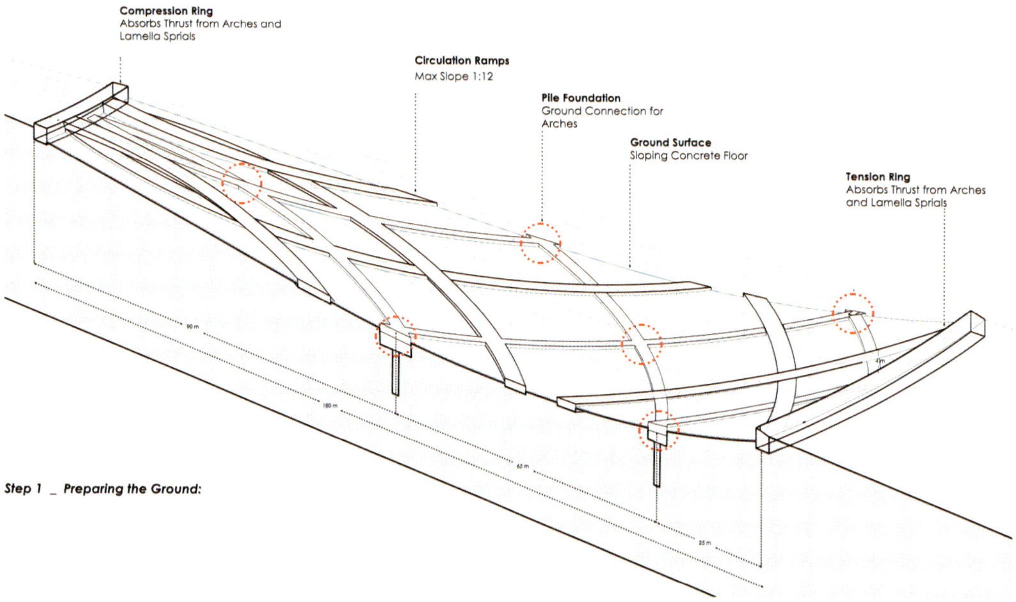

Step 1 _ Preparing the Ground:

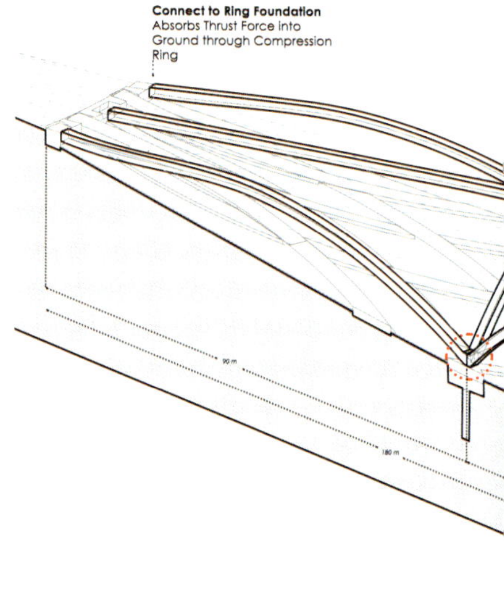

Step 2 _ Primary Tier of Arches:

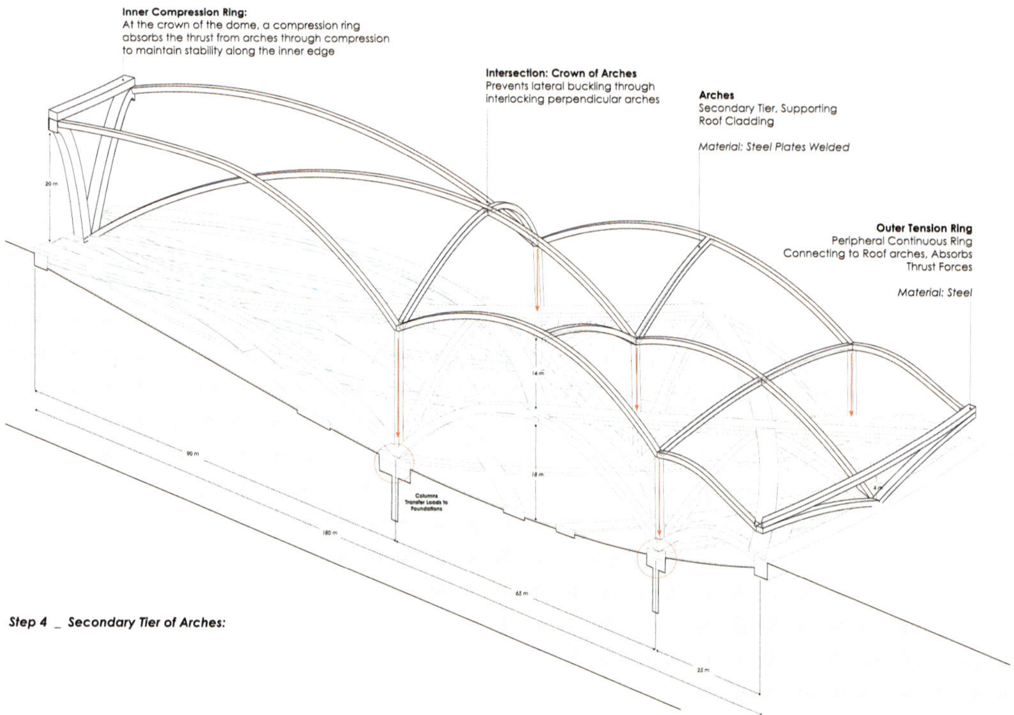

Step 4 _ Secondary Tier of Arches:

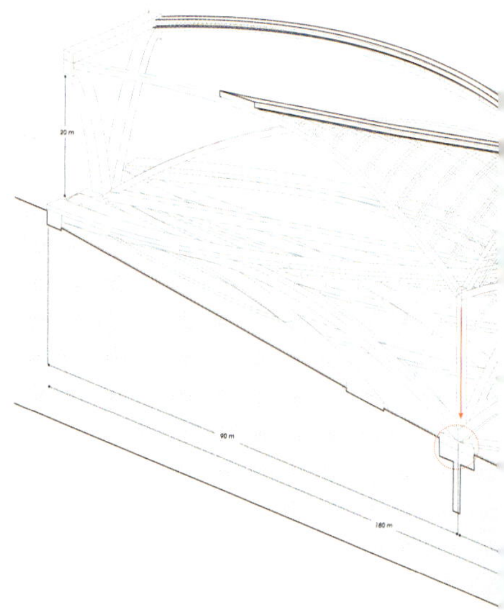

Step 5 _ Secondary Tier of Lamella Diagrid:

Model Construction:
3 Lamella Grid Layers Woven Together with an Arching System

1. Ground Beams and Circulation Ramps

2. Upper Ground Level: Spiral Diagrid, Arches, and Stressed Skin

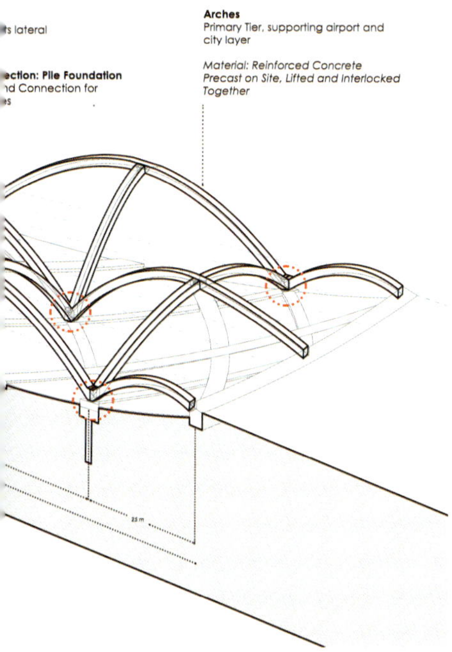

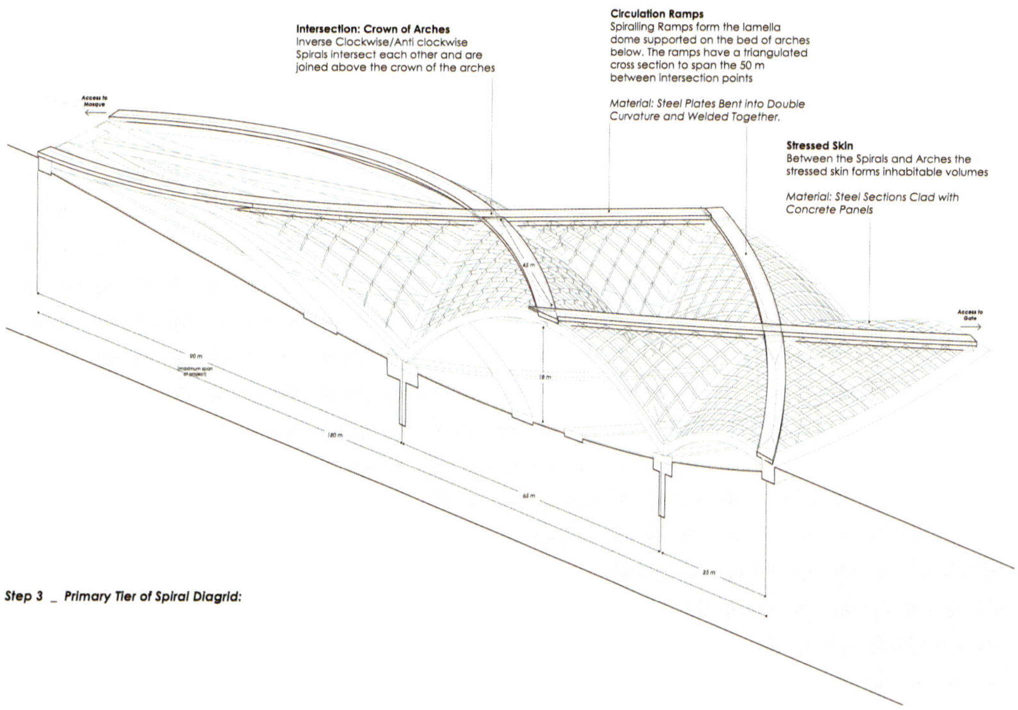

Step 3 _ Primary Tier of Spiral Diagrid:

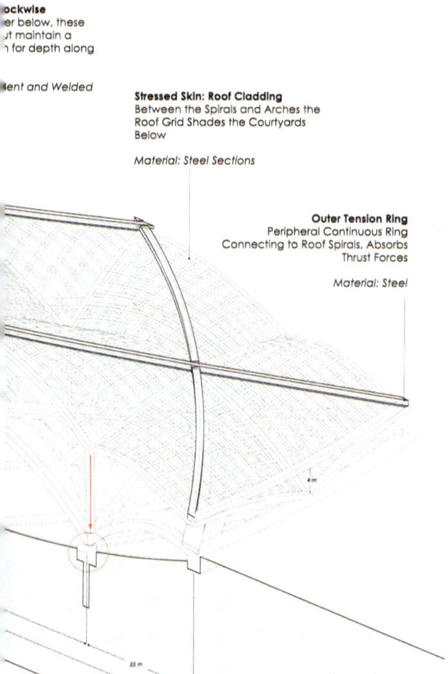

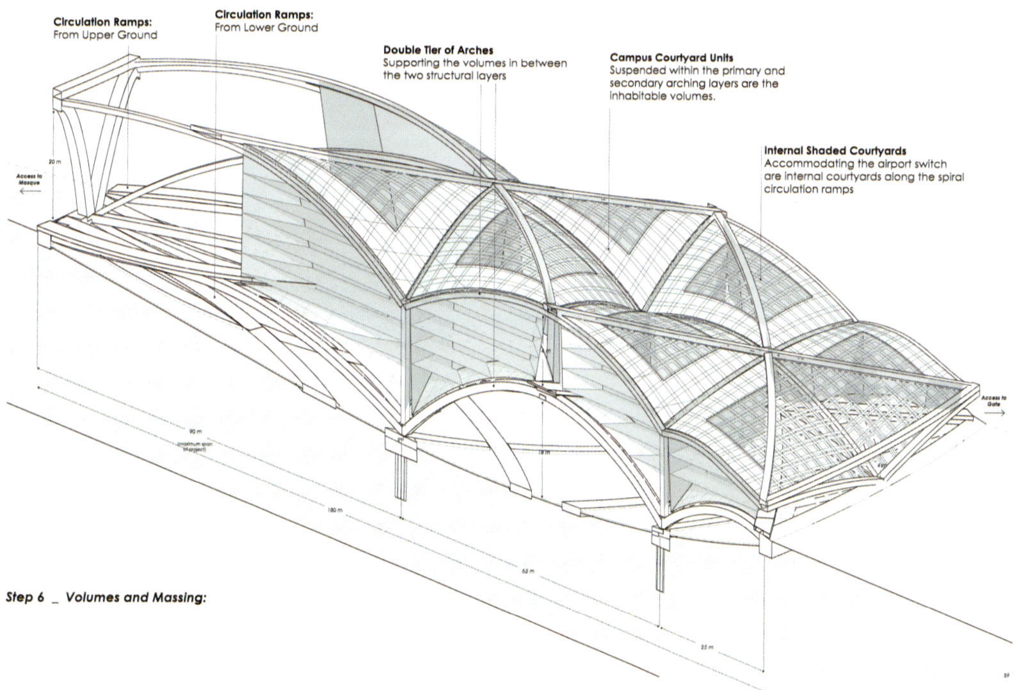

Step 6 _ Volumes and Massing:

3. Roof Cladding: Superimposed Arches, Support Diagrid and Stressed Skin

5.2 Airport to Islamic University

The airport functions are absorbed within a network of open space courtyards, where circulation is blocked off to form arrival and departure corridors. While the Campus activates the volumes framing the open spaces during the off-peak season for the majority of the year.

Seasonal Hajj Airport

1 Month: Courtyard Chains Activated

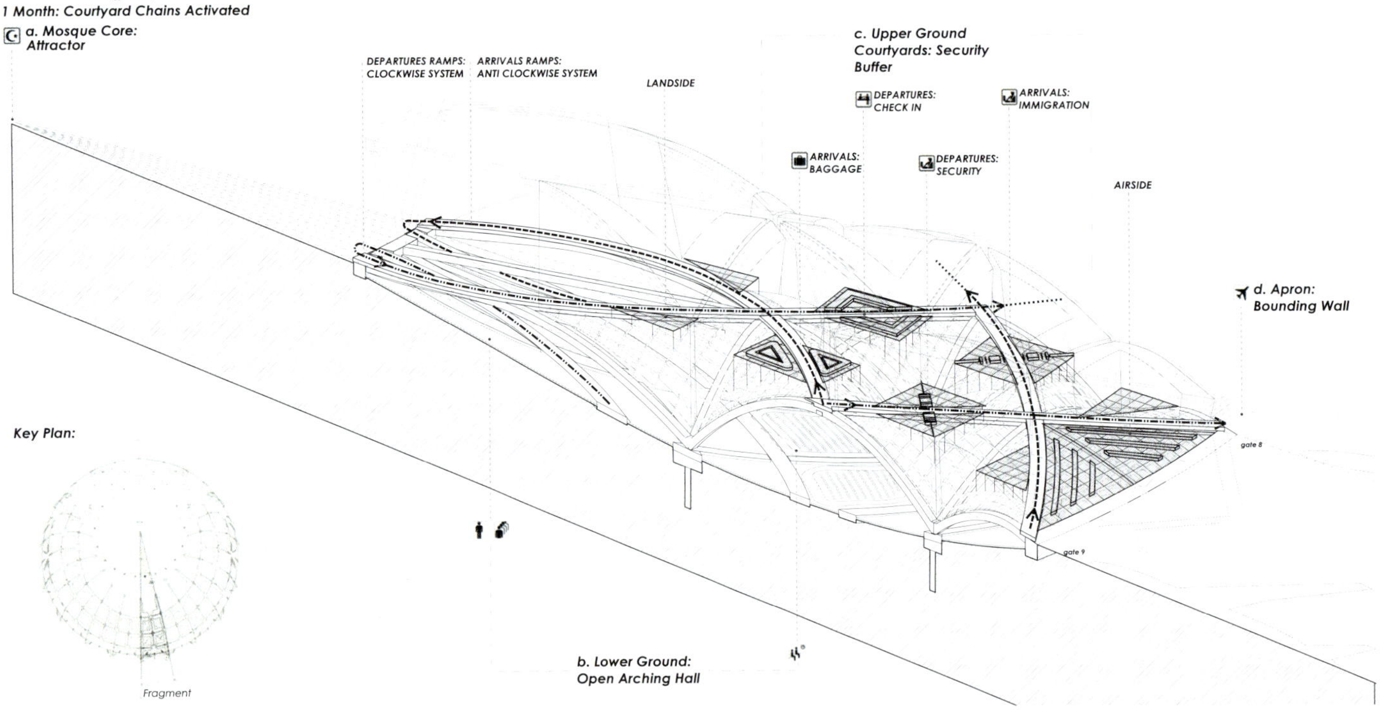

Courtyard Voids: One Way Arrival vs Departure Corridors

Islamic University

11 Months: Massing Units Activated

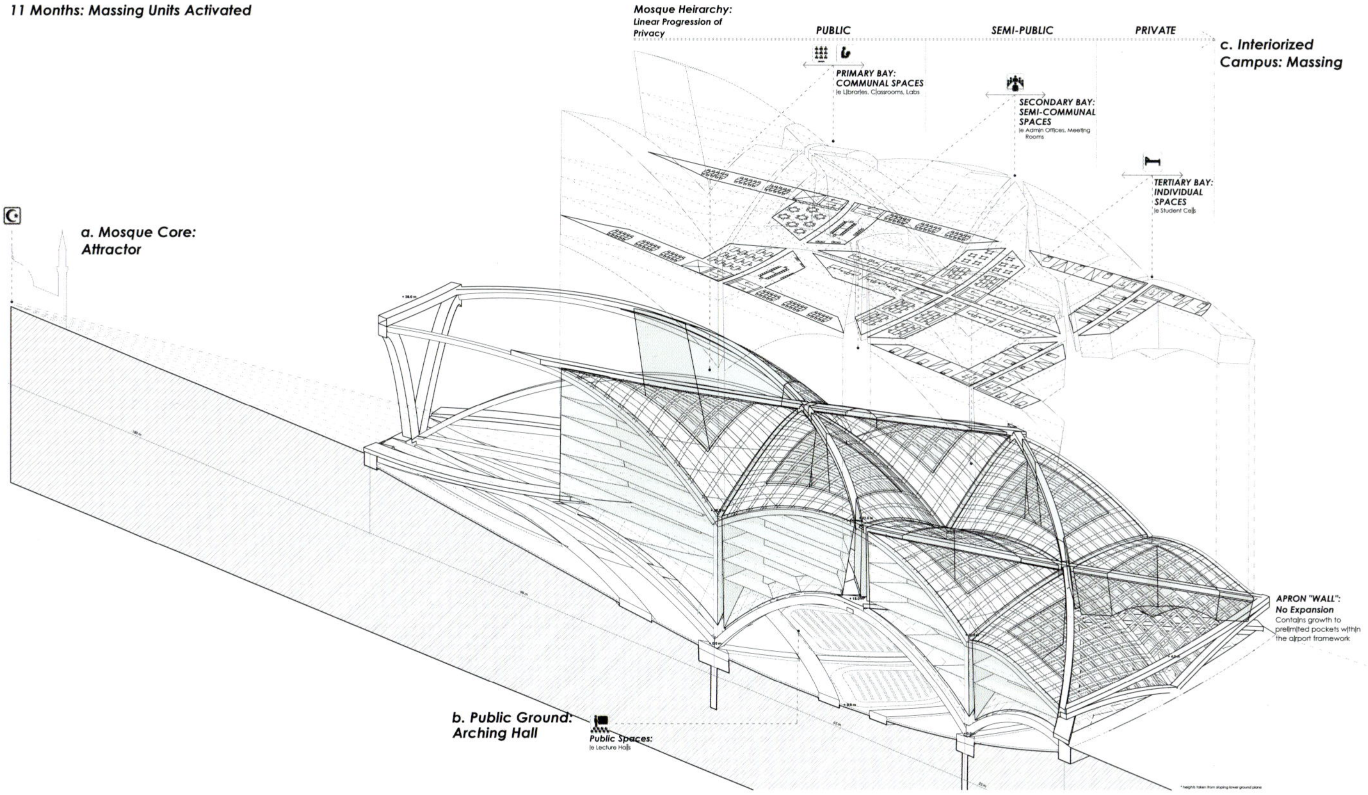

Courtyard Massing: Heirarchical Educational Programs

Laser Cut Fragment Model
Scale 1:200

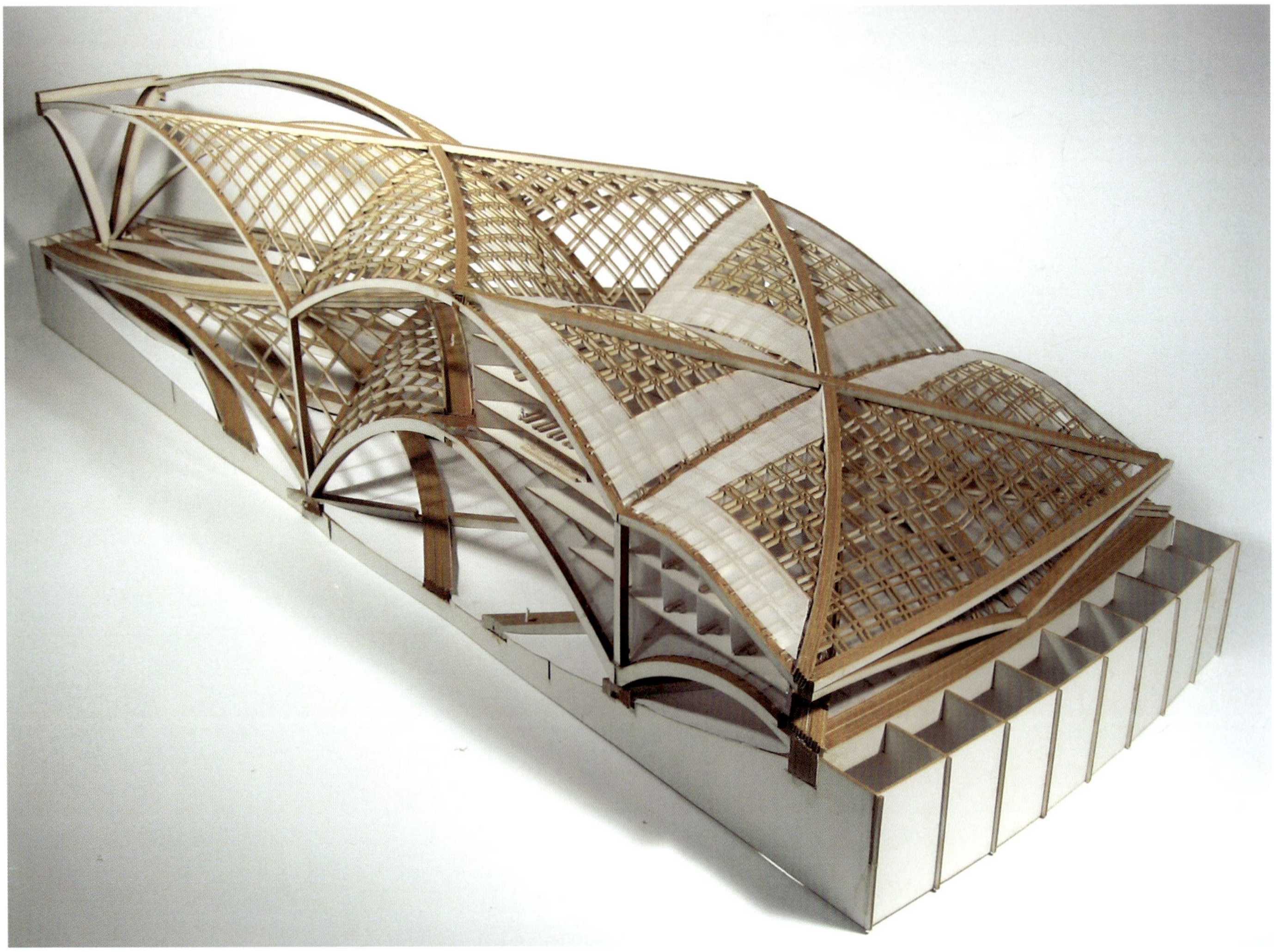

6. Gateways to the Holy City

6.1 Valley Gateway
6.2 An Autonomous Aerotropolis

The Manhattanization of Makkah:

In a move that has been dubbed "the Manhattanization of Makkah" by the Islamic Heritage Research Fund, in the next four years over £15 billion will be pumped into the city by local and foreign investors to fund the construction of over 130 skyscrapers – including the world's largest building to-be, Abraj Al Bait Towers. These gargantuan projects will overshadow and suffocate the sacred Haram Mosque at the heart of Makkah. They exacerbate issues of congestion through increasing centralization and density within the humble city's already threatened core. Furthermore, they continue to fuel Makkah's dependence on an economically draining, rather than profiting, Hajj pilgrimage. The provision of thousands upon thousands of hotel rooms to accommodate 3 million pilgrims for a fleeting six days will increase problems of redundancy. Although a comparatively small percentage of facilities will be always be occupied by Umra (the lesser pilgrimage) pilgrims year round, it is hardly enough to sustain this overprovision of Hajj infrastructure.

In contrast to the escalating centralized congestion of Makkah today, the project's proposed polynodal Gateways disperse the Hajj entry-congestion away from one crippled infrastructural corridor, distributing it along multidirectional routes that pass through valley entries. Moving away from overdependence on the Hajj, the Gateways allow for specialized (knowledge-industry based) and diversified satellite growth around Makkah. As autonomous entities, they do not compete with the Holy City's historic Haram core but instead rehabilitate it during pilgrims lulls by encouraging new growth within the airport nodes and along infrastructural corridors.

Valley Gateway:

Celebrating the holy threshold, the autonomous airports nestle into Makkah's surrounding valleys, above sunken highways. Framed and cradled by the mountainous landscape, they filter both vehicular and airplane access into the Holy City. As distinctly contained and clearly demarcated entities, they symbolize arrival upon and entry into a divine land protected by sacred boundaries. The ancient relationships of gateway to city, and pilgrim to pilgrimage are resurrected.

An Autonomous Aerotropolis:

During the month-long airport season, pilgrims circulate towards the core where they can access ramps to their gates. Here, the mosque serves purely as a visual orientating landmark guiding pilgrims as they move through the airport spaces – their motions alluding to the once-in-a-lifetime pilgrimage on which they are about to embark. While during the 11 month off-peak season, the mosque instead becomes a thriving magnet attracting programs around it and revitalizing the otherwise redundant, stagnating infrastructure. On the upper ground level are the series of linked open spaces for the airport, that are framed by the educational volumes and all contained between the landside mosque core and the growth-controlling wall of the airside apron.

Makkah 1850

Makkah 2004

Makkah 2010

6.1 Valley Gateways

Restoring gateway entries to Makkah, the airports are framed by valleys. Aligned with the Qibla direction, they highlight a crucial axis towards the Holy City.

Regional Plan: Madinah Aerotropolis Gateway

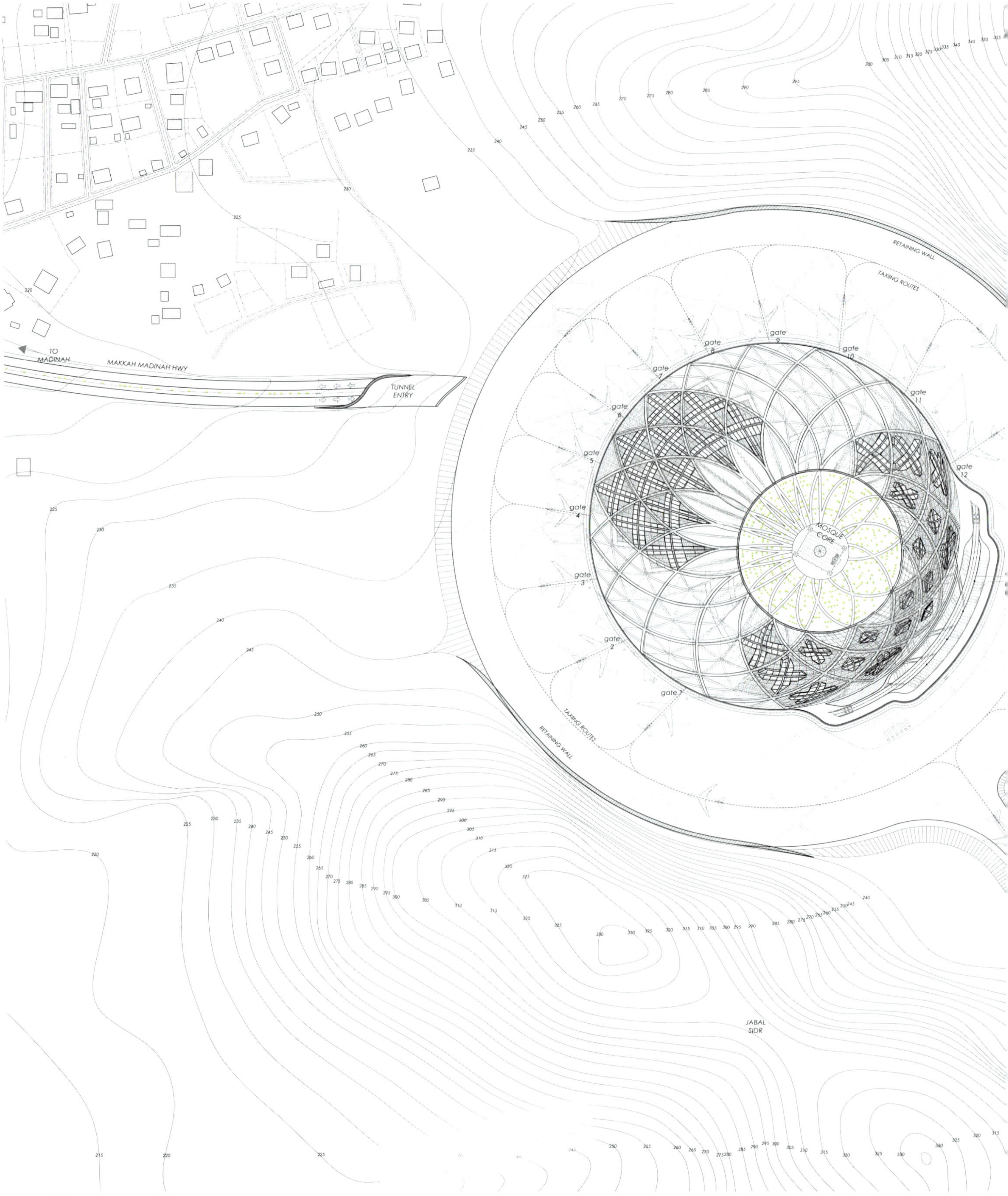

MAKKAH MADINAH HWY
(BELOW GRADE)
RUNWAY (NW-SE ORIENTED)
TUNNEL
ENTRY
TO
MAKKAH
0
100 m
N
Qibla

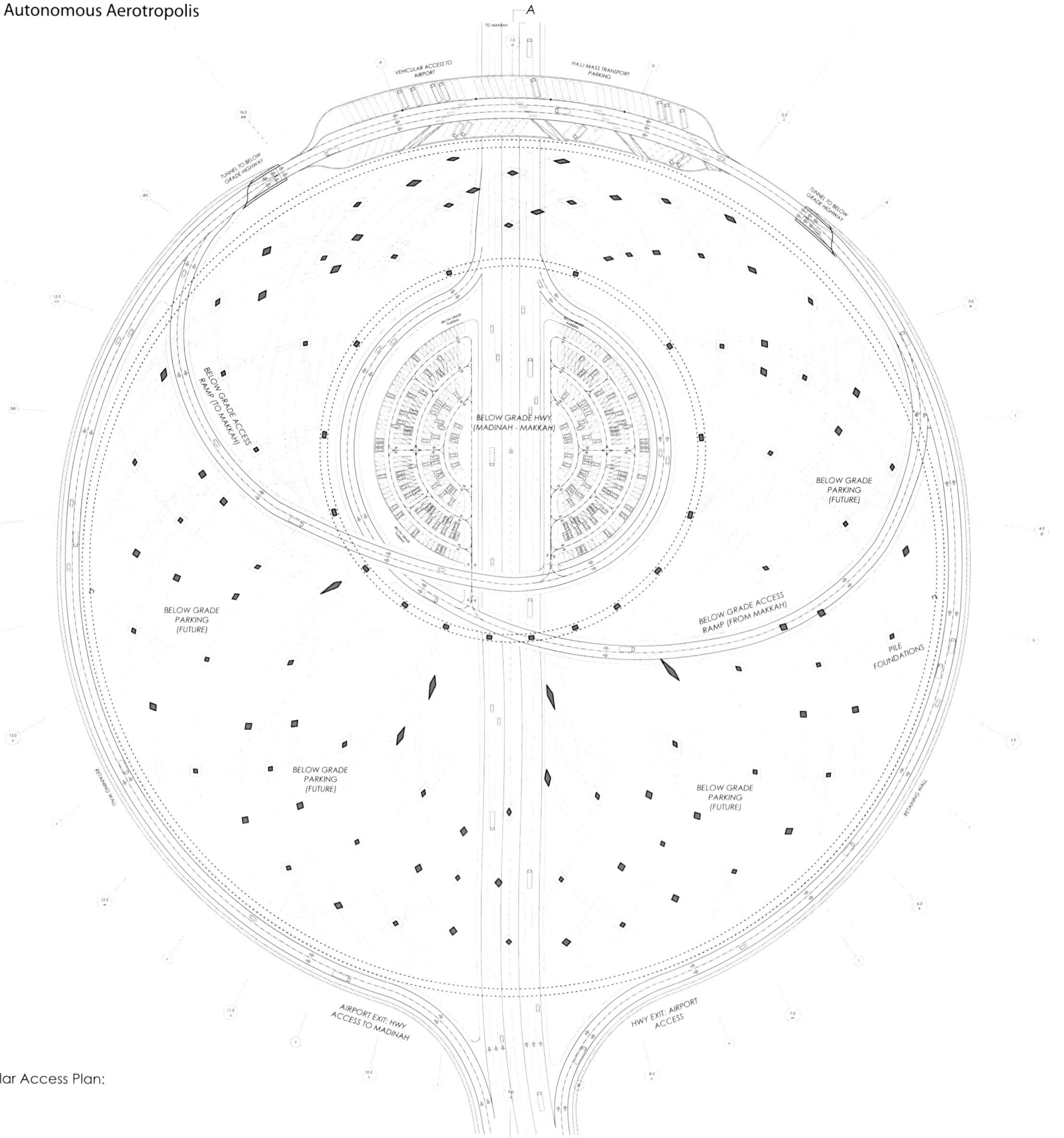

Vehicular Access Plan:

Section A-A':

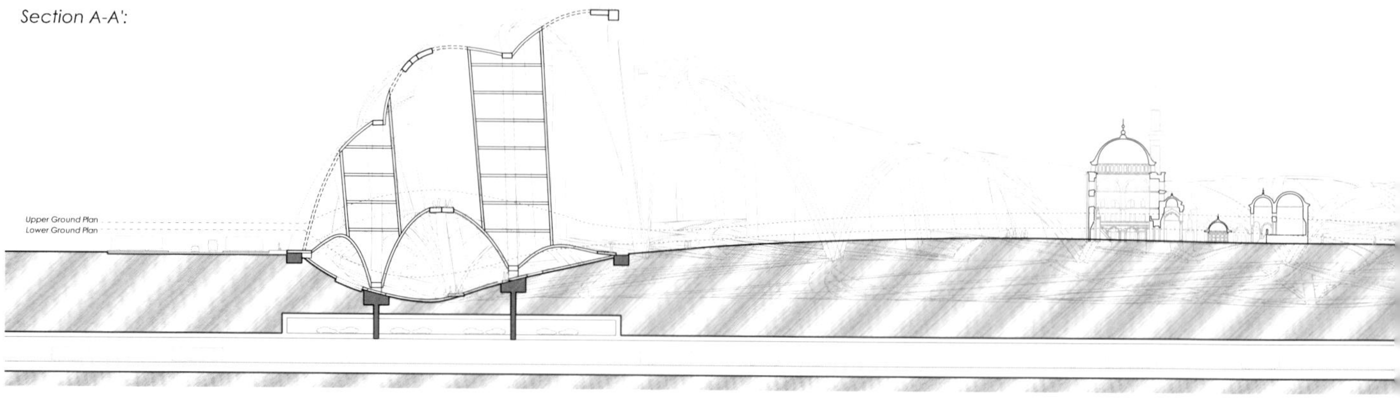

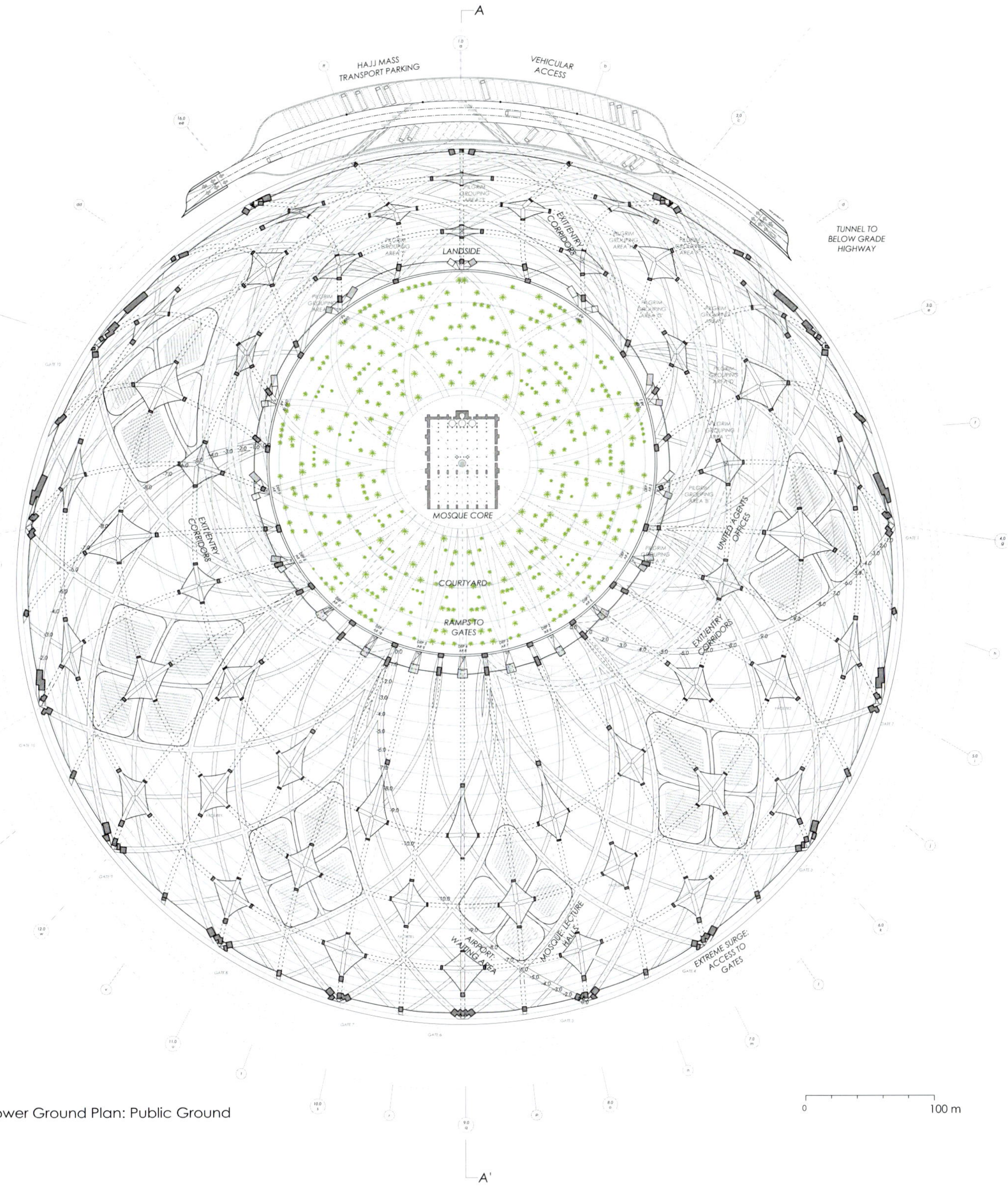

Lower Ground Plan: Public Ground

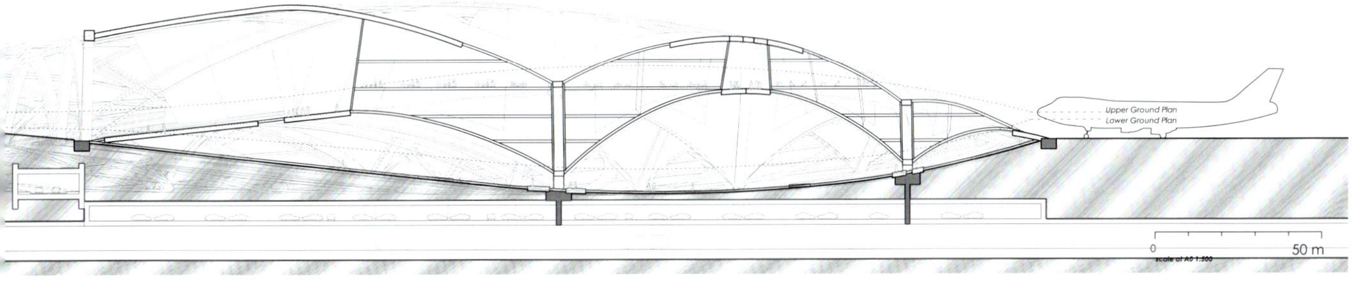

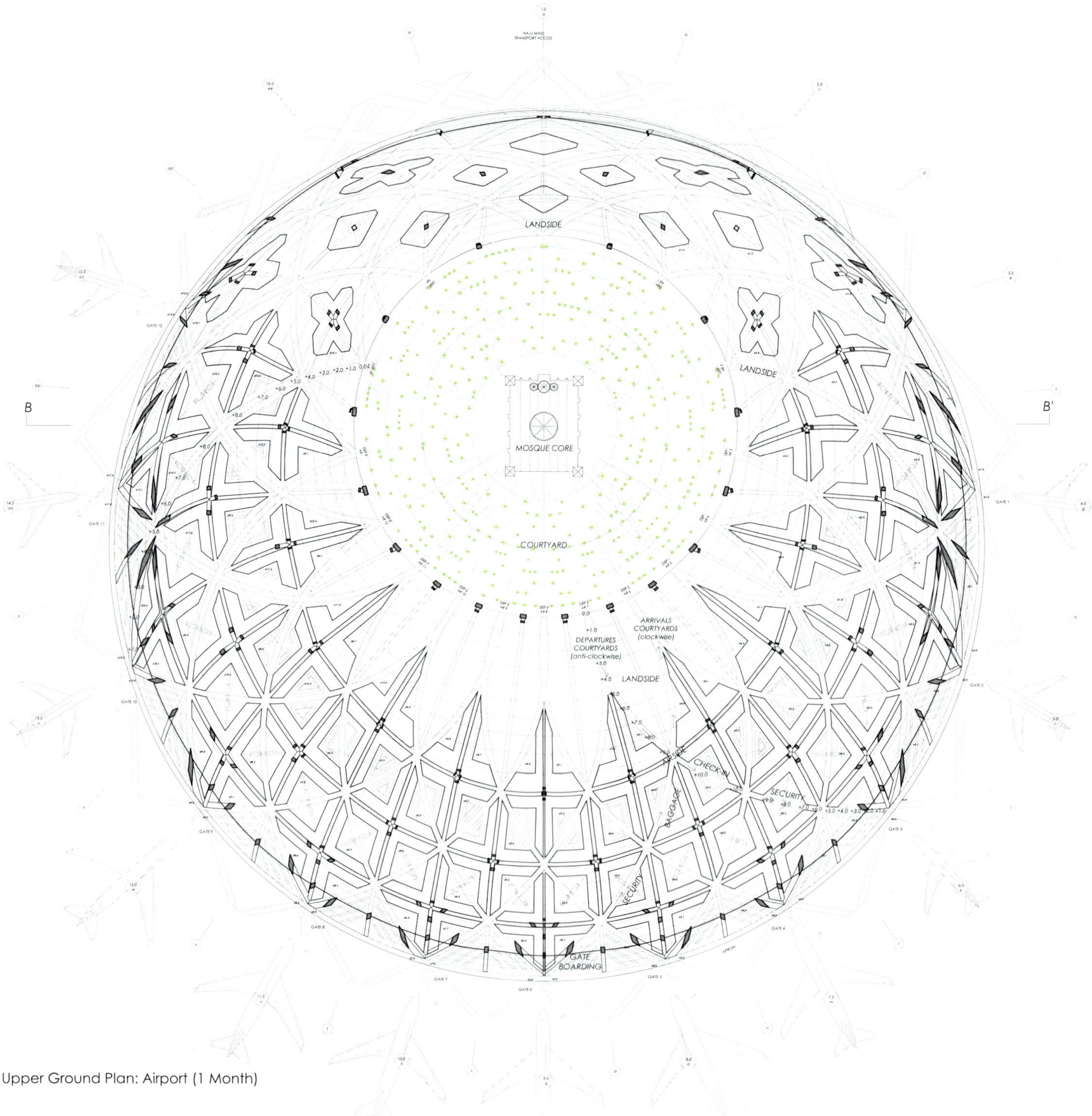

Upper Ground Plan: Airport (1 Month)

SECTION B-B'

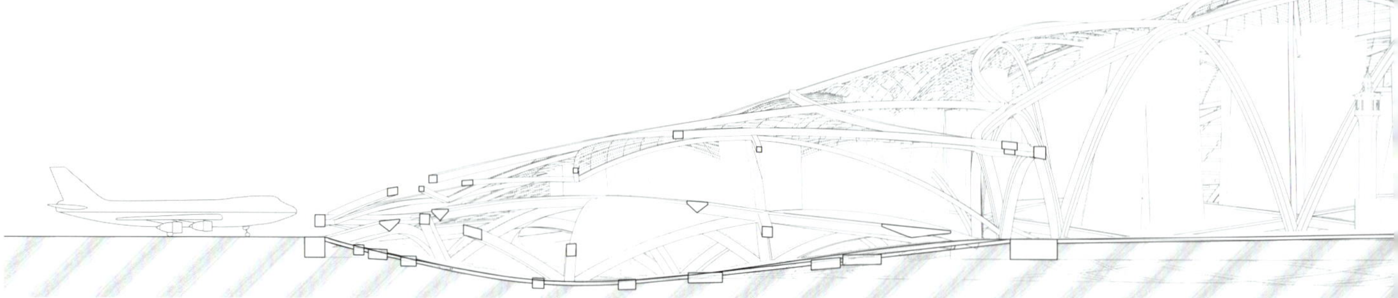

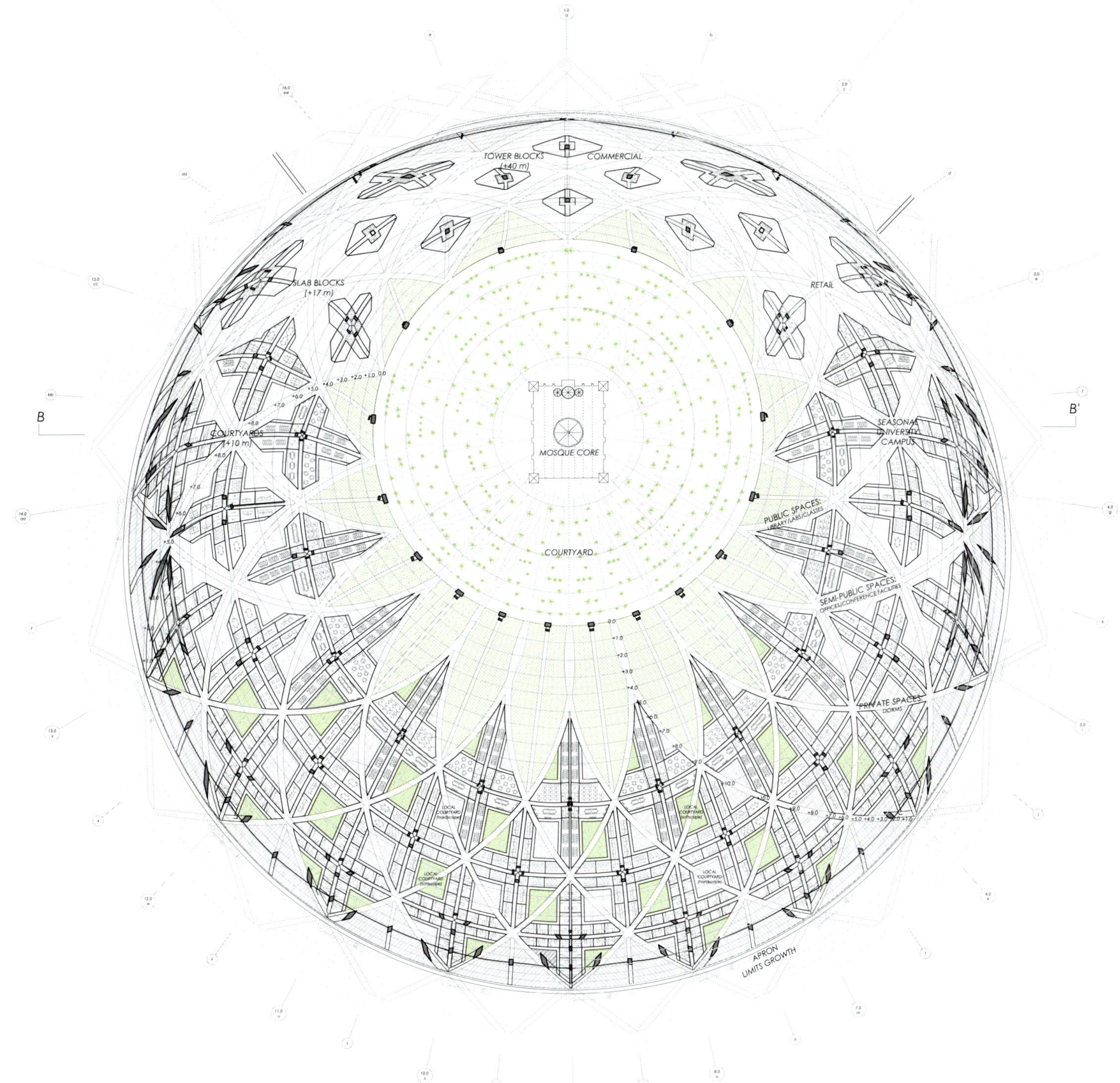

Upper Ground Plan: Mosque (11 Months)

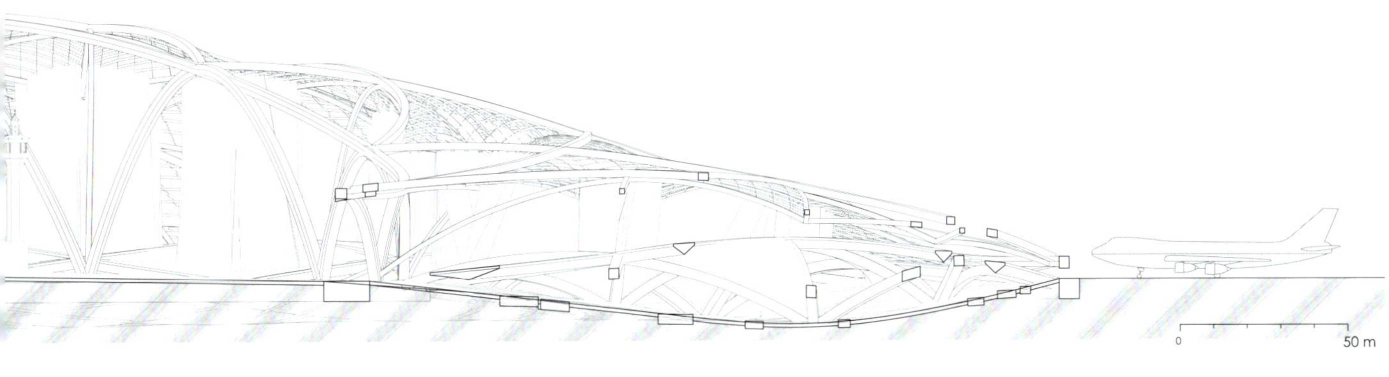

Airport: The Mosque as a Visual Orientating Landmark

Islamic University: The Mosque as a Programmatic Magnet Revitalizing the Redundant Airport

Airport: Activating Open Space Corridors

Islamic University: Contained Within Landside and Airside Boundaries

7. Conclusion

Idea for the City:

Thus the airport type that has both equally nourished and crippled the Holy City is now re-imagined as an airport, a mosque, a small city and a gateway.

As multiple gateways surrounding Makkah, they resurrect the traditional pilgrimage rite of entering the Holy City through its valleys.

These city-gateways disperse the pressure of growth and surges away from the increasingly congested centre, and maintain the legibility of the Holy City by preserving its natural limits and ultimately - the very Idea of its Sacredness.

Redundant Airport vs Opportunistic Airport-Mosque

Insufferable Bottleneck vs Provocative Gateway

Typological Changes Summary:

a. Unlike any other in the world, the Hajj airport's flows are completely predetermined and highly controlled. To accommodate the capped pilgrim numbers and one directional flows, Tegel's circulation model is twisted and overlaped to reduce redundant floor area - maximizing the plan solely for arrivals to the Hajj or departures from it. While multiple accesses to each gate, acting as emergency valves, relieve extreme bottlenecks and stampedes.

b. Although circular airports are extremely efficient and democratically organized, their symmetry and a lack of heirarchy disorientates users. Through offsetting the central core a linear axis is inscribed that highlights a clear direction within the airport to the Holy City. This directionality establishes a crucial relationship between pilgrim and divine destination.

c. The greatest burden for the Hajj Airport however, is its fleeting and seasonal activity. The massive amounts of infrastructure needed to process the pilgrims inevitably remains mostly redundant and financially draining. Yet, overlaying a captive and regenerative mosque core within the airport counters redundancy. An opportunistic Islamic University, also seasonally active with stretched vacations for students, is then able to coexist within the airport's framework.

a. Distributing Pilgrim Surges: Spiral Framework

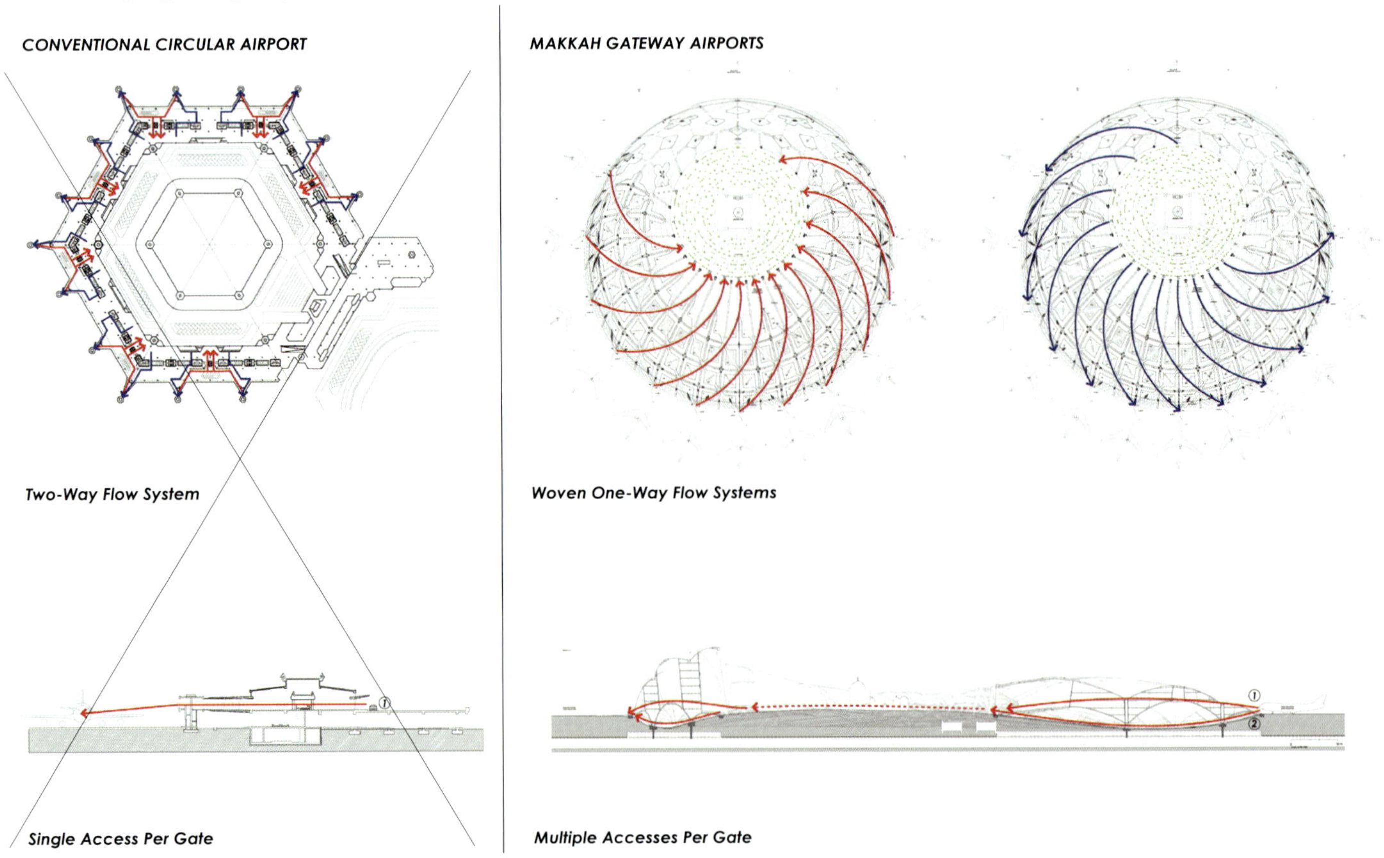

b. Gateway Orientation to City: Offset Core

CONVENTIONAL CIRCULAR AIRPORT

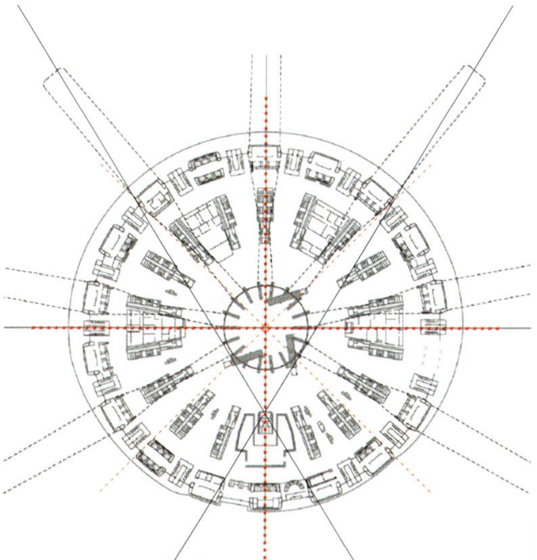

Disorientating: Total Symmetry

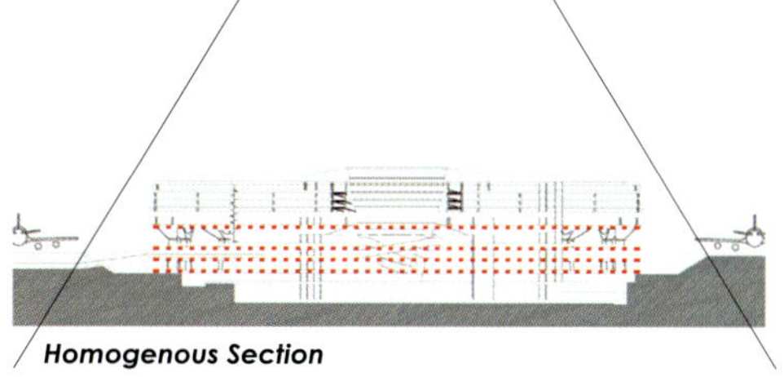

Homogenous Section

MAKKAH GATEWAY AIRPORTS

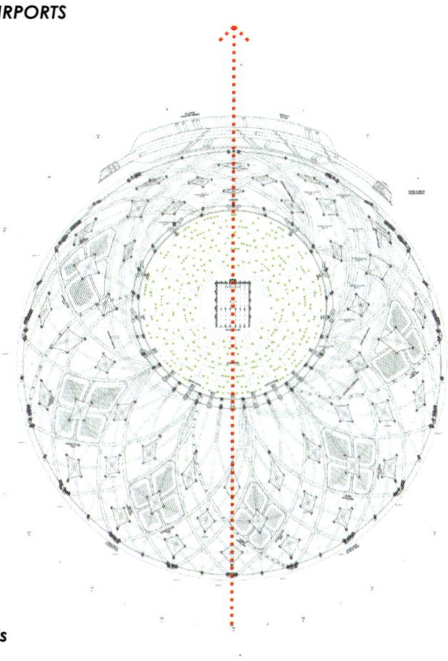

Orientating: Linear Axis

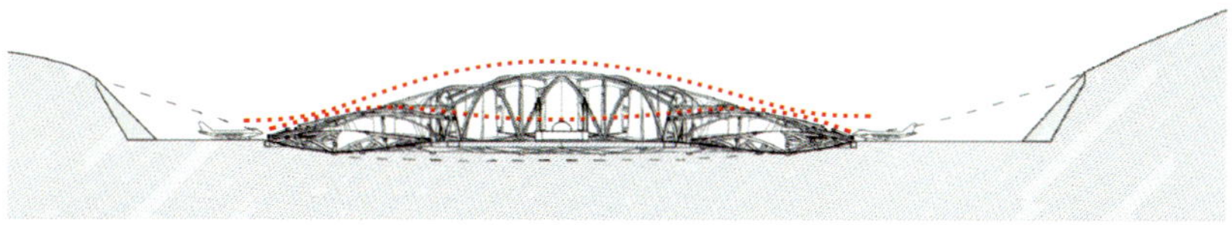

Varied Section

c. Seasonal Redundancy: Overlaid Mosque

CURRENT HAJJ AIRPORT

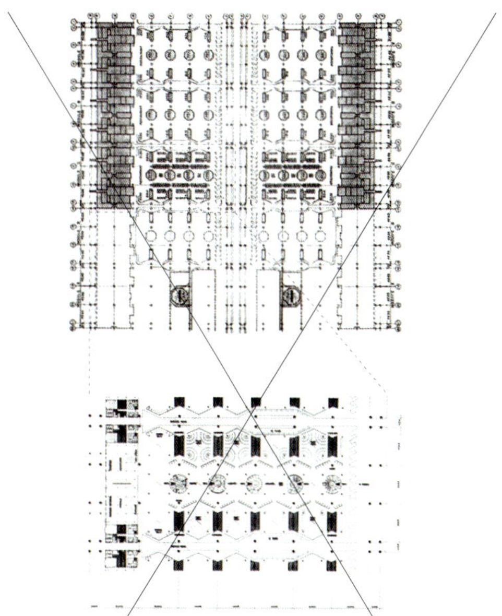

Redundant, Temporary Plan

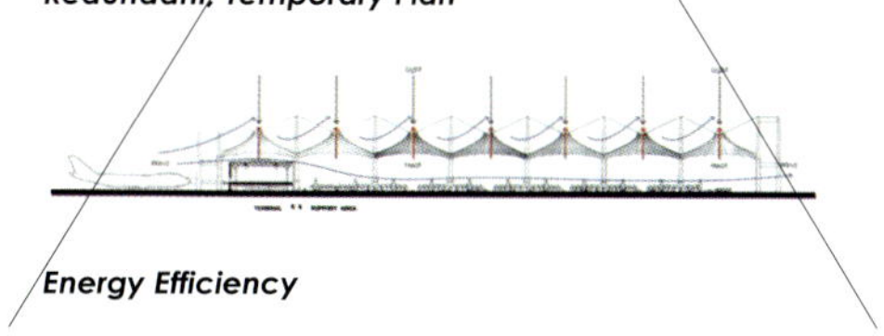

Energy Efficiency

MAKKAH GATEWAY AIRPORTS

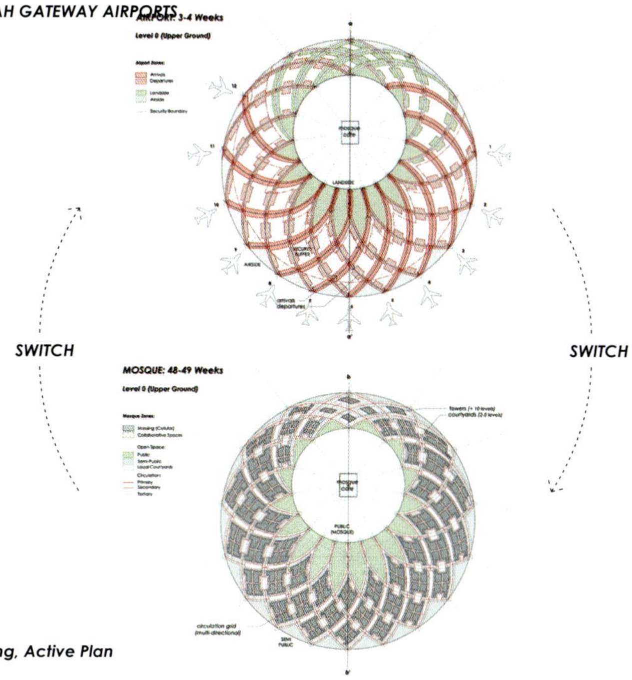

Switching, Active Plan

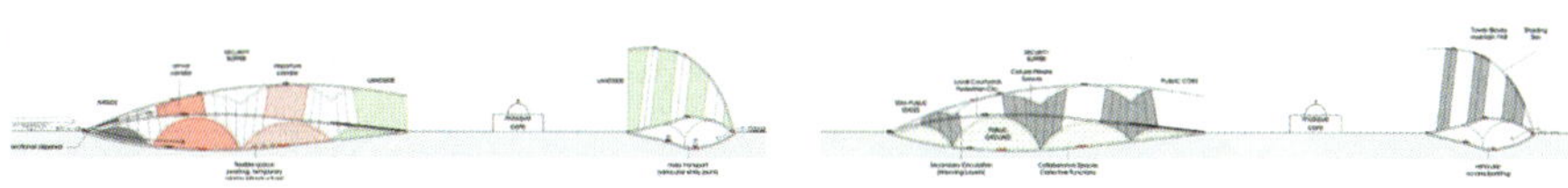

Programmatic Flexibility

An Airport, A Mosque, A Small City and a Gateway.

References

"A New Era of Visionary Expansion." In Saudi Arabia: A 21st Century Kingdom. PM Communications Reporting. 30 Sep 2006. 7 Nov 2007 <http://www.pmcomm.com/pdfs/saudiarabia2.pdf>

Abdul Ghafour, P.K. "Plans Under Way for SR10 Billion Makkah Knowledge Economic City." Arab News. 21 Aug 2006. 5 Nov 2007 <http://www.arabnews.com/?page=6&section=0&article=78229&d=21&m=8&y=200>

Afsaruddin, Asma. "The Philosophy of Islamic Education: Classical Views and M. Fethullah Gulen's Perspectives." Fethullah Gulen. 12 Nov 2005. 12 Nov 2007 <http://www.fethullahgulen.org/conference-papers/the-fethullah-gulen-movement-i/2131-the-philosophy-of-islamic-education-classical-views-and-m-fethullah-gulens-perspectives.html>

Akeel, Maha. "Survey Reveals Alarming Facts of Life in Makkah." Arab News. 22 Mar 2005. 1 Nov 2007 <http://www.arabnews.com/?page=1&section=0&article=60820&d=22&m=3&y=2005>

Al-Baar, Dr Osama. "Report on the Number of Hajj and Umrah Pilgrims." The Institute of the Custodian of the Two Holy Mosques for Hajj Research. 2003. Saudi Railways Expansion Program. 10 Oct 2007 <http://www.saudirailexpansion.com/saudirailexpansion/slp/HAJJ.pdf> (Arabic)

Allawi, Ibrahim. 1988. "Some Evolutionary and Cosmological Aspects to Early Islamic Town Planning." In Theories and Principles of Design in the Architecture of Islamic Societies. Margaret Bentley Sevcenko (ed). Cambridge, Massachusetts: Aga Khan Program for Islamic Architecture.

Al-Saban, Dr Sahal. "Transport for Hajj and Umrah Pilgrims: Current and Proposed." Ministry of Hajj. 1 Jan 2003. Saudi Railways Expansion Program. 10 Oct 2007 <http://www.saudirailexpansion.com/saudirailexpansion/slp/ministry.pdf> (Arabic)

Butt, Riazat. "Islam's Holiest City Set for 130-skyscraper Redevelopment." The Guardian. 29 May 2008. 29 May 2008 <http://www.guardian.co.uk/world/2008/may/29/saudiarabia.islam>

Fadaak, Tarek and Hatem Merdad. "An Airport for Makkah." Department of Urban and Regional Planning, King Abdul Aziz University. 1990. 11 Oct 2007. <http://www.kau.edu.sa/engineering/Journal/Volumes/Vol_14_2/ARCH%20-%20%20matar1a.pdf>

Gerkan, Meinhard Von. Architecture for transportation : Von Gerkan Marg und Partner. Basel : Birkhauser, 1997.

Kasarda, Dr John. "Airport Cities." Aerotropolis. Apr 2007. 7 Oct 2007 <http://www.aerotropolis.com/aerotropolis.html>

"King Abdul Aziz International Airport (JED/OEJN), Jeddah, Saudi Arabia" Airport Technology. 8 Oct 2007 <http://www.airport-technology.com/projects/abdul_aziz/>

Long, David E. "The Hajj and Its Impact on Saudi Arabia and the Muslim World." Saudi-US Relations Information Service. 19 Jan 2005. 10 Oct 2007 <http://www.saudi-us-relations.org/articles/2006/ioi/060106-hajj-long.html>

Rahman, Maseeh. "Among Many, Many Believers." TIMEasia. 2006. 12 Oct 2007 <http://www.time.com/time/asia/photoessays/mecca_hajj/letter_hajj.html>

Serageldin, Ismail. 1985. "Planning and Housing for the Hajj." In Large Housing Projects: Design, Technology, and Logistics. Margaret Bentley Sevcenko, (ed). Cambridge, Massachusetts: Aga Khan Program for Islamic Architecture.

"Towards a Knowledge Economy." 1st Annual Conference of the Arabian Knowledge Economy Association. 23 Oct 2008. 3 Nov 2008 <http://www.xenel.com/pdf/AKEA_PA-newdate_Oct23.pdf>

"Traffic and Transportation Study in Makkah Province." The High Commission for the Development of Makkah Province. 2003. Saudi Railways Expansion Program. 10 Oct 2007 <http://www.saudirailexpansion.com/saudirailexpansion/slp/Makkah.pdf>

Wolfe, Michael, ed. One Thousand Roads to Mecca: Ten Centuries of Travelers Writing about the Muslim Pilgrimage. New York: Grove Press, 1997.

Thanks to,

At the AA:

I would like to express my heartfelt thanks to my tutors Chris Lee and Sam Jacoby for their endless energy, patience, guidance and support.

Thanks to Lawrence Barth and Michael Weinstock for their contributions towards this project and their encouragement.

At home:

To my family, I am truly grateful for your unfailing support throughout my five years at the AA. And thank you especially for tolerating the eccentric workaholic behavior typical of any architecture student.

Thanks to the Crown Prince's Scholarship Program, not only for their financial support, but for encouraging me to pursue this path.

Micro Urban Landspace

Keita Tajima
Diploma unit 11

AA Diploma Honour presented, 2008
High Pass in Technical Design Thesis, 2008

Keita Tajima

Keita Tajima studied at Tama Art University in Tokyo and the Architectural Association in London where he was nominated for AA Honours and received a High Pass in his technical studies submission which researched techniques of sustainable construction and the re-use of materials within the construction industry. He worked in Barcelona and London, most recentry with Peter Beard_LANDROOM on building and Landscape design.

Table of Contents

Introductions

This project is based on the brief of Diploma unit 11 – Micro City.
"Decollaging the city, Inspired by merz, and in particular the merzbau as an interior reconstruction of the city, the unit continued its preoccupation with developing strategies for observing the postindustrial urban landscape..." 1."Micro city" (text next page) is a main topic of Dip 11, the idea of a micro city (city within a city) was first explored through several group studies such as Urban sampling in Battersea and micro city collage. I will begin the first chapter by talking about those group studies and extracts in relation to my idea of micro city and its site in Battersea. Chapter-2 covers the development of spatial tools and possible spatial qualities of an urban micro junction. The idea of a micro junction was explored through anatomical drawings of Carlo Scarpa's architectural insertions into the fabric of the Castel Vacchio in Verona. The notion of a small architectural insertion to old fabric of the city, was interpreted and developed as responsive architectural tools to intervene the transitional aspect of post industrial context. Initial strategy of architectural intervention was tested by using transformative objects and drawings. In chapter-3, further detailed observations and technical investigations on site's textural details, behaviour of operations and structure of architectural interventions, allowed me to develop urban micro junction in different scales with precision as well as landscaping strategies in relation to existing materials and techniques. Final chapter explores proposal for renewal of post-industrial context though series of drawings that synthesizes contextual, technical knowledge and architectural artefact, and concludes my vision of micro city.

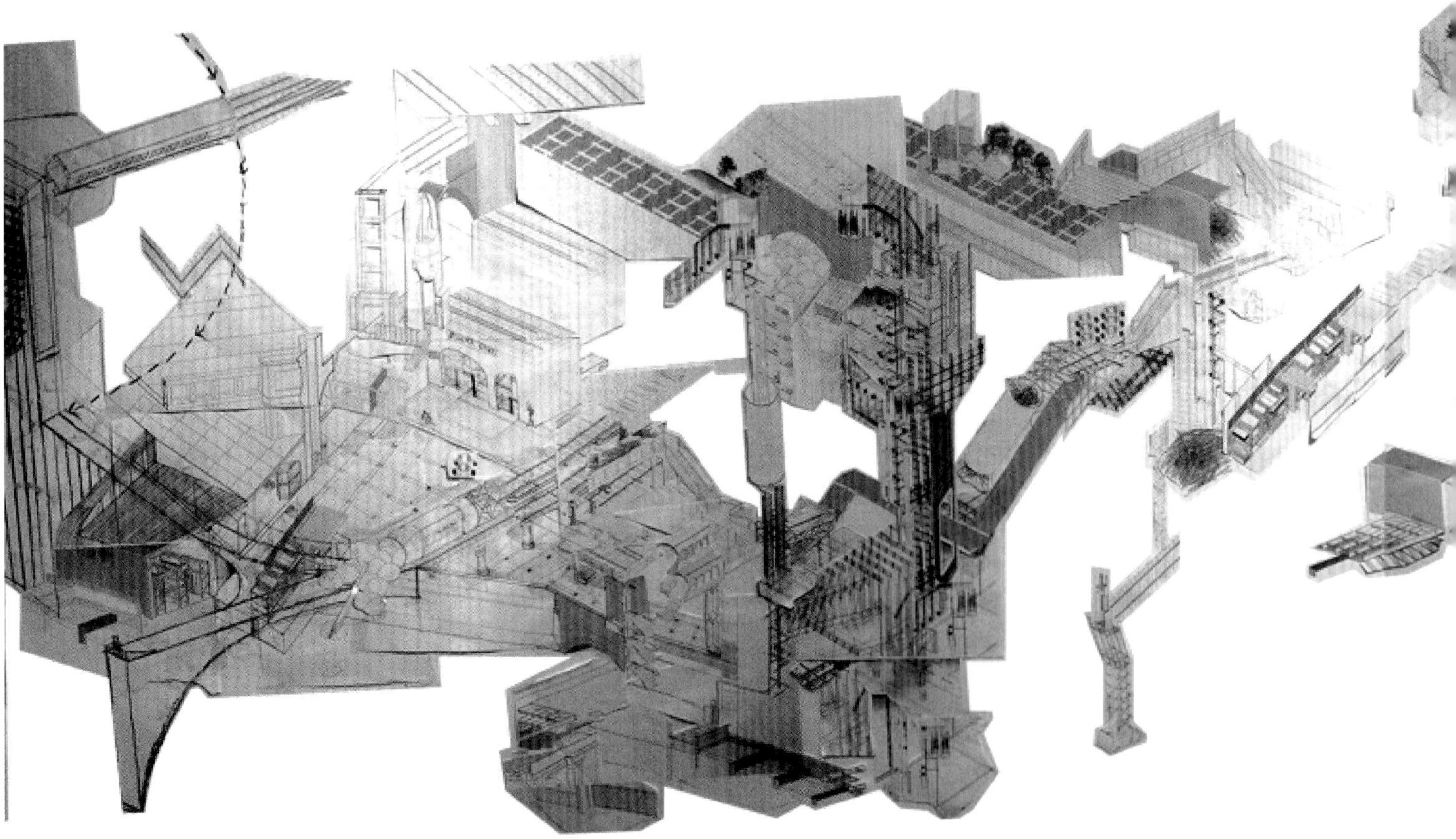

Micro city

micro city is big and small it's a city inside the city interiority behind the urban fabric

microcity is recursive seeing from the reverse view angle it is a flip side of the coin of the city as a map of unmapped

its autonomy expands and shrinks it has an operational field that links between a series of incomplete objects

microcity associates fragments, textures, gaps, events and memories its system utilises the gap between larger systems

microcity that we wish to design is a model unit of urban environment that facilitates network of relationships between objects, non-space and left over matters

and transforms them into a series of micro public space and a proximate network of social facilities

it should have a beautiful set of relationships between machine organizations and a material detail organizations

it may have a set of scale to balance between the volume of excess and rooms for the marginal architecture of the microcity

is the structure for behaviour that gives our body a sense of time and direction inside the city of constant state of flux

microcity is a collection of exceptions every city should envisage

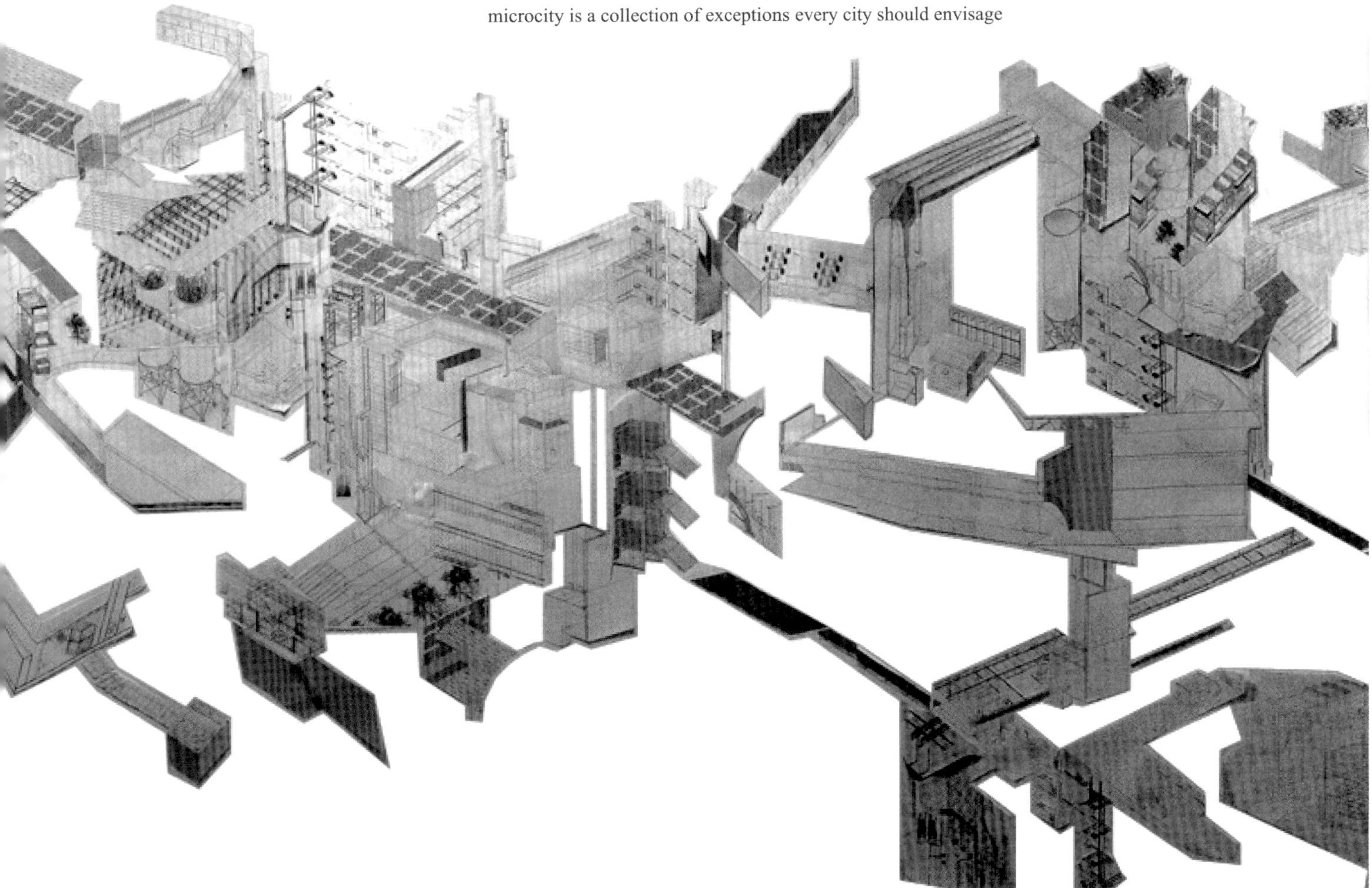

1. Urban sampling and Micro city

"I began to construct pictures out materials I happened to have at hand, such as streetcar tickets, chips of glass, cloakroom checks, bits of wood, wire, twine, bent wheels, tissue paper, tin cans etc. These things are inserted into the picture requires. They lose their individual character, by being dematerialized they become material for the picture."
Kult Schwitters

Inspired by Kult Schwitters's Merz, we made our picture out of the materials from Battersea, as he made Merz bau or any collage pieces by cutting and connecting everyday objects or trash, we began collecting urban samples from Battersea, located inner-city periphery of London, post – industrial brown fields, 60's rapid housing development fragmented by 19 century's railway arches and recent new riverside regenerations, different historical layers are all juxtaposed to each other. We collected urban elements instead of any trash or found objects, that make up the city, excessive textural details, strange adjoining of programmatic spaces in railway arches, left over gap spaces between residential area and industrial area etc... Once these city's elements are sampled, they are assembled again to create Micro city. Micro city collage thus contains urban elements of Battersea, but at the same time these elements are no more recognized as they were, rather creating new set of urban spaces and infrastructures that become our creative strategies to re-inhabit post - industrial urban peripheries.

Landscape of recycling waste
The site was chosen after sampling field work and it is located within residential area of Battersea, the fragmented brown fields in urban periphery where currently occupied by several recycle waste managements that accumulating and moving tons of recycling wastes every day. The transitional movement of recycling wastes and vast occupation of those industries create desolated landscape within urban condition and disconnects public activities within the area, but on the other hand, its textural detail, transitory landscape and existing industrial operations can be seen as an creative opportunity for change.

More urban samples.
Taking advantage of our sampling method, we continued to catalogue urban samples in Haneda, Tokyo, Japan. Small town emerged by forced eviction by expansion of US Air base in post war time in Tokyo. As it's strange fate of the town, the configurations of city structure is random and complicated. Our group's samples are mainly narrow paths within residential blocks that can be potentials of being public shared spaces or already used as shared spaces. The interiority of narrow paths is explored further in relation to its spatial volume, in order to develop spatial quality of micro city.

Collective works

Micro-city collage, Battersea, Tokyo sampling and reconstructing Time machine are all done by groups and each of us extract our own idea, views and techniques from them and developed and elaborated through each one's criteria of the project.

images
top left - Merz bau, top right - Merz, Bottom right - site bird's eye view, page left - time machine

Keita Tajima

1.1 Time machine

Our Time Machine is constructed based upon the text "How to construct Time Machine" by Alfred Jarry. The direct translations of speculative text into physical object requires imaginative interplay between practical aspect of "machine" and impossibility of the idea of time travel. This was an introduction or sort of exercise in beginning of the year, however it was still meaningful for later stage to explore the poetics within pragmatic role of architecture of the city in time.

Commentary and Instrvctions for the Practical Constrvction of the Time Machine

I. The Nature of the Medium

It is no more difficult to conceive a machine to travel in Time than one to travel in Space, whether one consider Time to be the fourth dimension of Space or a place differing essentially in its contents.

One usually defines Time as: the taking place of events, just as Space is the place taken by bodies. Or more simply: Time is succession, whereas Space — whether it be a question of three-dimensional or Euclidean space; four-dimensional space, implied by the intersection of several three-dimensional spaces; Riemannian spaces, where spheres may be turned inside out, since the circle is a geodesic line on the sphere of the same radius; Lobachevskian spaces, where planes cannot be flipped over; or any non-Euclidean space, distinguishable by the fact that one cannot construct two identical figures in them as one can with Euclid — Space is simultaneity.

Any parcel of Time that is simultaneous has extension, and may therefore be travelled in by means of Space Machines. The present has Extension in three directions perpendicular to one another. Whether one transports oneself to some point in the past or in the future, that point will be present and extended in three directions during the visit.

Reciprocally, Space and the Present extend into the three dimensions of Time: the space that has been covered, or past space; the space to come; and the present proper.

Space and Time are commensurable: navigating by the sighting of points in Space may only be onwardly pursued through the medium of Time; and to measure Time quantitatively, one reduces it to the Space of dials and chronometers.

211

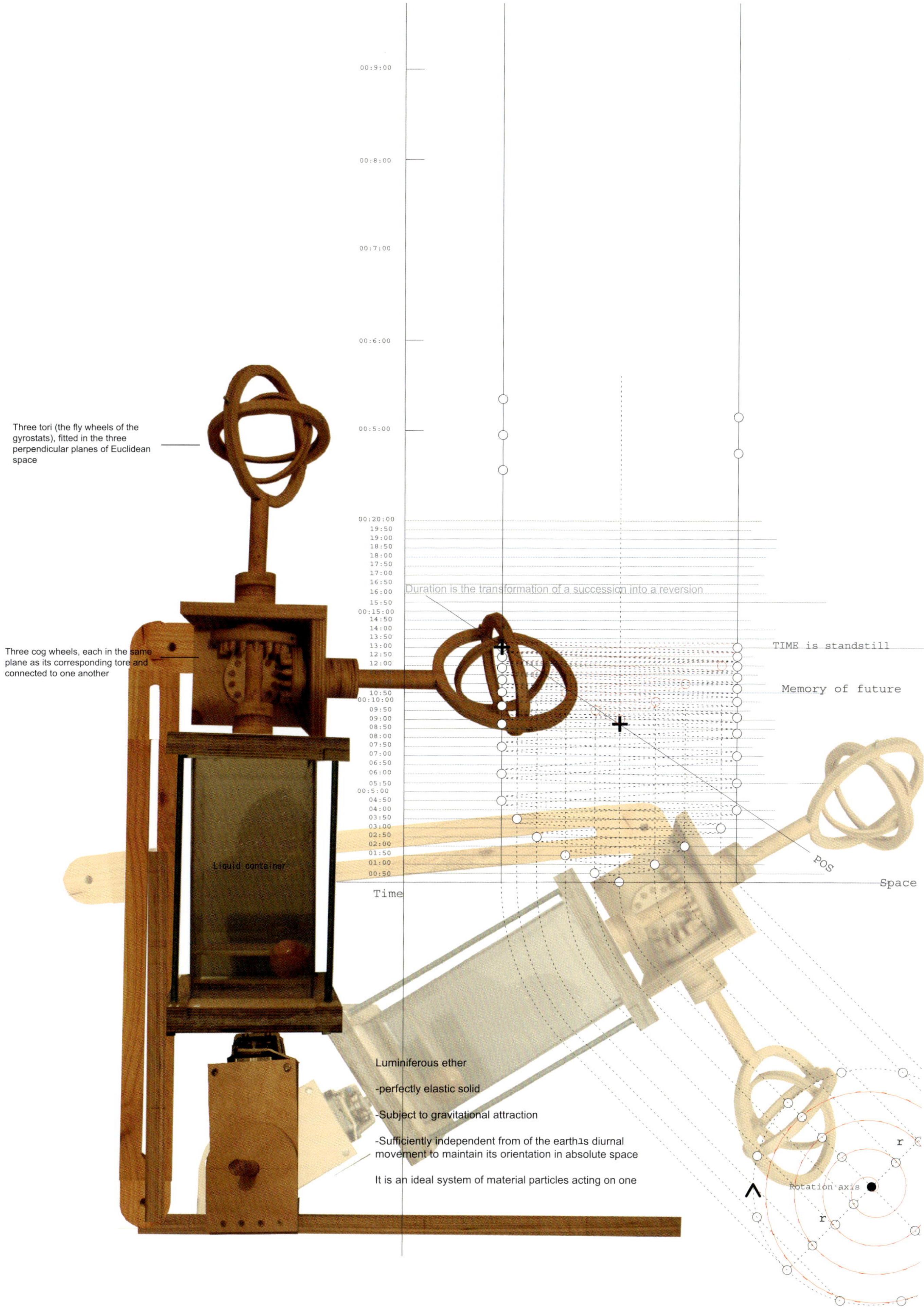
Three tori (the fly wheels of the gyrostats), fitted in the three perpendicular planes of Euclidean space
Three cog wheels, each in the same plane as its corresponding tore and connected to one another
Liquid container
Duration is the transformation of a succession into a reversion
TIME is standstill
Memory of future
POS
Space
Time
Luminiferous ether
-perfectly elastic solid
-Subject to gravitational attraction
-Sufficiently independent from of the earth1s diurnal movement to maintain its orientation in absolute space
It is an ideal system of material particles acting on one
Rotation axis
r
r

1.2 site introduction - Battersea

Batteasea is located on the inner –city periphery of London, From late 18 century to comparably recent times, Battersea was established as an industrial area together with arrival of railway and post war housing development, and that demarcates motley assortment of different territories and programs. The site is on the edge of Battersea area, north of wandsworth road where the railway arch cuts through 60's residential blocks and recycling wastes industries are wedged into this gap. The pile of metal scrap, car tyres, timber and construction debris occupied this desolate brown field site together with operations of busy industrial machineries and other small cottage industries. This brown field site leaves strange gap and discontinuity within residential area and its transitional aspect of industries (specially those waste managements) and riverside's recent luxurious developments emphasize unstable development of the area. Therefore the project intends to find the way to facilitate smooth transition from post industrial brown field site and strong integration to fabric of the city.

view from the railway. on the left side of railway - residential area. on the right side - industrial estates

Wandsworth Road

1.3 Urban sampling - Battersea

The aim of urban sampling is to observe and extract specific detail or behaviour of post-industrial urban phenomenon in Batttersea. Specially, textural detail and related industrial operations or infrastructure were sampled from our site as unresolved urban resources of the city and the landscape. Each samples are also used to compose Micro city collage together with other groups samples.

metal scrap yard occupies and operates large area within industrial estates, creates desolated landscape.

construction wastes recycling industry

sorting waste on the pavement. Coucil housings behind over look entire field of recycling wastes.

local car breakers

mountain of ferrous metal wastes

local car breakers yard. Broken cars are manually seperated into body and other parts. Car body is then brought to metal scrap yard.

railway arch devides residential area and industrial area

local cafe parasetically built within railway arch

building sand inside railway arches. Railway arches are used by several small work shops, storages and car washing / fixing companies.

1.3 Urban sampling - Battersea

EMR. Metal scrap yard

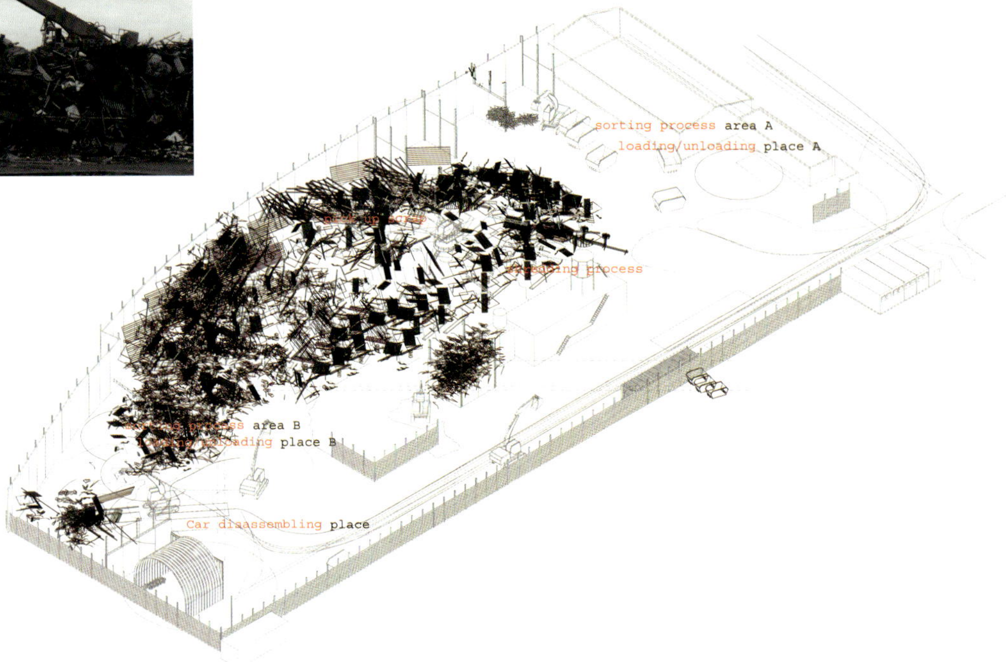

Materials

ferrous metals
primary metals
non ferrous metals
cast iron
alminium
brass
secondary metals
nickel
heavy iron
stainless steal
zinc
steel
light iron
alloys
tin
borings
boturnings

Facilities

a - media separator
b - shredder

Rigid spatial elements

A - short fences
B - tall fences
C - car disassembling place
D - shredding place
E - office
F - vehicle store
G - management office

SITA W aste Management

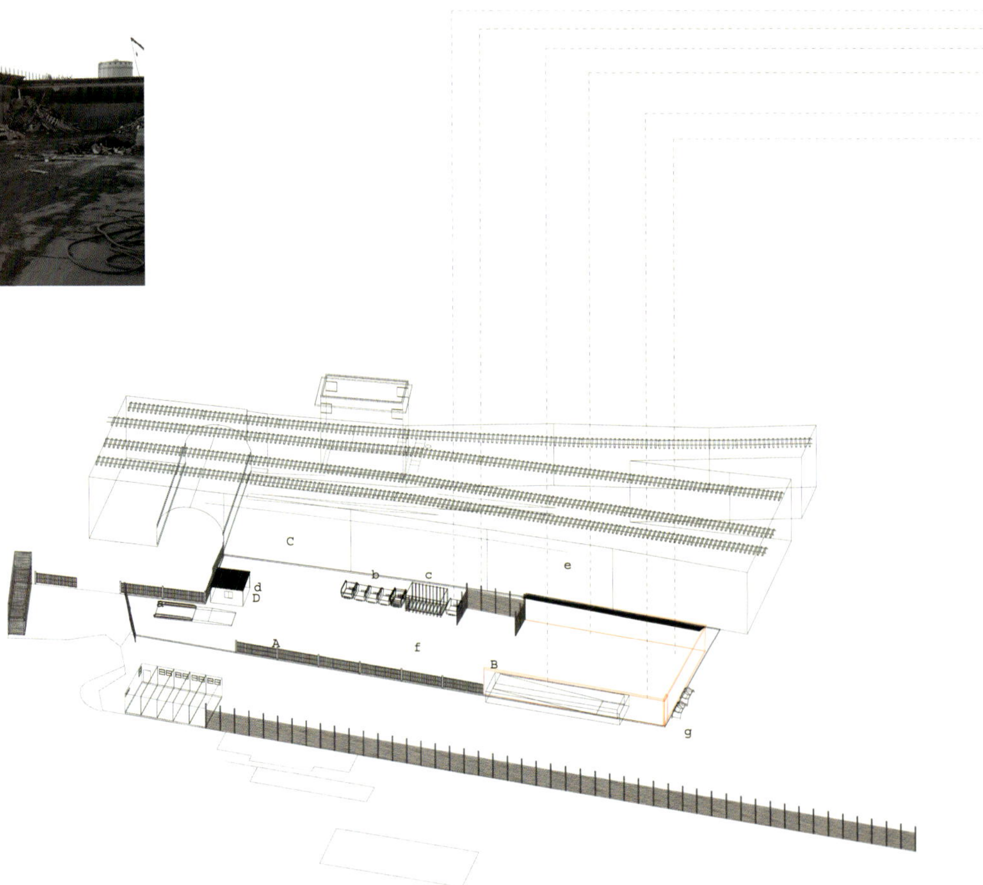

Materials

Aluminium
Aluminium
Aluminium
landfill
concrete/soil/sand
wood

Facilities

a - scale
b - skip
c - container
d - management office
e - sprikler
f - loader
g - water tank

Rigid spatial elements

A - short fences
B - tall fences
C - viaduct
D - management office

short constructural

cast iron

fragmentised

HMS

light steel

ferrous

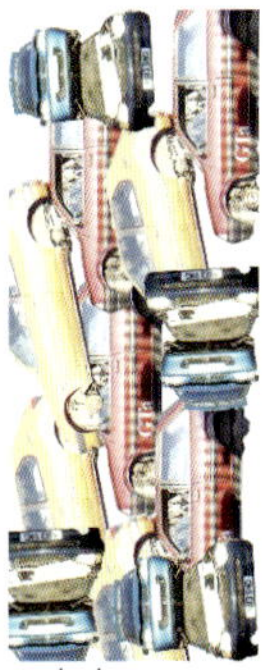
car body

copper wire

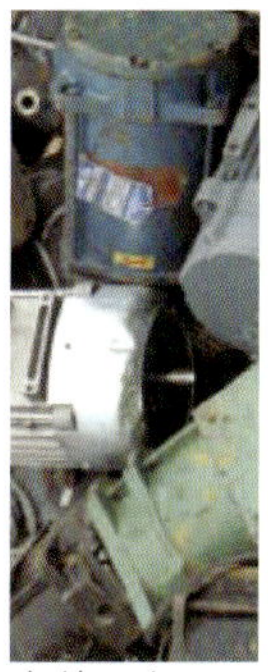
electric mortors

non ferrous

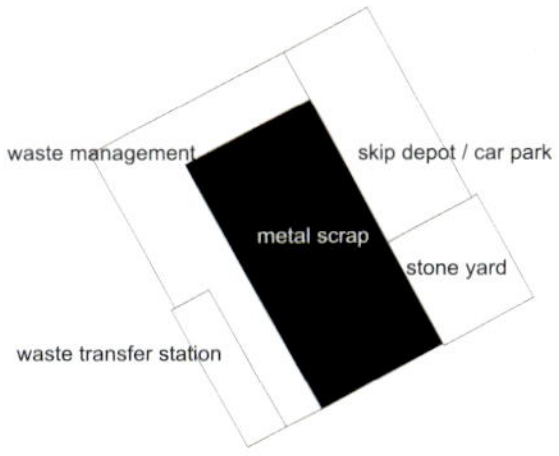

neighbouring program

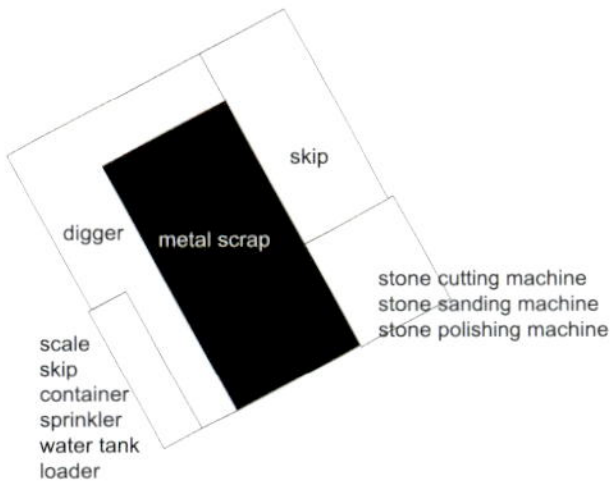

neighbouring facilities

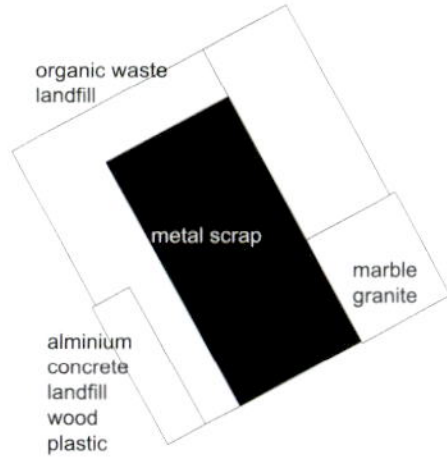

neighbouring materials

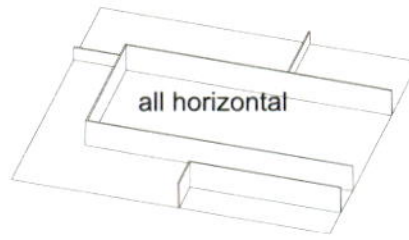

Spatial condition

Materials

Wood

Aluminium

Landfill

Concrete/soil/sand

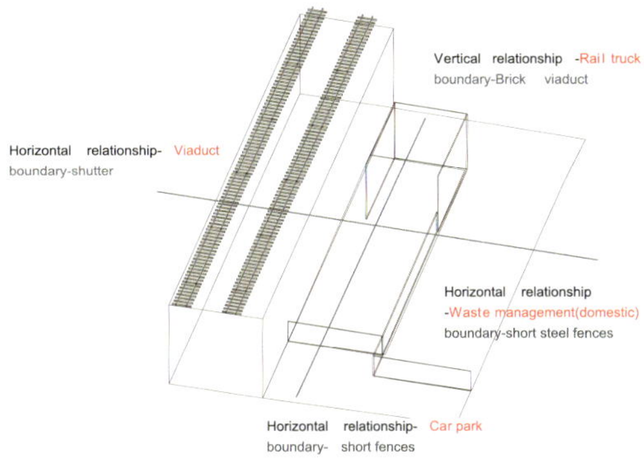

Spatial conditions

Neighbouring program

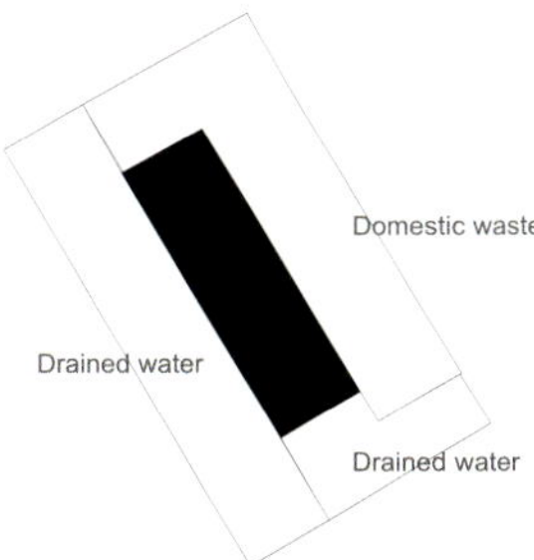

Neighbouring materials

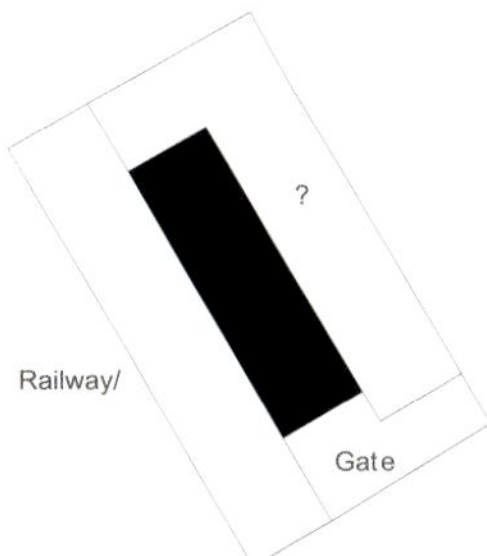

Neighbouring facilities

1.3 Urban sampling - Battersea

Car repairing place extension

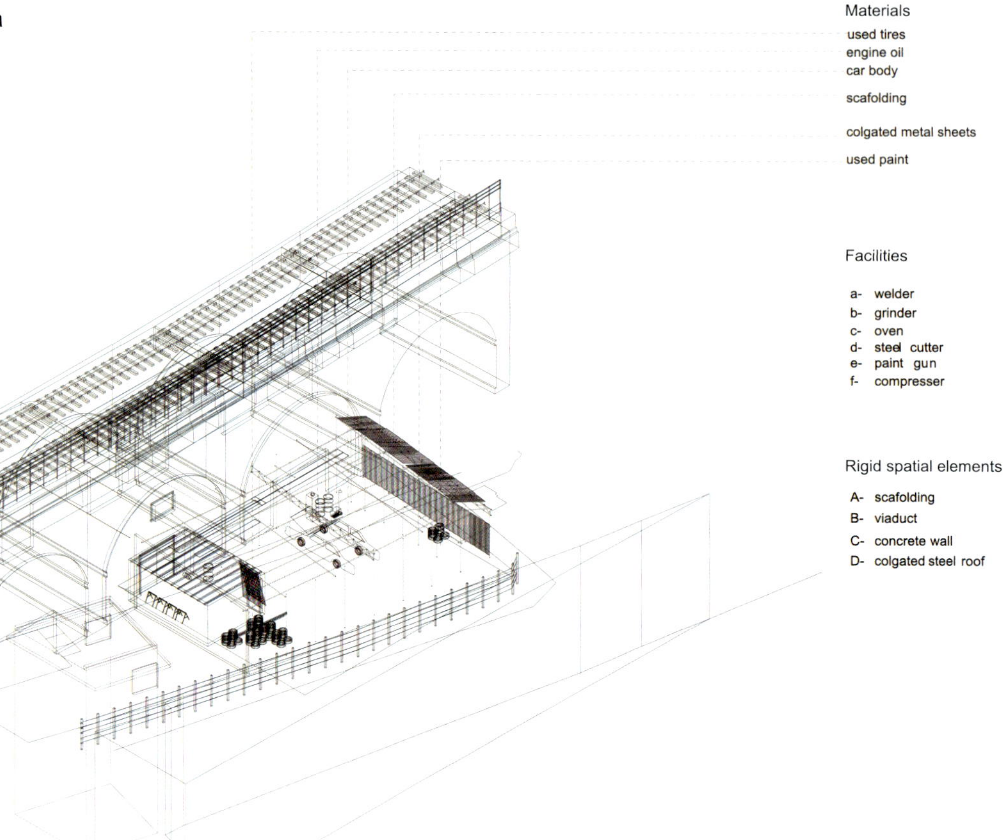

FM Conway Surfacing / Civil engeneering

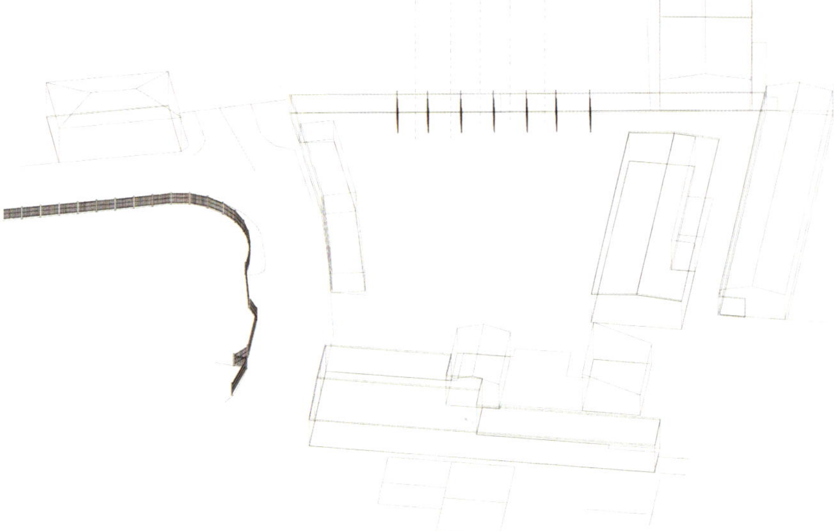

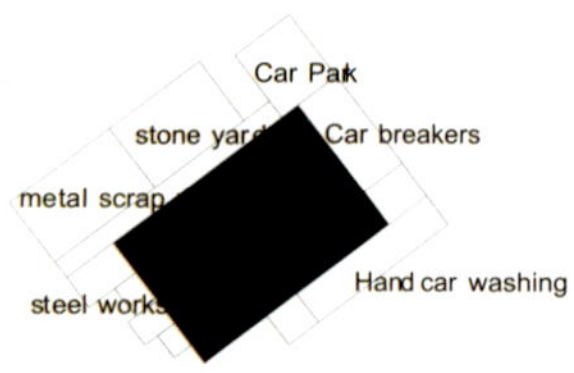

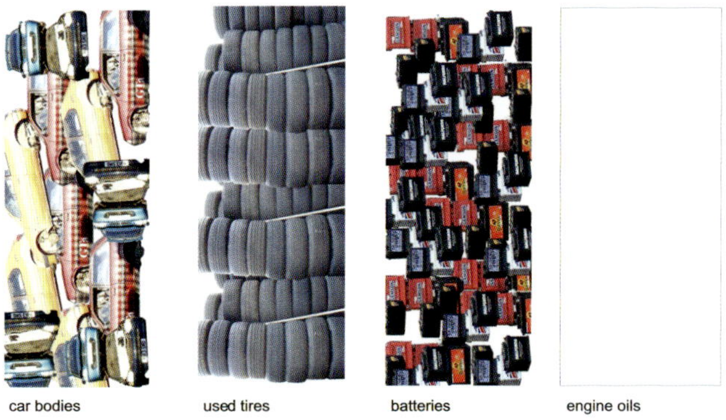

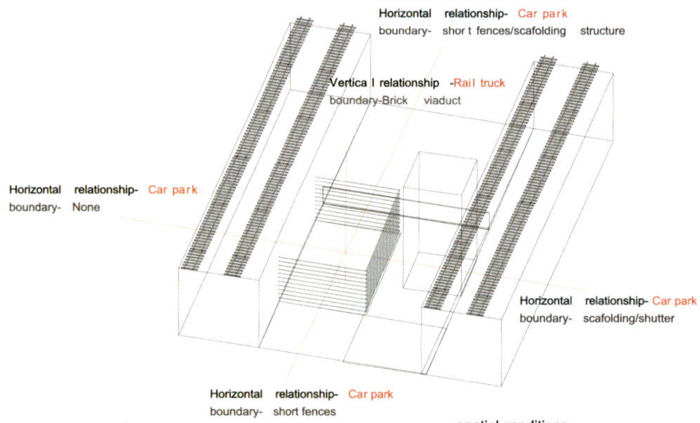

spatial conditions

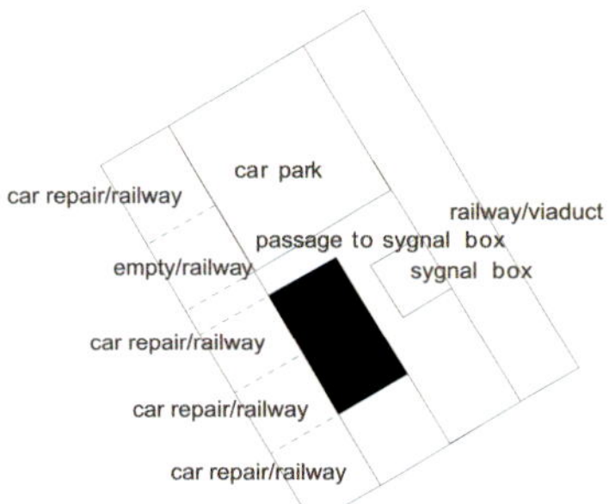

Neighbouring program

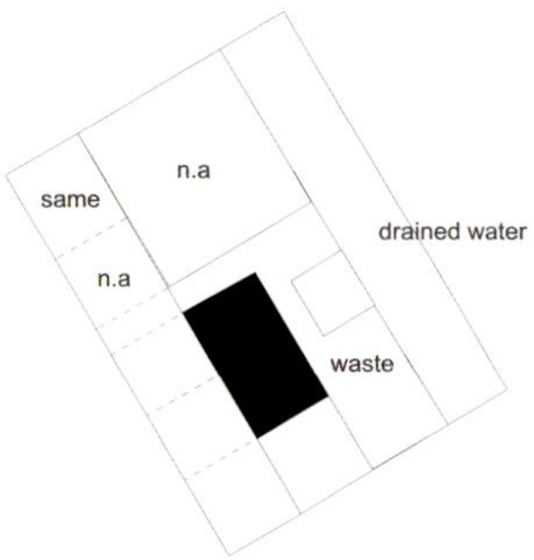

Neighbouring materials

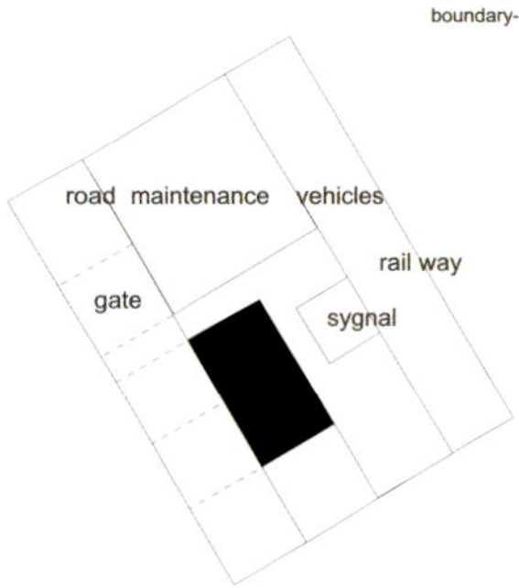

Neighbouring facilities

SITA Waste management

Materials

Organic wastes

landfill

Domestic wastes

Facilities

a- scale
b- skip
c- container
d- management office
f- loader

Rigid spatial elements

A- short fences
B- tall fences
C- viaduct
D- management office

Keita Tajima

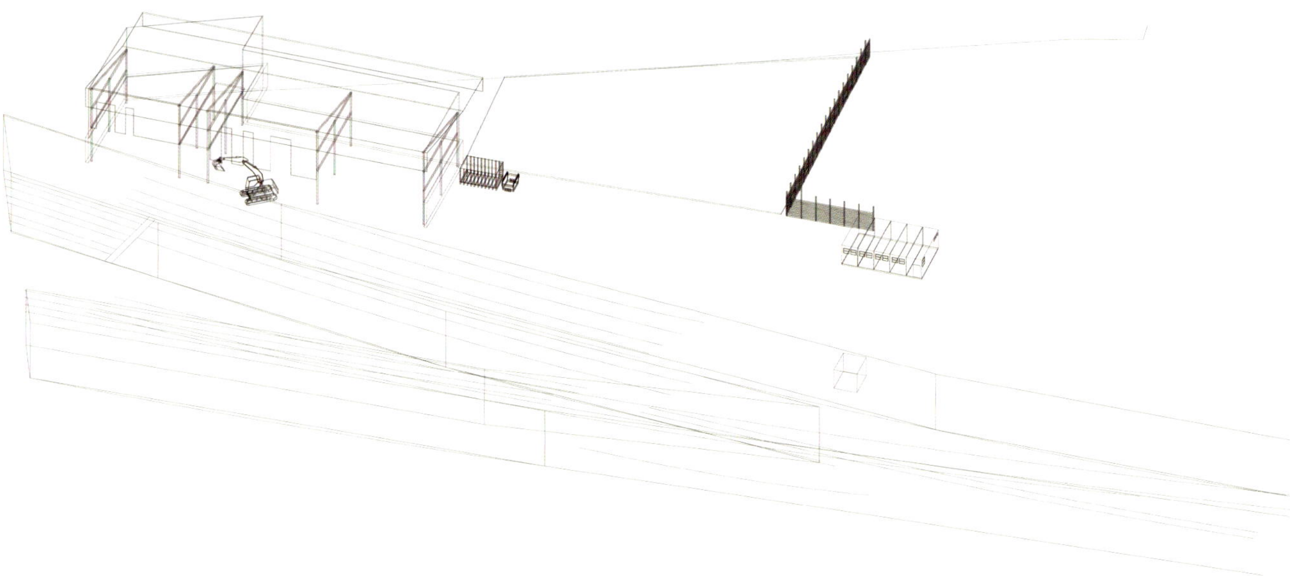

1.4 Urban sampling - Tokyo

Urban sampling was continued in Haneda, Tokyo where random urbanization process had affected on its built environment, resulting to produce unique spatial phenomenon. Our group was particularly interested in narrow passage within residential block that eventually being perceived as interior space rather than out door space by its exposed domestic appliances in relation to size of the space that provides opportunity to share within community. Exploring and sampling this phenomenon feeds to our idea of "micro city" in terms of its detailed contents of spaces as well as transfiguration of micro built environment of the city. Later, this sampling was explored more on its spatial quality as inverted space...

The town of Haneda

The current condition of the town of Haneda has been established through its unique history as it locate itself next to Tokyo international Airport (Haneda Airport). The town used to be located next to Haneda Airfield which established in1931, in 1945 the airport became Haneda Army Air Base under the control of the United States army and its expansion of Air base forced people who had been settle in the town to be evicted from surroundings within 48 hours. The people evacuated to the land across the river which now almost divides Airport and the town, then settled randomly and that formed original configurations of the town. The town has been changing over the decades by implementations of new infrastructures that makes the contrast between old and new more obvious.

Interiority of narrow path

The drawing next page shows one of the samples from our research of narrow paths, that explores the interiority of the narrow paths by indicating locations of exposed domestic appliances within a residential block and orientations of faces of each houses that defines use of space as public, private and semi public. Cul-de-sac type of external space is used as both communal shared space and public passage, ambiguity between public and private is modestly formalized by surface conditions, plants and stepping stones that makes buffer zone as semi public territories

exposed domestic appliances

Sample 01 – extremely narrow spaces

The minimum width of the path is less than 80cm, very intimate space has been emerged between private houses, the space resulted as a shared space or extension of each private spaces by exposing their domestic appliances and garden plants. However, the space is still function as passage, infrastructure for localised fire brigade by placing small water tanks, the journey through this passage is almost going through someone's living room.

sample - 01

Sample 02 – micro public square

As a consequence of random urbanization of this small town, the contrast between old and new sometime produces strange space within dense residential fabric. The small square appeared in the middle of dense residential block where used to be occupied by three small houses creating typical narrow labyrinth. After land owner demolished them and built one house in the middle, left over land became private garden of this newly built house but still being used as public passage. The exposed private appliances and plants contribute this place as communal shared square rather than private garden.

sample - 02

Sample 03 – localised Laundromat

This is a typical example of spatial use of cul-de-sac in this town. The space is equipped with washing basin and washing machine (placed in outside), shared with four or five houses in order to do their laundries. All the house's entrances are facing to this area and treated it as their own space as well as communal space.

sample - 03

sample - 1

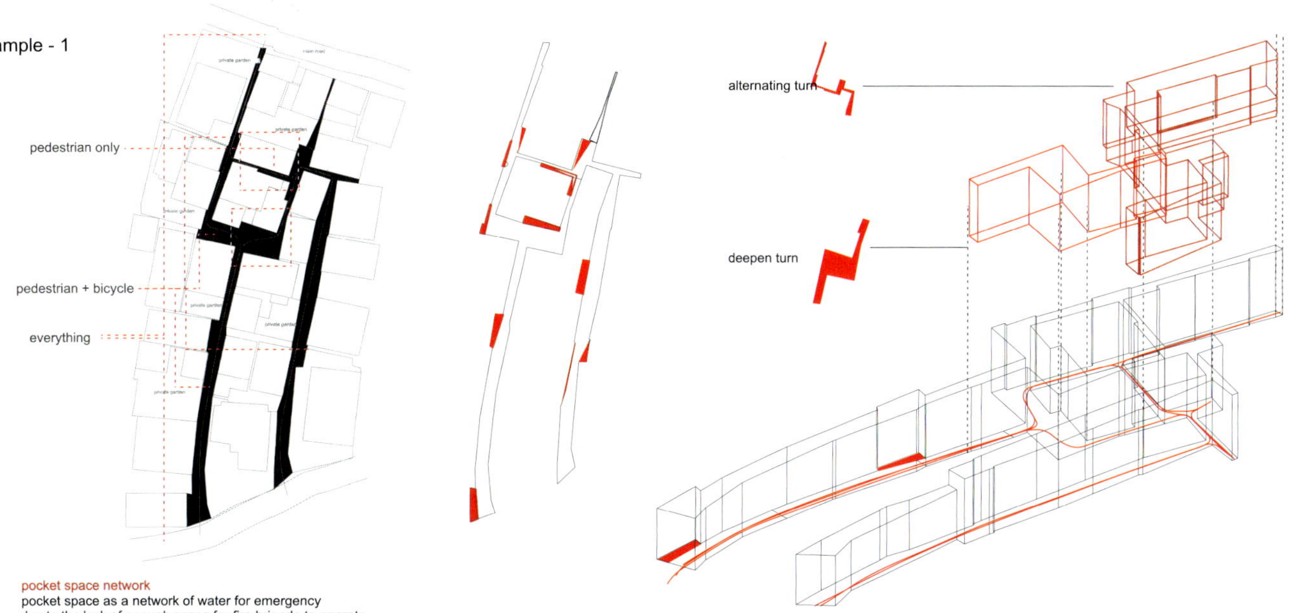

pocket space network
pocket space as a network of water for emergency
due to the lack of enough space for fire brigade to operate.

minimum width of the space 52cm
localized infrastructure - shared water tank network

sample - 2

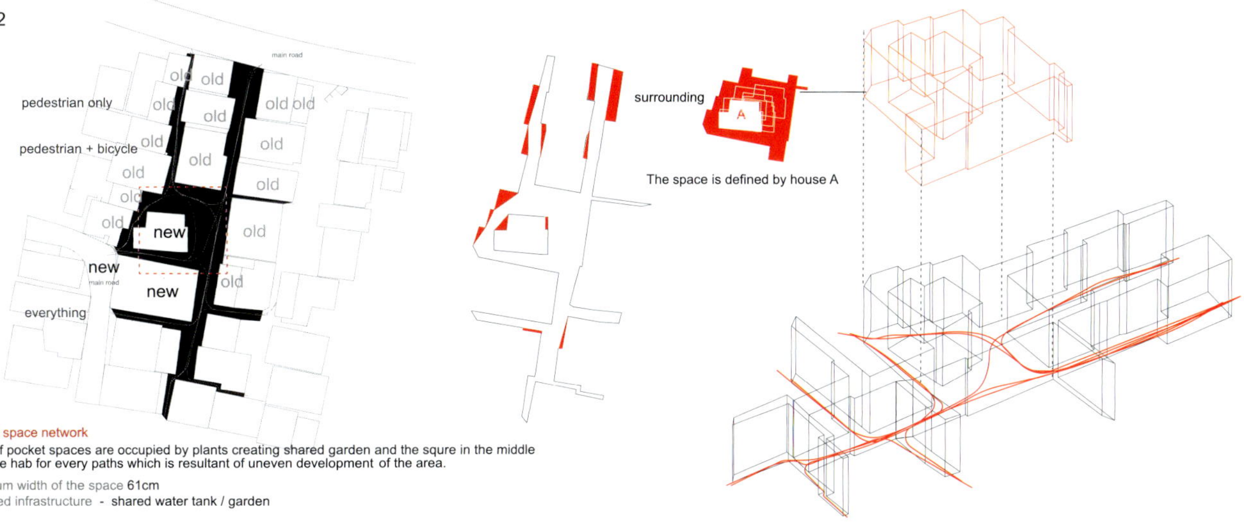

pocket space network
Most of pocket spaces are occupied by plants creating shared garden and the squre in the middle become hab for every paths which is resultant of uneven development of the area.

minimum width of the space 61cm
localized infrastructure - shared water tank / garden

sample - 3

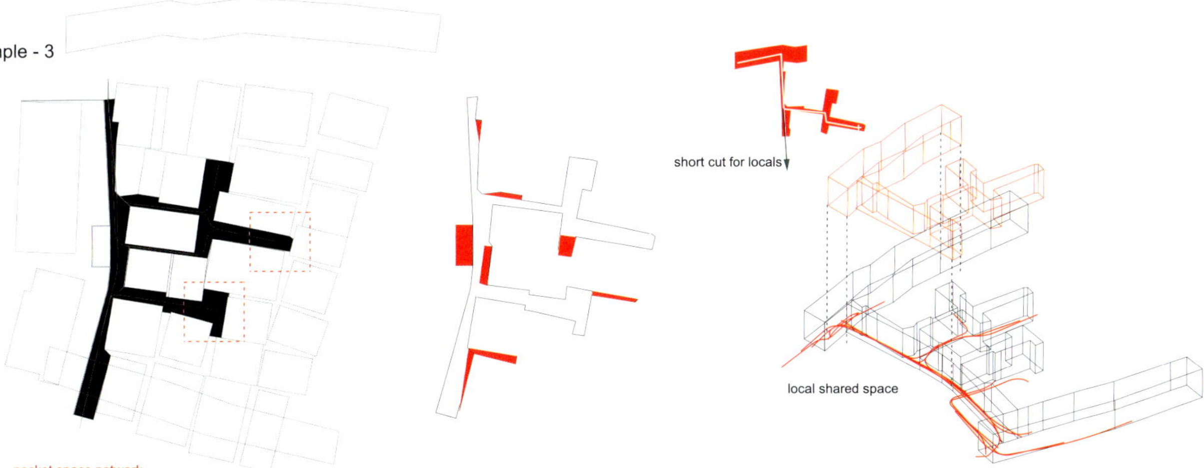

pocket space network
There is local recycling network, not only by using pocket space.
but one of the recycling box is located on the pocket spaces.

minimum width of the space 65cm
localized infrastructure - shared washing / drying area

1.5 reading micro city collage

The part of a micro city collage was extracted to analyse further by individually. Each part has its own story of micro city that each of us invented, but also there are coinciden-tal urban units or spaces emerged when it was assembled all together. My part of collage suggests something between urban landscape and large storages connected to infrastructures, and has own sorting mechanisms and different state (temporally stored and consolidated) and different type of materials. Public shared facilities such as kitchen units and bath rooms are found in between those accumulated volumes, utilizing their excess to feed urban pocket gardens. The collage next page is a re-construction of original collage which to speculate my own idea of a micro city with building elements of the site.

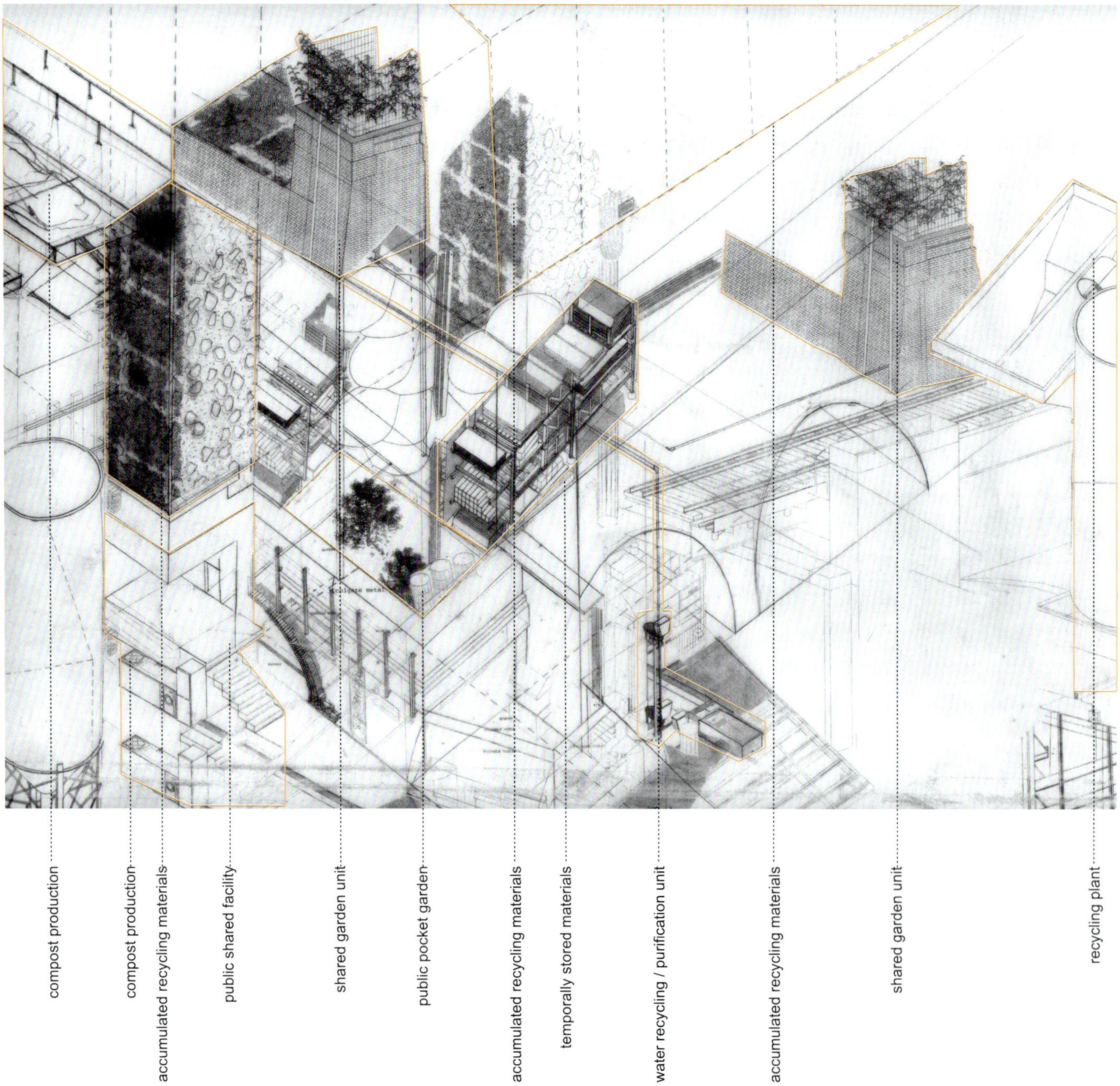

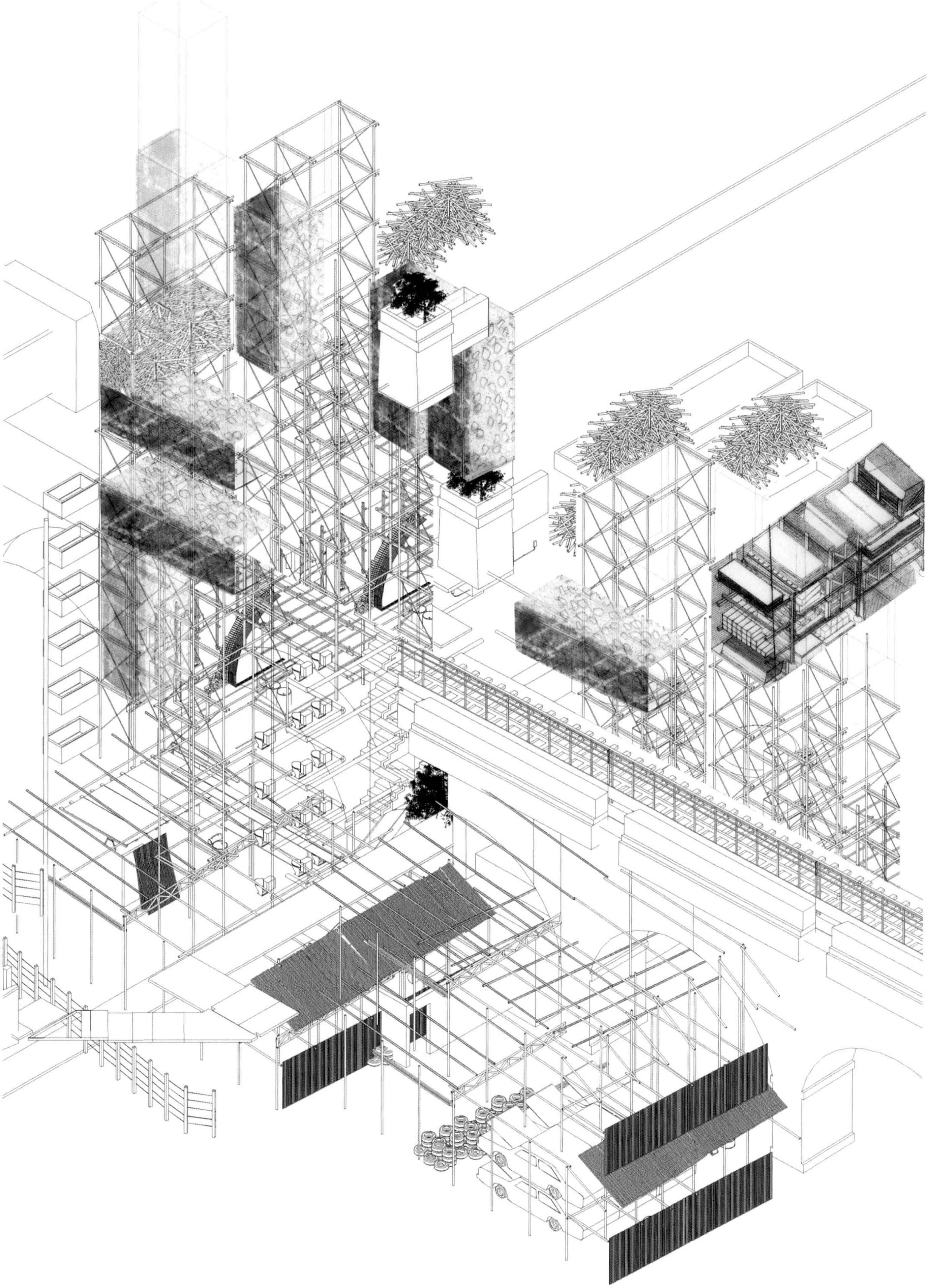

2.0 Initial development of Tools and strategy

2.1 urban micro junction
2.2 insertion model
2.3 insertion model – spatial exploration
2.4 exploration of initial strategy

Urban micro junction - Castel Vecchio, Italy

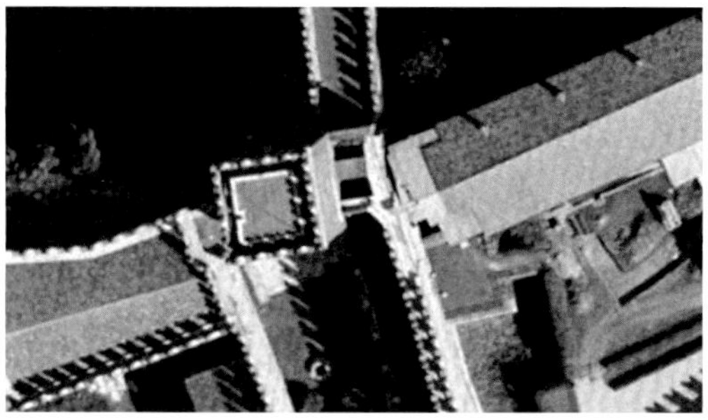

Micro urban junction is a space which composes different historical layers of material, series of small architectural insertions and activities. Taking Castel Vecchio's Cangrande space as an example of junction space within urban fabric, the idea of space is explore through drawing which takes Cangrande space apart into different spatial elements. However, the space is composed in such a small scale in relation to the fabric of city, Scarpa's re-insertion of architectural elements and opens up the space into the city as a form of museum as well as infrastructure which physically connects over the river. The idea of micro urban junction suggests how small architectural space can be strategic point which functions and coordinates surroundings within the city over the time.

Insertion model study

Inspired by Scarpa's architectural insertion into Castel Vecchio, the idea of small architectural insertion into urban fabric is initially explored through abstract model which constructed in series of cut, addition and re-insertion. Taking the behaviour of model as both strategic and spatial organization device, idea of urban junction is developed as coordinating point for current industrial landscape of the site (urban bunker). Urban bunker is defined as landscaping device which coordinates the verticality and horizontality of sediments and orchestrate landscape and existing industrial operations while providing safe space for public access. The idea of urban bunker was examined by series of sketch models and drawings in relation to the context, simultaneously the initial model was de - collaged into series of detail model of junctions to explore further potential as spatial device.

Initial strategy

The notion of micro urban junction is a result of certain process as its space eventually composes intersections of layered materials, public activities, micro infrastructure and re-establishes the relationship between interiority of micro urban space, new textural ground and the city. By developing the idea of urban micro junction, the project speculates time based strategy of architectural insertion (urban bunker) into existing landscape and how it would gradually transforms the existing condition into new type of landscape and Micro Urban Junction within. At this stage, the strategy was explored in abstract level, testing the idea of urban bunker as a series of strategic points within the landscape to speculate the possible transformation of existing condition.

images
top left - Castel Vecchio
top right - insertion model
bottom - insertion model components

2.1 Micro urban junction - Castel Vecchio, Velona, Italy

The idea of urban micro junction is taken from the case study of Cangrande space in Castel Vecchio, Italy.
Cangrande space locates itself in conjunction with two old buildings that form museum and foot bridge across the river. Drawing explores its spatial quality by each architectural elements that have been built and inserted in different time. At the same time, it reveals how series of small architectural insertions integrates different materials and functions into one small junction space, and at the same time it opens up into old fabric of city as an infrastructure.

Urban scale - insertion to old fabric of the city as a part of infrastructure

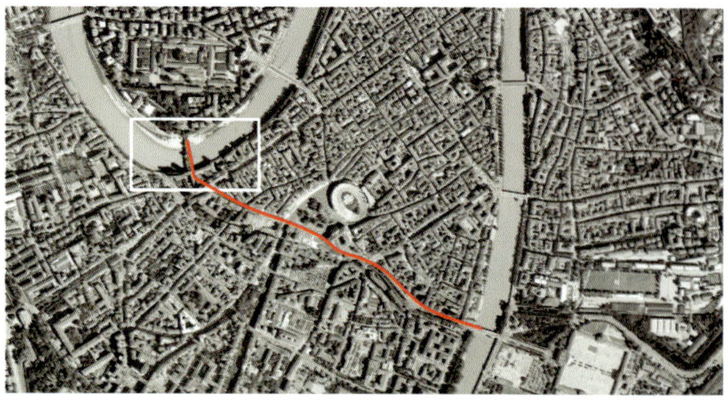

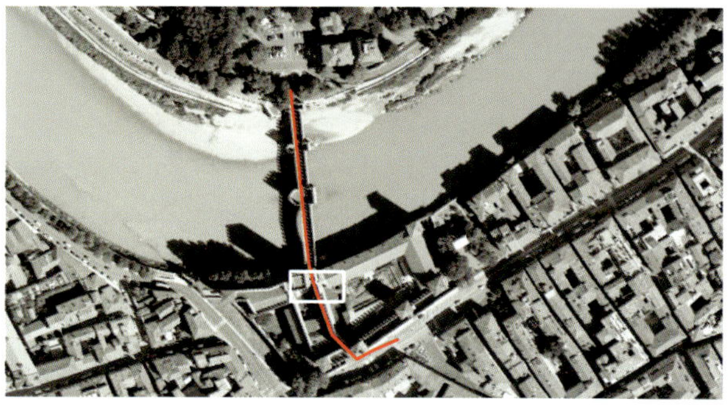

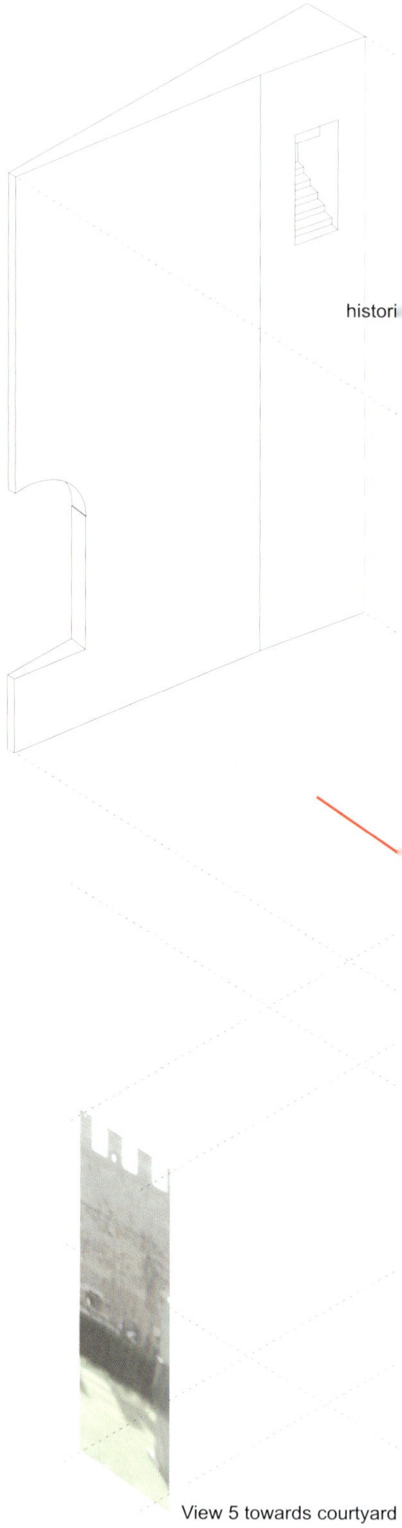

Urban Junction -architectural insertion to the old part of the castle as a part of museum

detail sample

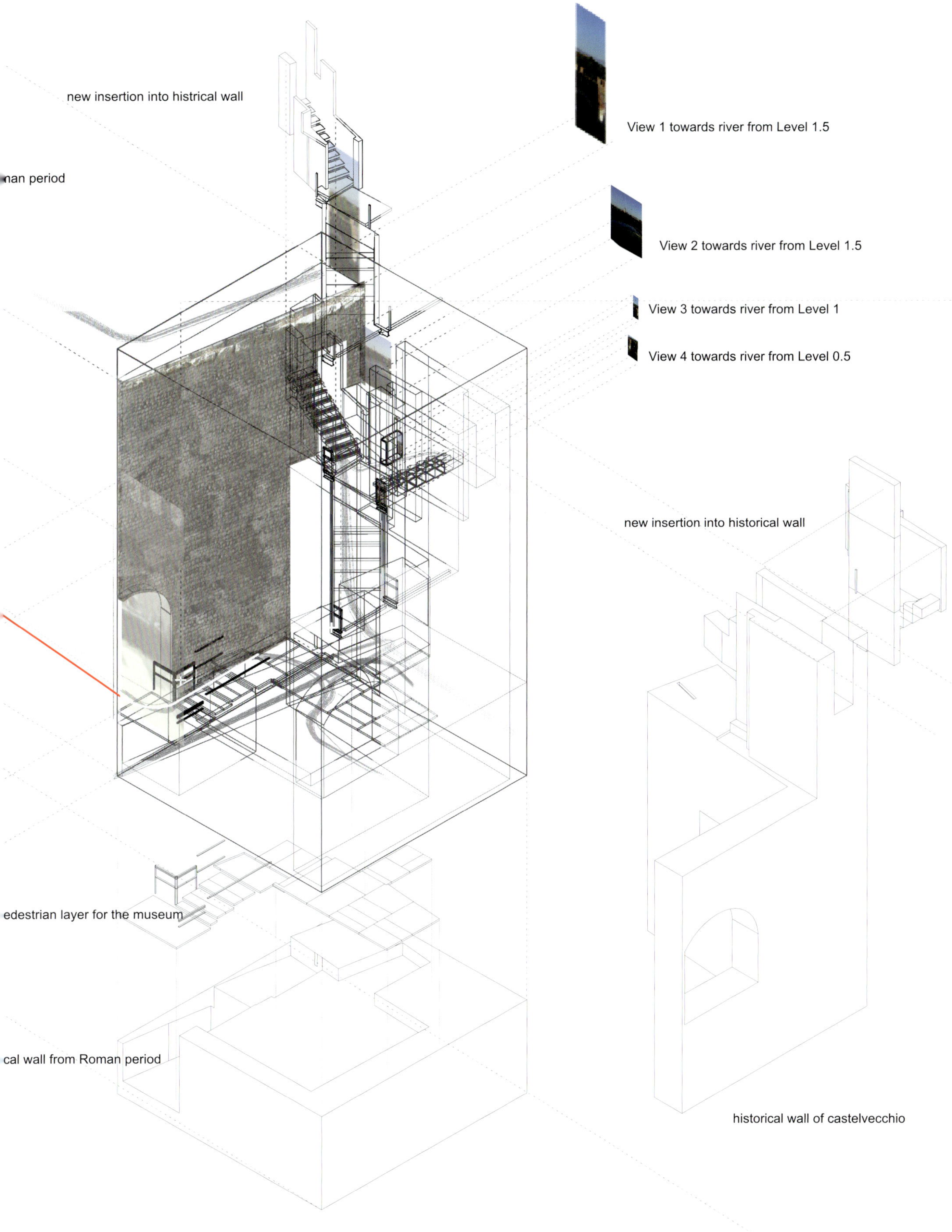
new insertion into histrical wall
nan period
View 1 towards river from Level 1.5
View 2 towards river from Level 1.5
View 3 towards river from Level 1
View 4 towards river from Level 0.5
new insertion into historical wall
edestrian layer for the museum
cal wall from Roman period
historical wall of castelvecchio

2.2 Micro urban Junction - model study - 1

By following the idea of Scarpa's architectural insertions into Castel vecchio by means of its material combinations as well as its construction process in time, abstract model was constructed. Form of the model represents itself as a result of continuous insertions and cutting processes. By considering this process as operational tool which to explore the strategy of architectural interventions into the site in Battersea where industrial operation takes place, each fragment of the model was analysed further in terms of their performances as spatial devices.

assembled junction model

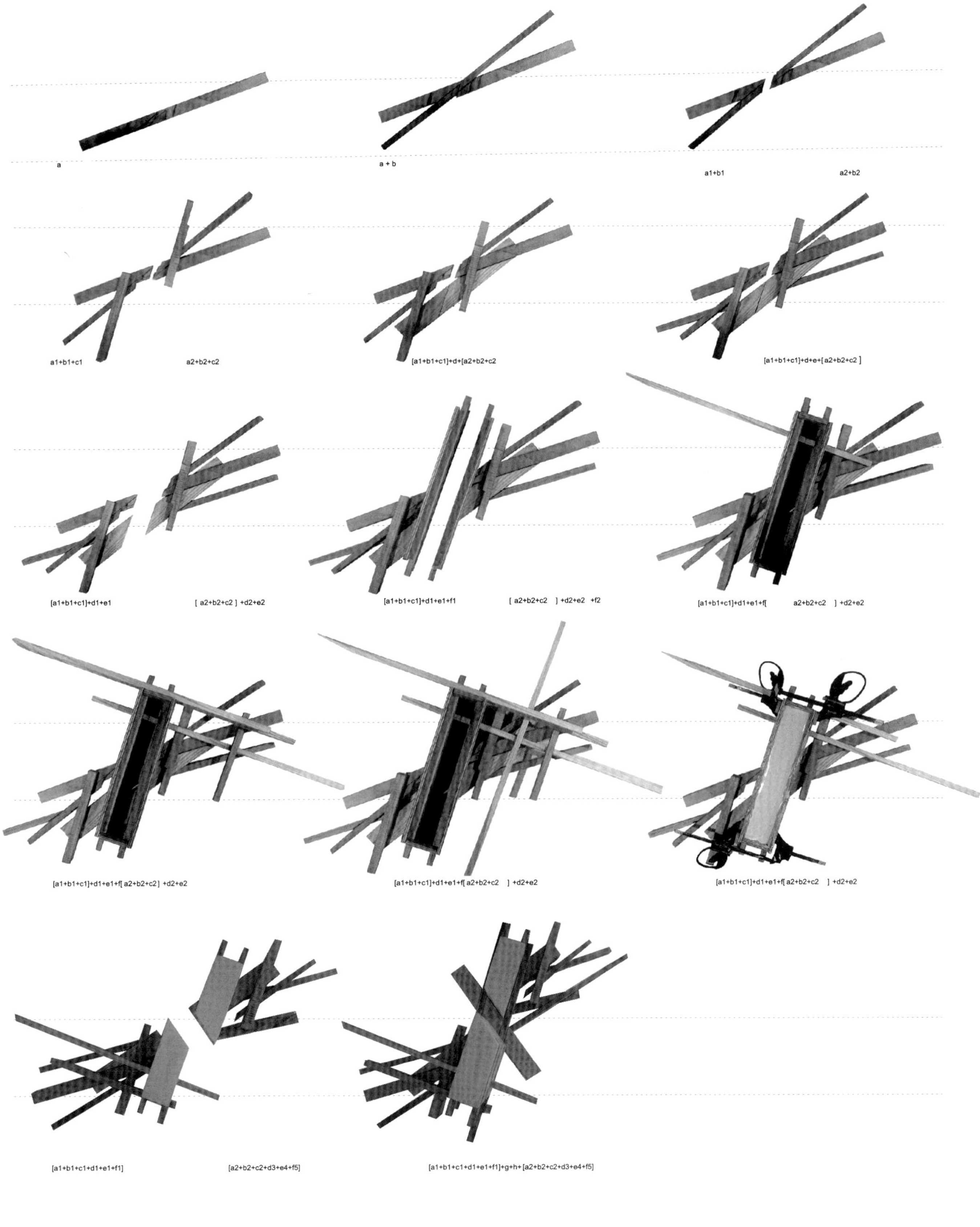
a
a + b
a1+b1
a2+b2
a1+b1+c1
a2+b2+c2
[a1+b1+c1]+d+[a2+b2+c2
[a1+b1+c1]+d+e+[a2+b2+c2]
[a1+b1+c1]+d1+e1
[a2+b2+c2] +d2+e2
[a1+b1+c1]+d1+e1+f1
[a2+b2+c2] +d2+e2 +f2
[a1+b1+c1]+d1+e1+f[
a2+b2+c2] +d2+e2
[a1+b1+c1]+d1+e1+f[a2+b2+c2] +d2+e2
[a1+b1+c1]+d1+e1+f[a2+b2+c2] +d2+e2
[a1+b1+c1]+d1+e1+f[a2+b2+c2] +d2+e2
[a1+b1+c1+d1+e1+f1]
[a2+b2+c2+d3+e4+f5]
[a1+b1+c1+d1+e1+f1]+g+h+ [a2+b2+c2+d3+e4+f5]

2.3 Micro urban Junction - model study - 2

The junction model is once taken apart into series of detail fragments, and each fragments are analysed in series of horizontal sections to understand the spatial territories in every different level. The evolution of those territories in horizontal plane becomes hint to explore how spatial volumes change in relation to the continuous insertion of solid objects.

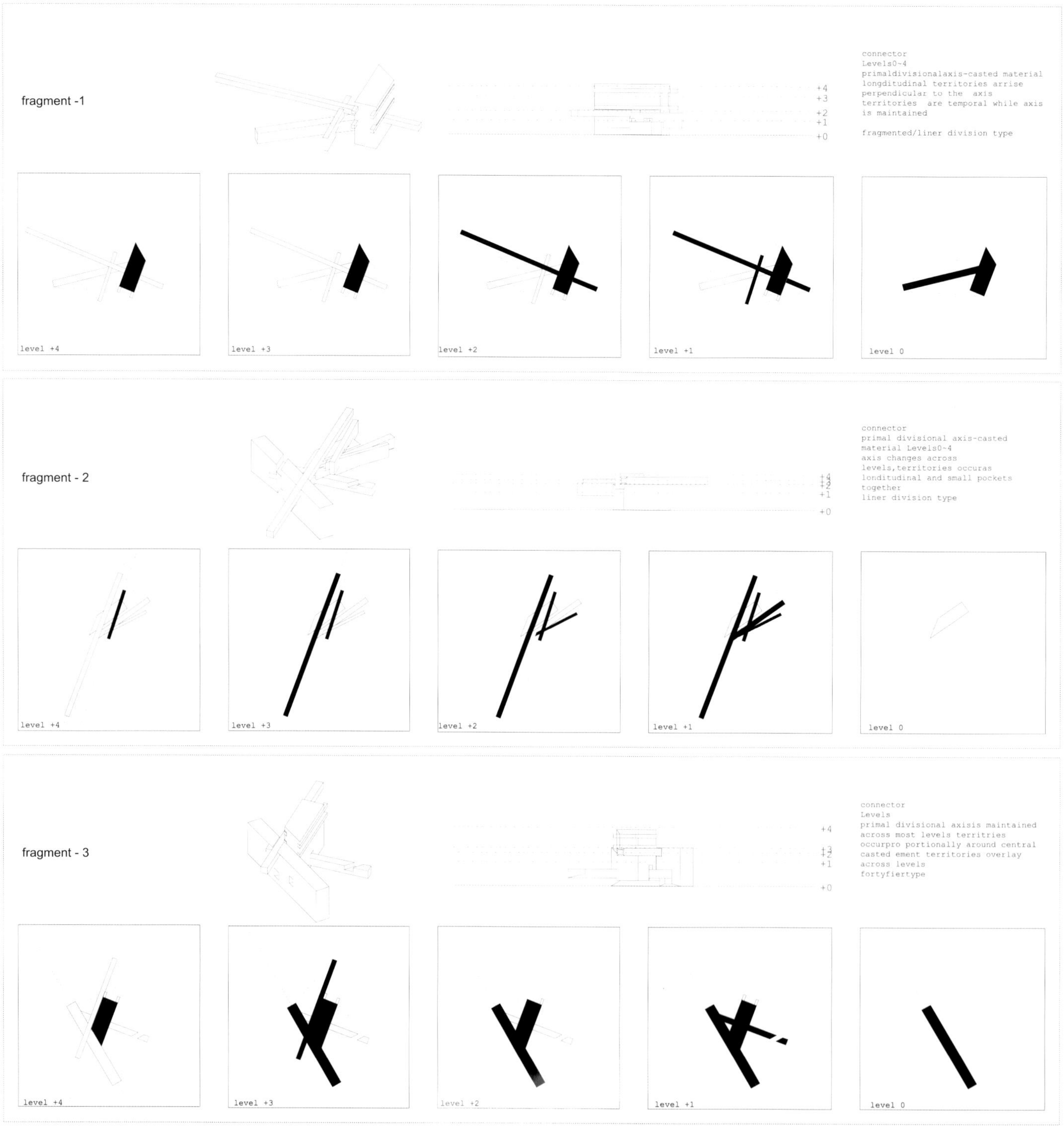

*fragment -3

fragment - 4

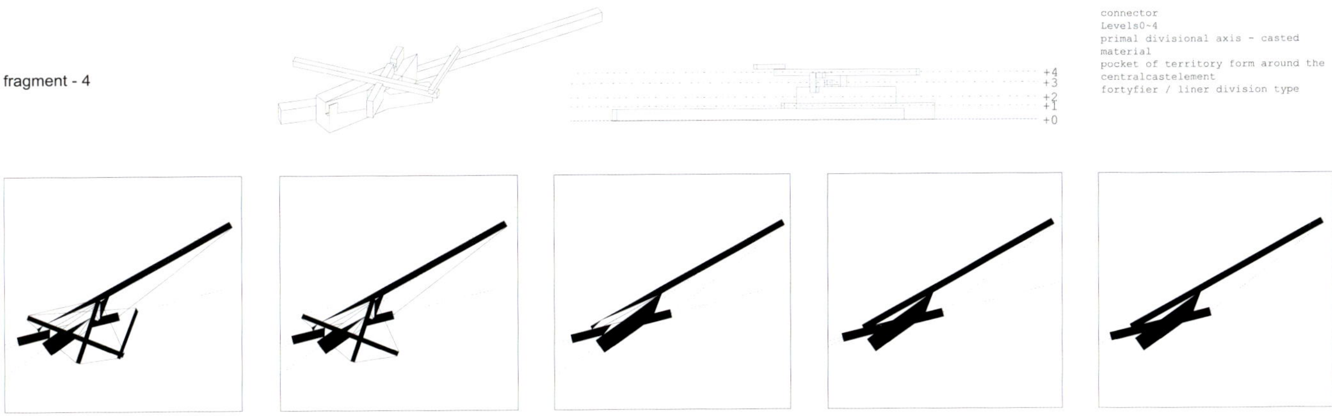

fragment - 5

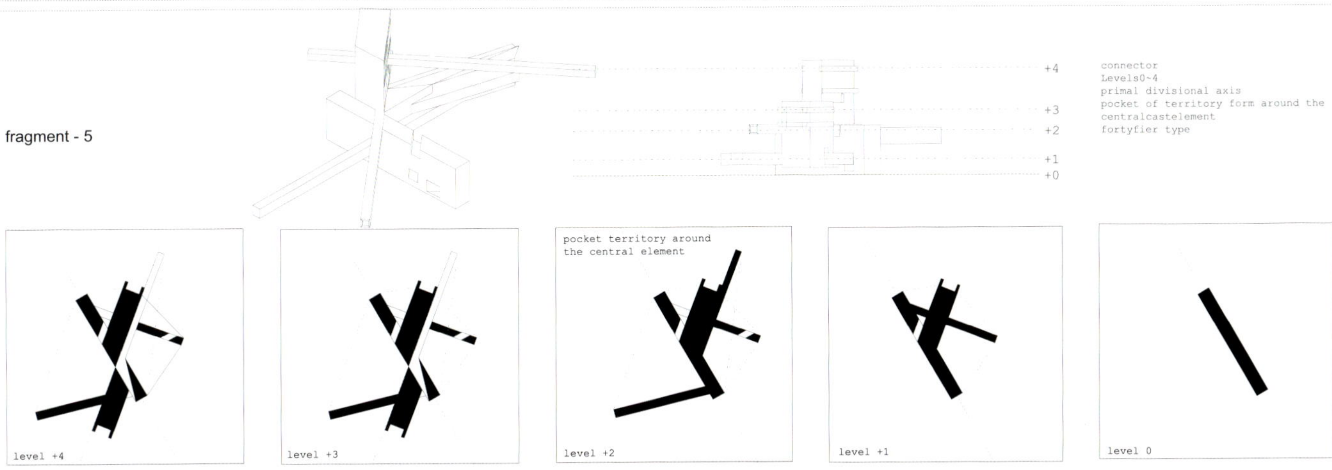

fragment - 6

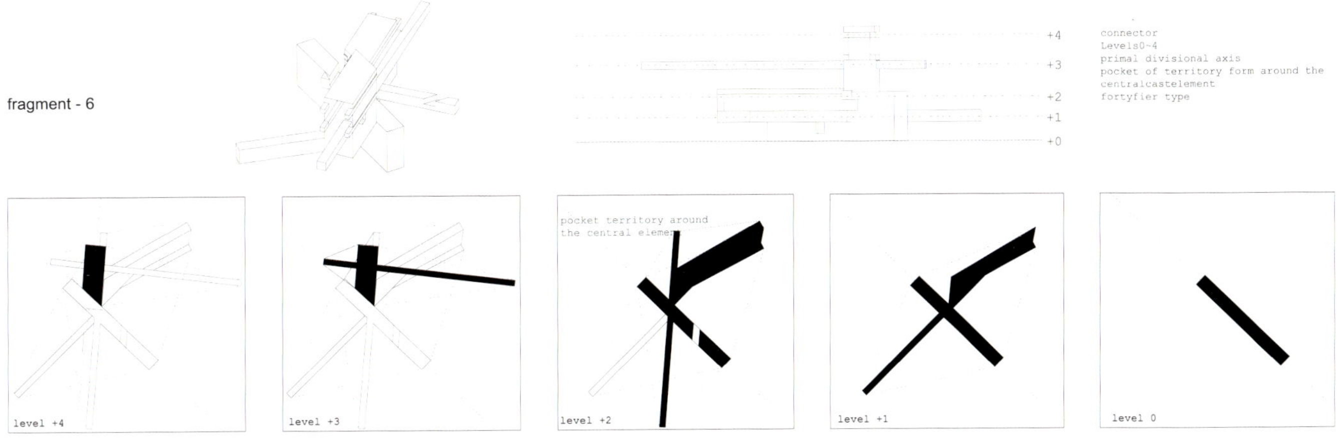

Keita Tajima

2.4 exploration of initial strategy

With understanding of junction model as a spatial coordination device, I developed the strategy for re - organizing recycling waste industries and their occupied materials. The idea of coordination device is once simplified as assembled wall elements as land forming devices, and phasing process of continuous insertions are simulated by sketch models.

sketch proposition of land forming process by inserting cordination points

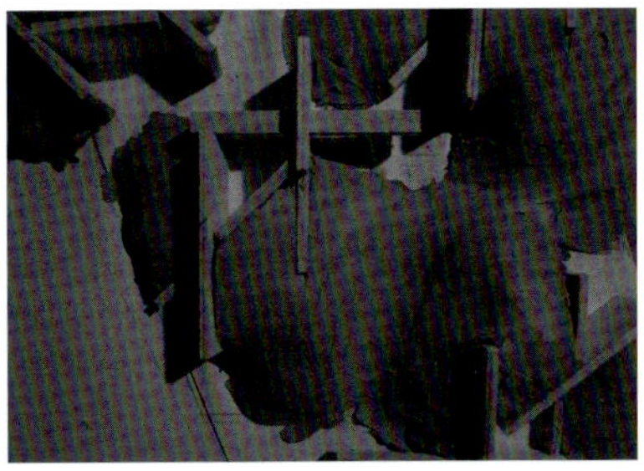

stratified land forming strategy diagram

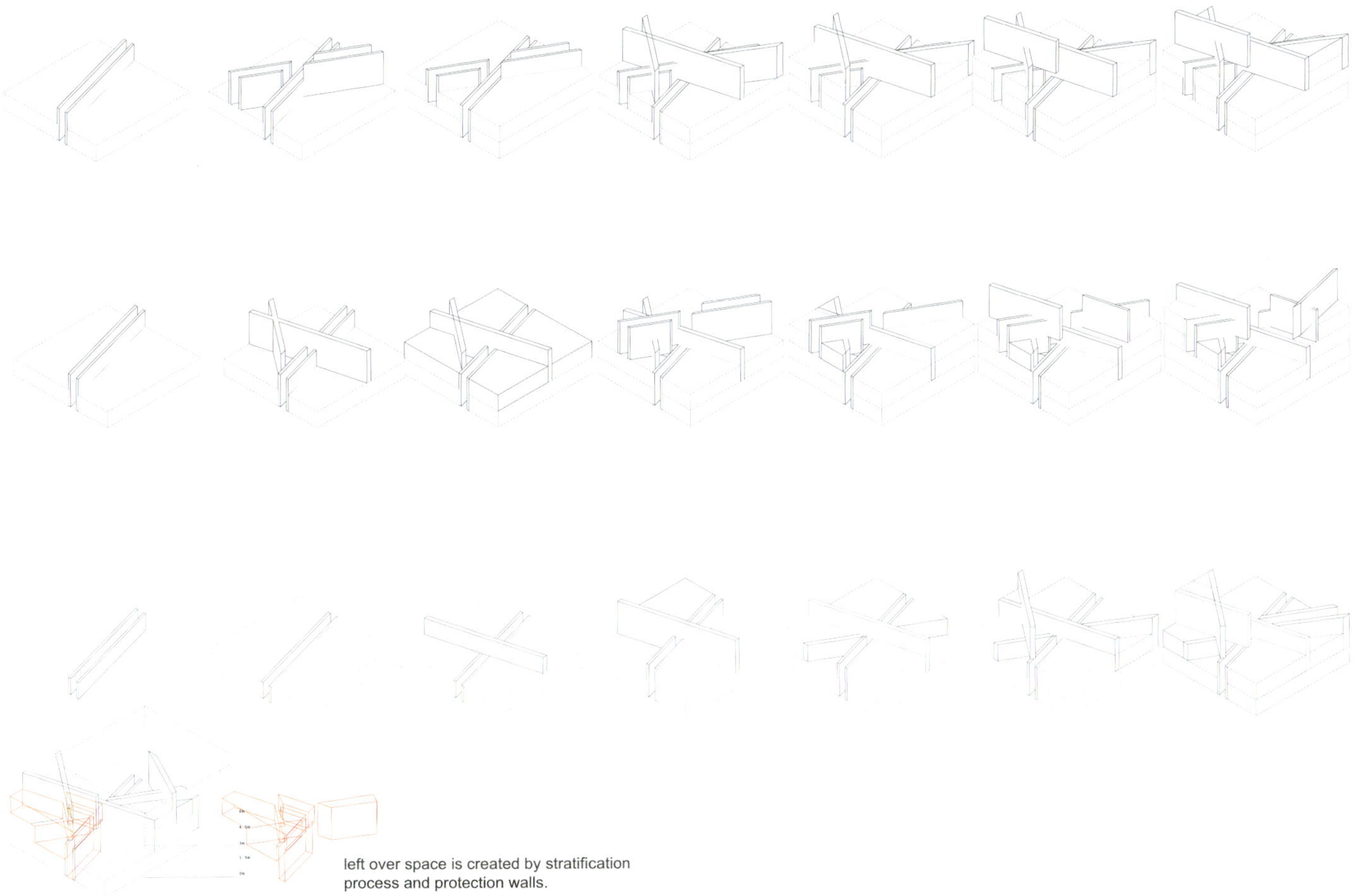

left over space is created by stratification process and protection walls.

3 Design Research

3.1 Layers - operations, topography, texture
3.2 Volume and topographic change
3.3 Operations
3.4 Texture - material evaluation
3.5 texture - material evaluation2
3.6 Re-organizing the land of recycling industries
3.7 Inverted space
3.8 Material experiments

Detailed observation

The existing industrial landscape of recycling waste was observed in three different layers, topographic change of the volume of wastes, existing operations and textural details. The analysis of overlapping layers reveals the potential of re-organization of existing operations, utilization of waste materials, transient movement pattern of topography, which informs possibilities for strategic transformation of post-industrial site by re-inserting architectural elements as urban components of micro city.

Existing operations / volumetric change of wastes

One of dominant force within current waste management site is a series of industrial sorting operations that occupy and prevent other activities to intervene. In order to propose urban strategy of re-inhabiting and alternating process of existing industries, understanding of existing operations and climate of recycling industry are essential part of project.
Existing operations are re-organized according to their territories, handling materials and frequencies that influence the change of sedimentary landscape. Each operations are carefully observed and separated into industrial components that make up small units. The masses of wastes are compacted either processed and re-distributed to each small unit in relation to their operations and sediments.

Textural details

Huge amount of metal scrap, layered car bodies, tires, pile of timber waste and construction debris, all those wastes are the elements of urban bricollge and have possibilities to create new ground within transient working landscapes. The current landscape is firstly re-organized according to their economic values and differentiated into immediate, temporal and permanent landscaping materials. The quality of materials thus suggests different registration to surface conditions that invites various activities to accommodate. Therefore new landscape is always in transition, working landscape while gradually providing public activities within industrial territory. Immediate materials stays in moment, always on demand of recycling industry while permanent materials are regarded as useless or hopeless materials to recycle and are normally transported to landfill site in London which already exceed its capacity to be filled. Each material has its own time scale of sedimentary processes influenced by frequencies of wastes collections as well as its own natural process. Permanent materials such as construction debris, car tires and timber waste are re-evaluated in order to be utilized as fill materials, filtering material and surfacing materials while immediate and temporary materials such as non ferrous and ferrous metals are compacted in order to reduces their occupied space. Several case studies of existing landscaping techniques and physical experiments for filtering ability of materials inform the individual quality and its industrial process as well as natural process that suggest advantage of each material and articulating spatial conditions and programs.

Inversion of sedimentary volumes

Sedimentary process of material facilitates excessive volume of materials to be perceived as inverted space rather than volume of materials. The inversion change the relationship between original ground level and sediments, sediments are no longer perceived as volumes and starts to provide connections to surrounding as new ground, while leftover original ground surface become inverted space to be inhabited. The principle of inversion is studied through narrow spaces in Tokyo as analogical spatial experience. And Chinese court yard house is a example of inverted architectural space in terms of programmatic relationship between ground surface and space beneath.

Meterial experiments

The idea of Urban bunker is developed through physical material experiment along with series of related researches. Material experiment of concrete retaining wall is to find out principle of proposals in small configuration which to explore further development of urban components.
The sectional model of urban bunker demonstrate the process of casting concrete wall, inlaid timber structure and retaining resource materials. The further exploration of this structural principle is supported by understanding of geotectonics of retaining walls in relation to different pressure arise from the density of each materials and their processed form, which defines proportion of wall structures in relation to retained materials. The case study of Siza's swimming pool is to explore spatial quality of retaining wall structure which combines both architectonic and geotectonic by applying timber structures onto concrete retaining wall. And its structural principle suggests new relationship between retained materials and extended timber structure.

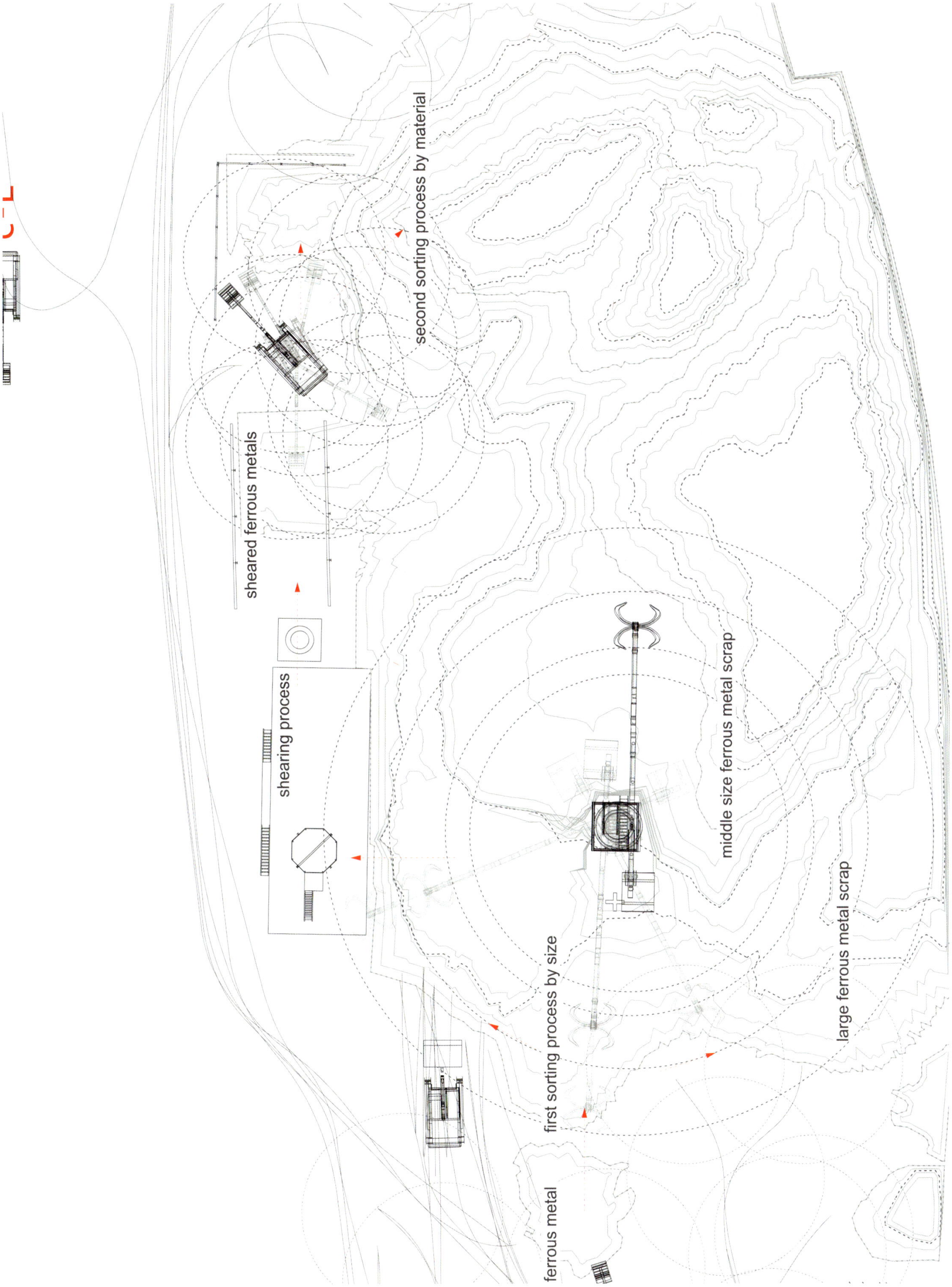

operations and topography in metal scrap yard

3.1 Urban layers - waste recycling industry

heap of resources

heap Z - nonferrous metal
heap Y - cast iron
heap X - ferrous (first area for ferrous metal to be accumulated)
heap W - large pieces of ferrous metal
heap U - smaller pieces of ferrous metal
heap T - second accumulation of shread ferrous metal
heap S - first accumulation of shread ferrous

heap R - light steel
heap Q - car body
heap P - construction debris
heap O - municipal waste
heap N - construction debris
heap M - timber
heap L - large refuse
heap K - ferrous metal

Facilities that process resouces
shearing ferrous metal
car depollution process

tipping area

area 1	non ferrous loading/unloading
area 2	ferrous loading/unloading
area 3	small stuff and mechanical stuff loading/unloading
area 4	scrap car and light steel loading/unloading
area 5	scrap car loading/unloading
area 6	large refuse, construction debris and ferrous loading/unloading
area 7	large refuse loading
area 8	municipal waste loading/unloading

operation teritory
movement of vehicle
coverage of cranes

texture (resources)

Circulation of waste in West London

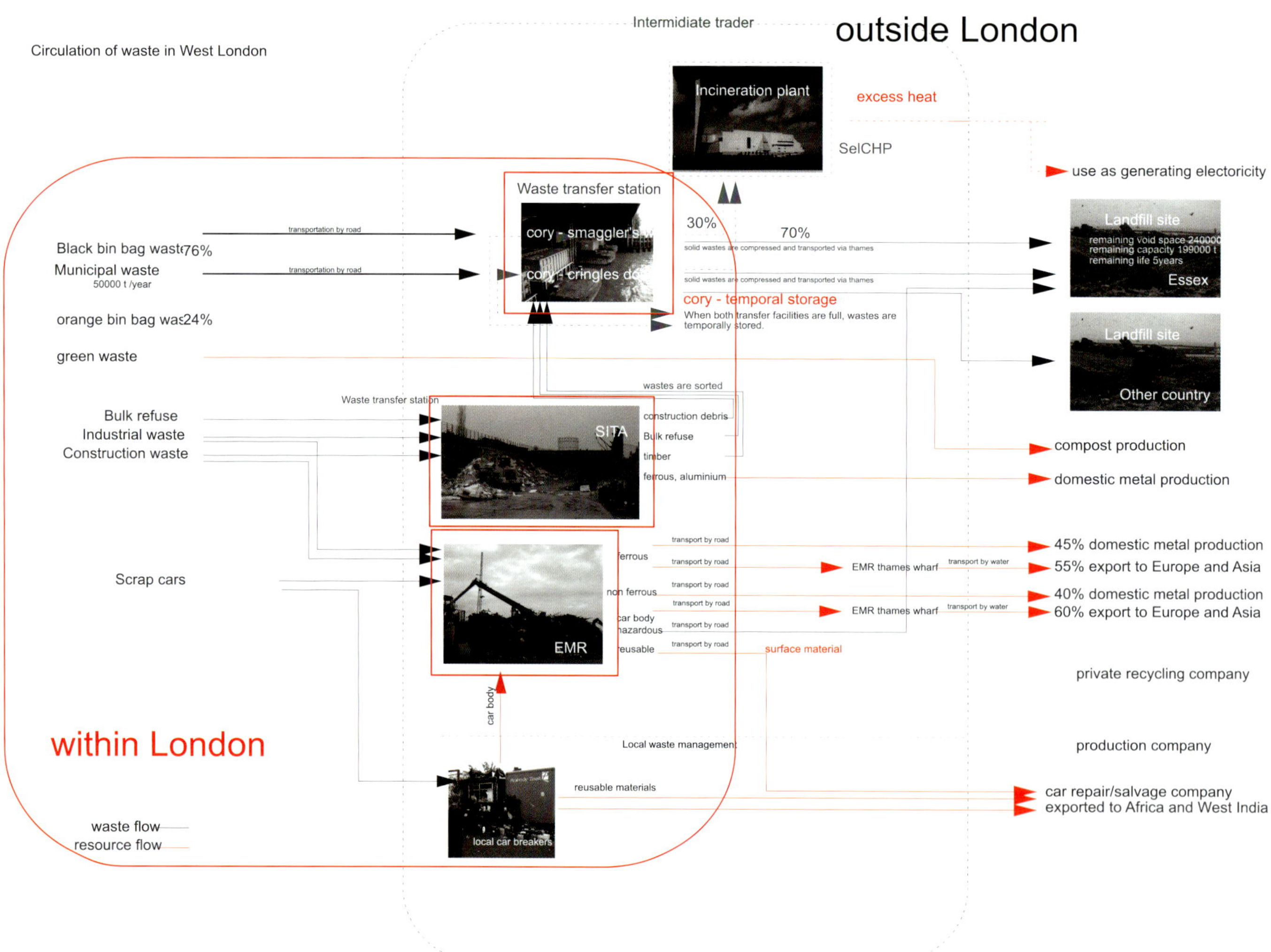

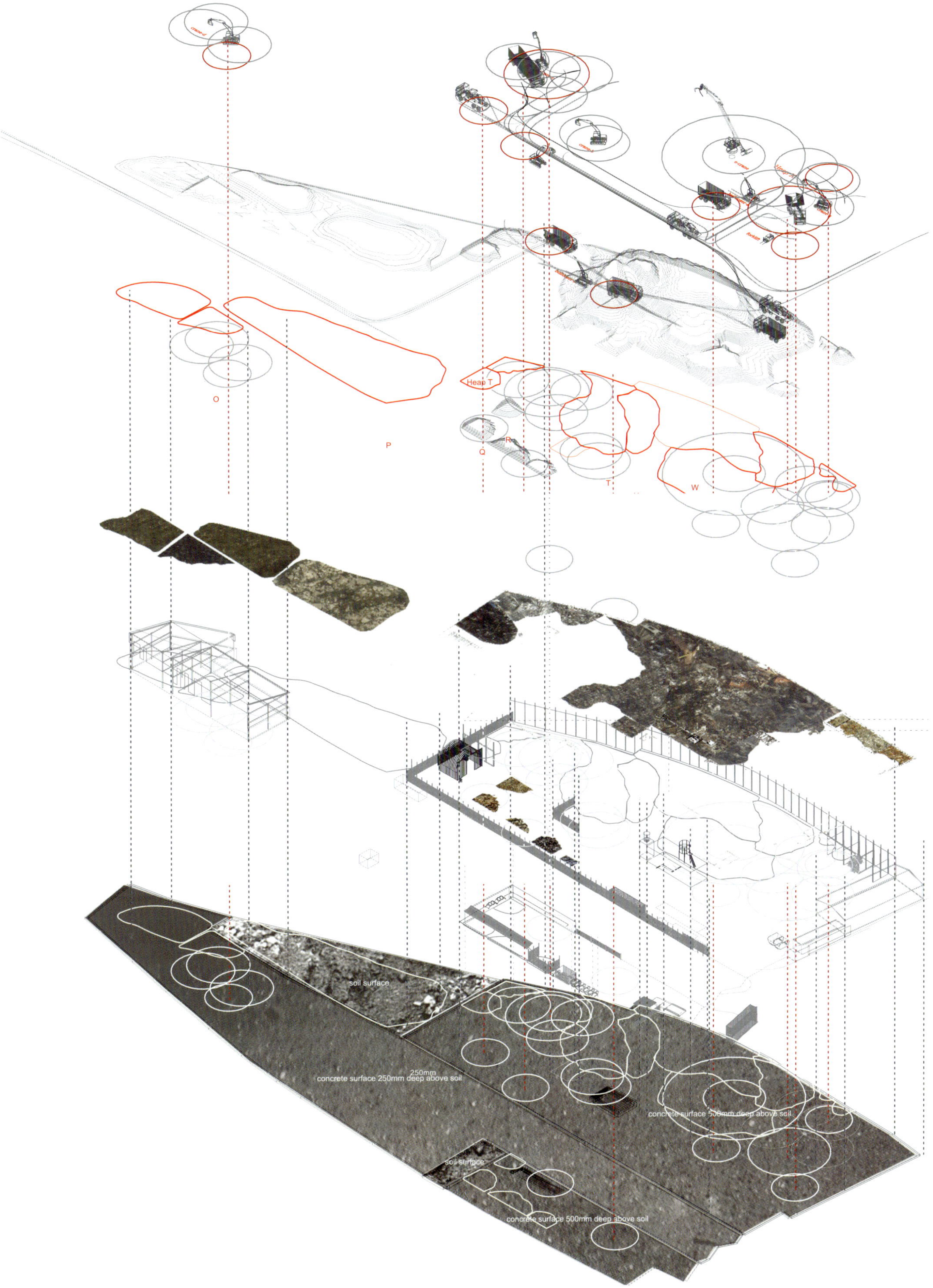
Heap T
O
P
Q
R
T
W
soil surface
250mm
concrete surface 250mm deep above soil
concrete surface 500mm deep above soil
soil surface
concrete surface 500mm deep above soil

3.2 volumetric analysis of transitional topography

Volumetric analysis of recycling waste shows the change of transitional landscape of the site. The frequencies and amount of wastes trading process was mapped, in order to identify topographic change of the site over time. The volumetric change of each different materials shows their territorial change within the site, and thus suggesting the possibilities of architectural intervention.

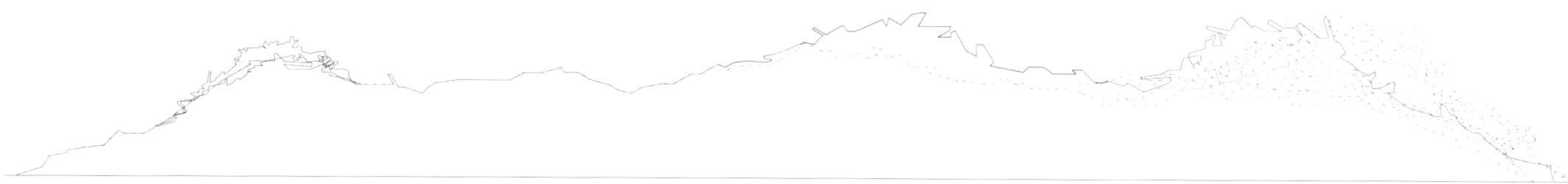

day-1 07:00-17:00

total in volume 1470m3
out volume m3 570
increased volume 900m3

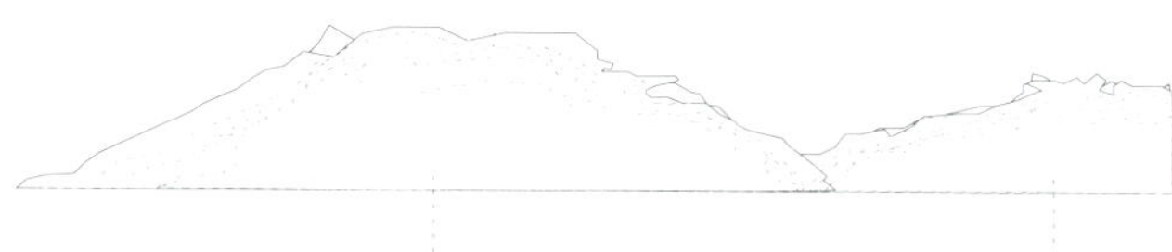

large refuse

total in volume 371m3
out volume m3
increased volume 900m3

timber

total in volume 166m3
out volume m3 570
increased volume 900m3

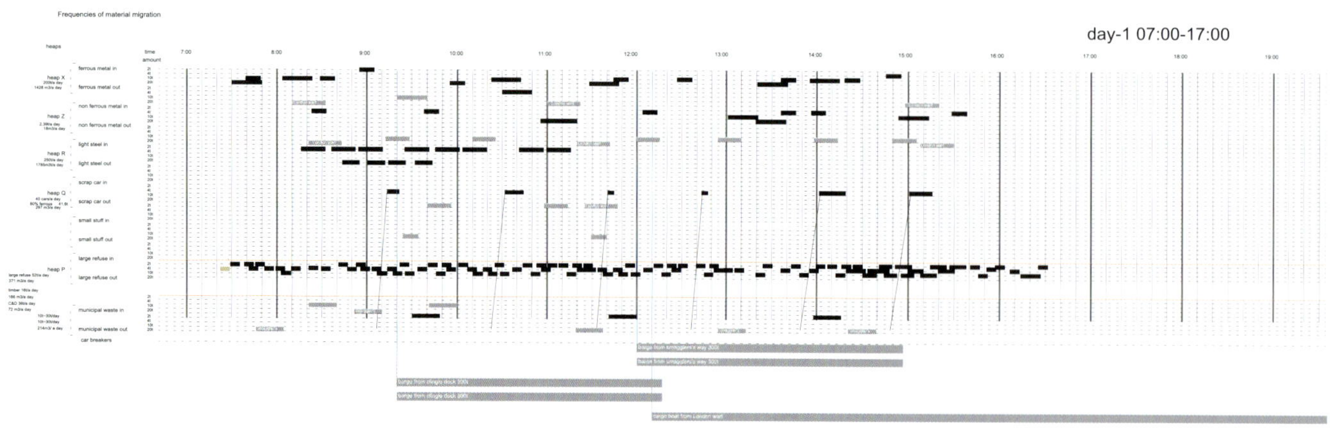

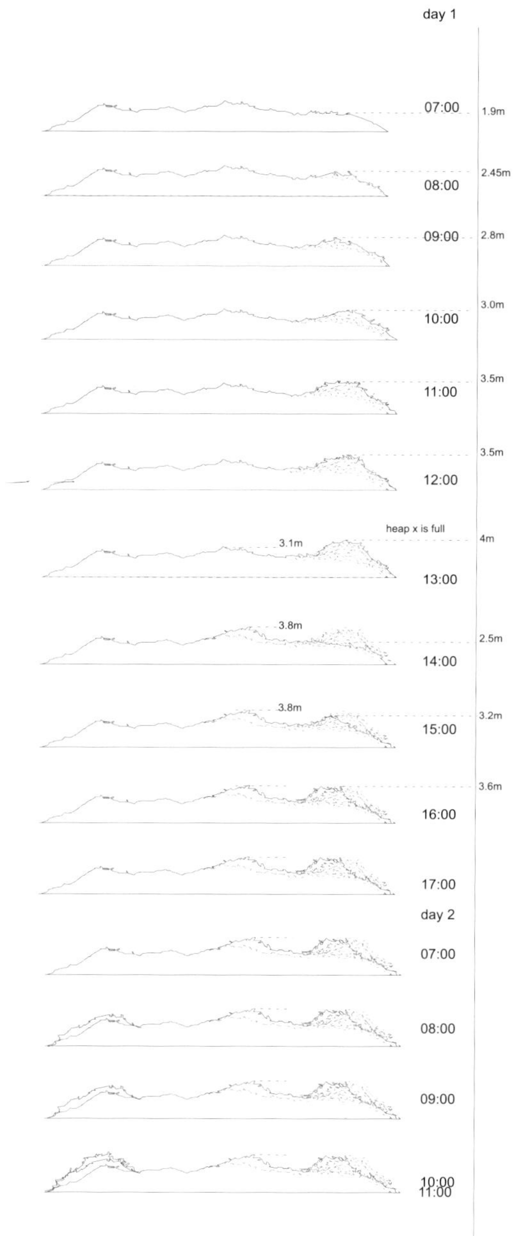
day 1
07:00
1.9m
08:00
2.45m
09:00
2.8m
10:00
3.0m
11:00
3.5m
12:00
3.5m
heap x is full
3.1m
13:00
4m
3.8m
14:00
2.5m
3.8m
15:00
3.2m
16:00
3.6m
17:00
day 2
07:00
08:00
09:00
10:00
11:00

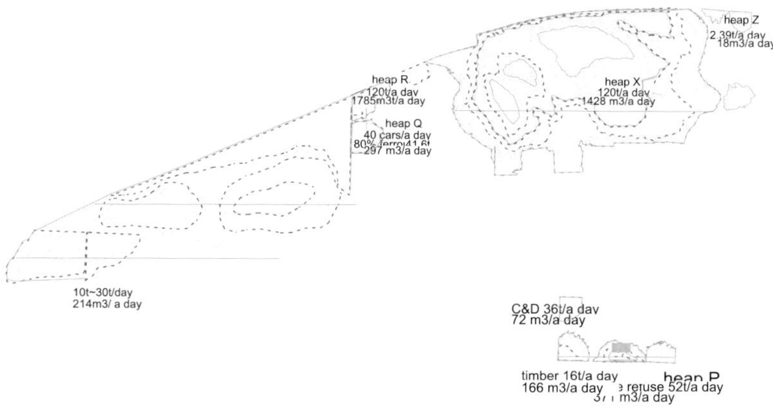
heap Z
2.39t/a day
18m3/a day
heap R
120t/a day
1785m3t/a day
heap X
120t/a day
1428 m3/a day
heap Q
40 cars/a day
297 m3/a day
10t~30t/day
214m3/ a day
C&D 36t/a day
72 m3/a day
timber 16t/a day
166 m3/a day
heap P

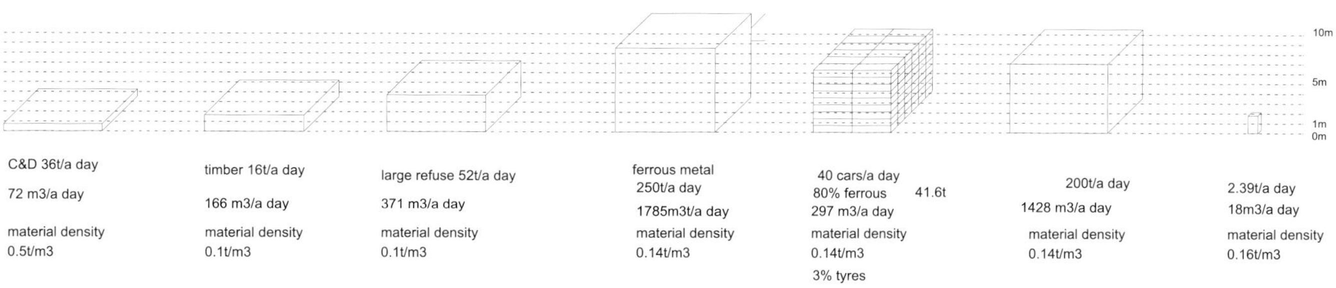
10m
5m
1m
0m
C&D 36t/a day
72 m3/a day
material density
0.5t/m3
timber 16t/a day
166 m3/a day
material density
0.1t/m3
large refuse 52t/a day
371 m3/a day
material density
0.1t/m3
ferrous metal
250t/a day
1785m3t/a day
material density
0.14t/m3
40 cars/a day
80% ferrous
41.6t
297 m3/a day
material density
0.14t/m3
3% tyres
200t/a day
1428 m3/a day
material density
0.14t/m3
2.39t/a day
18m3/a day
material density
0.16t/m3

3.3 Industrial operations - analysis

1

2

3

4

5

6

7

8

1

non ferrous loading/unlodaing

unloading
duration 00:08:00-00:15:00
loading
duration 00:12:50-00:18:00
materials
cast iron/copper/aluminium
pipe/brass/el cables and wires
operating vehicles
bulldozer/crane/manually by hand

2

ferrous loading/unlodaing

unloading
duration 00:12:50-00:18:00
loading
duration 00:12:50-00:18:00
materials
large ferrous metal
operating vehicles
bulldozer/crane/

3

small stuff/mechanical stuff store

unloading
duration 00:5:00-00:20:00
loading
duration 00:10:00-00:15:00
materials
motors/fragmetized metals/cables/fridges
operating vehicles
orklift/van with skip/manually by hand

4

scrap car/light steel loading/unlodaing

unloading
duration 00:12:50-00:18:00
loading
duration 00:12:50-00:18:00
materials
car bodies/light steel/ steel furnitures

5

scrap car unlodaing/depollution

unloading
duration 00:6:50-00:18:00
loading
duration 00:18:00-00:26:30
time include compressing car
materials
scrap cars
operating vehicles
folk lift/crane/

6

unloading commercial/industrial waste

unloading
duration 00:03:00-00:15:00
loading
duration 00:10:00-00:20:00
materials
domestic waste(black bin bag)/
construction debris
operating vehicles
forklift/van with skip/manually by hand

7

loading pit for 20/10 t truck

loading
duration 00:11:40-00:20:00
materials
aluminium/ferrous metal/
timber /construction debris/landfills

8

domestic waste loading/unloading

unloading
duration 00:4-50:12:00
loading
duration 00:11:50-00:15:00
materials
domestic waste(black bin bag)
construction debris

minimized and subdivided operations units

unit a

operations
ferrous bale store
shredding tyres
materials
ferrous bale
tyre shredds

unit b

operations
riquid removals
sorting
materials
hazadous riquid
from car
ferrous bale
other small stuff

unit c

operations
sorting
accumulating
materials
non ferrous metal
ferrous bale
tyre residues

unit d

operations
baling
accumulating
materials
non ferrous metal
ferrous bale
ferrous unprocessed
sand

unit e

operations
shearing
accumulating
materials
loose ferrous metal
ferrous bale
sheared ferrous metal
concrete crush

unit f

operations
disassembling
sorting
accumulating
materials
loose ferrous metal
ferrous bale
car body
sand

unit g

operations
sorting
accumulating
materials
no ferrous
ferrous bale
car body
sand

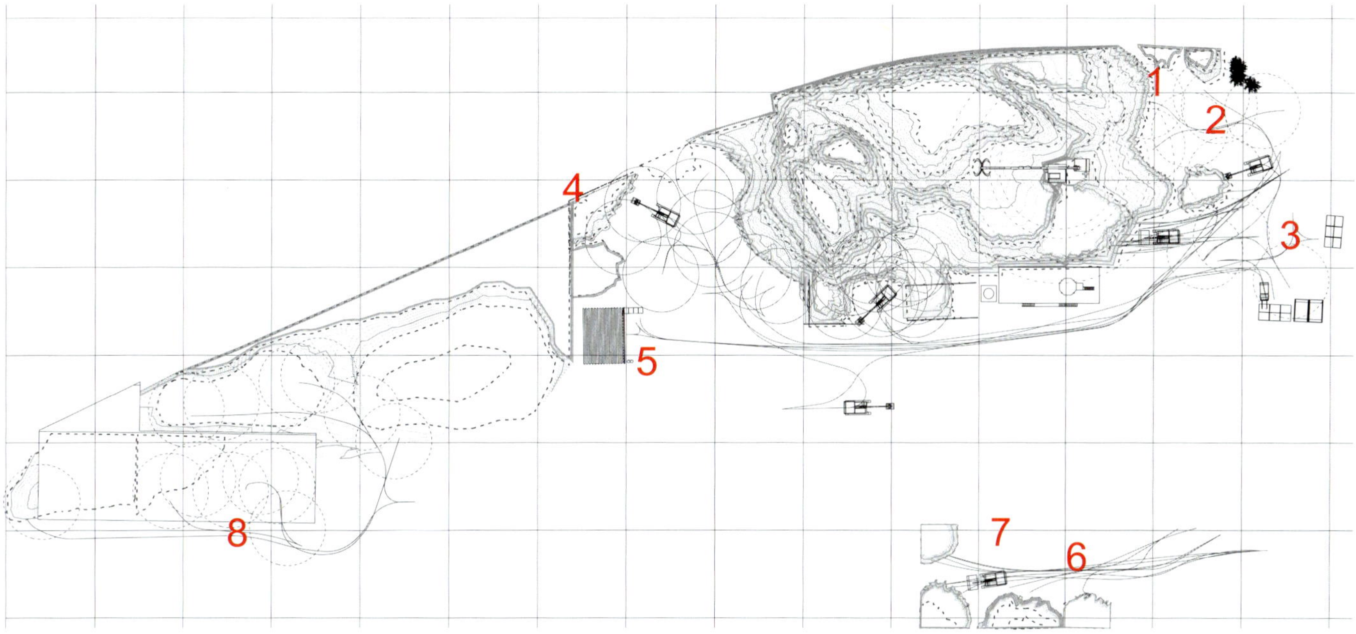
1
2
3
4
5
6
7
8

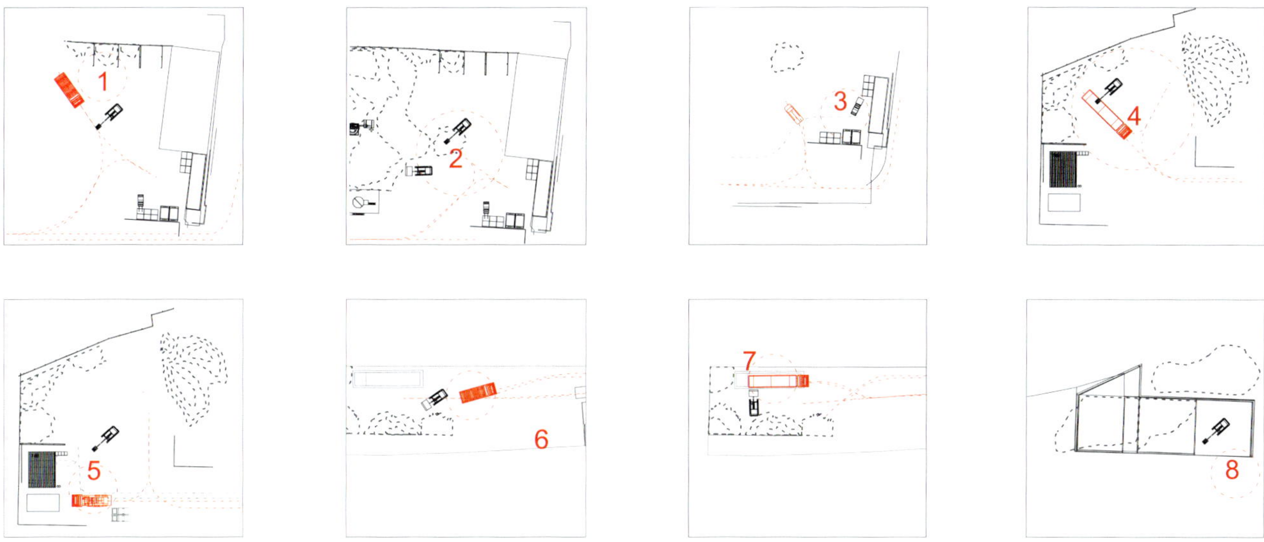
1
2
3
4
5
6
7
8

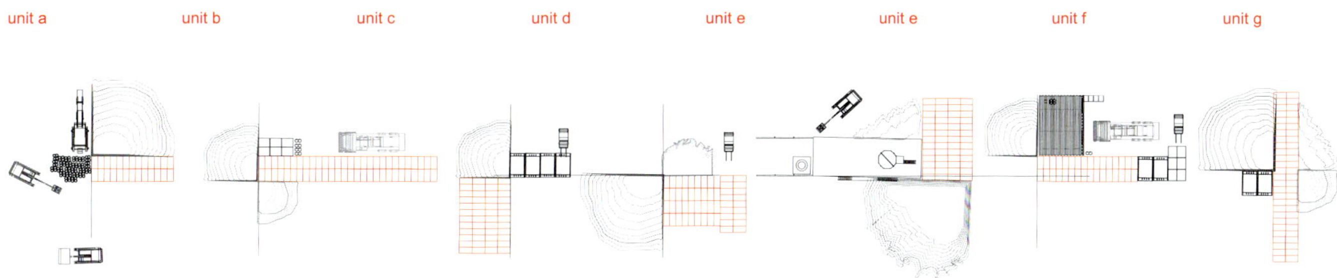
unit a
unit b
unit c
unit d
unit e
unit e
unit f
unit g

3.4 Texture - material evaluation 01

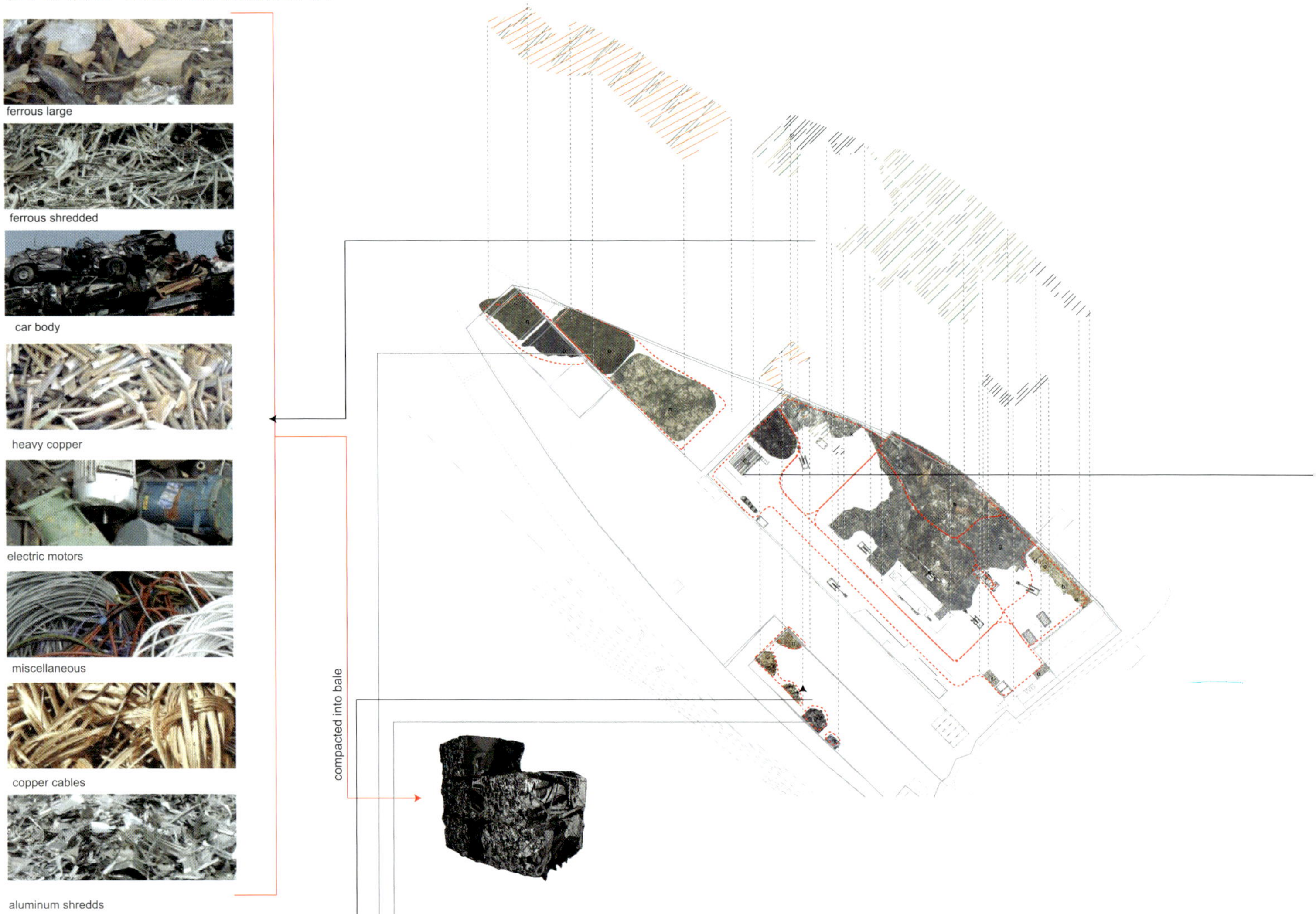

Material	Density		Process	Use	Remarks
concrete	2403kg/m3		crashed andsort into different sizes	aggregate for concrete gabion construction surfacing material	the use of crush concrete is depends on their sizes. but, as a compacted fill, it has strength comparabel to conventional fill materials
sand	1602kg/m3	filteration	washed	aggregate for concrete	good perfomance for utilizing as a filtering material such as sand bed
gravels	1746kg/m3	filteration draining	washed and sort into diferent sizes	aggregate for concrete filling as compacted	good perfomance for utilizing as a draining material,
soil	2002kg/m3			sufacing as top soil filling as compacted and composited with other materials	usefull for any compacted fill in combination with other materials, as well as surfacing for vagetation
saw dust	298kgm3	filtering	granurated	bioretention medium	good performance in filtering but takes long time to drain out takes long time to drain but bio-degradable
wood mulch	288kgm3(soft wood) *the density and colour are depended on the type of timber		mulched	surfacing for play ground/vegetation	good performance in shock absorption good for vegetation together with soil to feed nitrogen and later become sustanable soil

car tyres

process	material density	application	performance
shredded	 30mm~300mm	light weightfill drainfill	lighter than normal fill material such as soil. decrease vibration by its elasticity water permieable
rubber grnule 1522kg/m3	 0.5mm~5mm	surfacing pavement	decrease vibration by its elasticity, good for noise reduction. reduction rate - 6~8dB
recycle rubber products	 4mm~8mm	surfacing playground	decrease vibration by its elasticity, shock absorption, and the water permiable.

filteration saw dust or mulch + sand

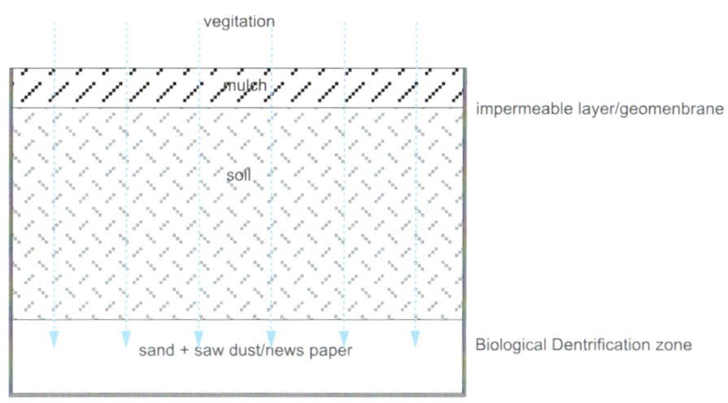

Surfacing recycle concrete + sand or earth(compacted fill)

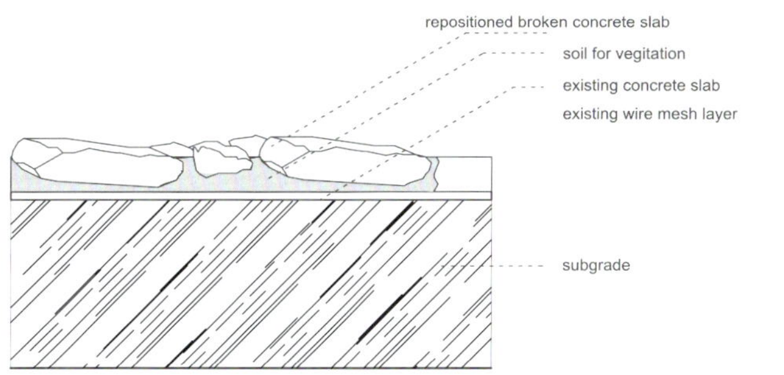

Surfacing recycled tire rubber + sand or earth(compacted fill)

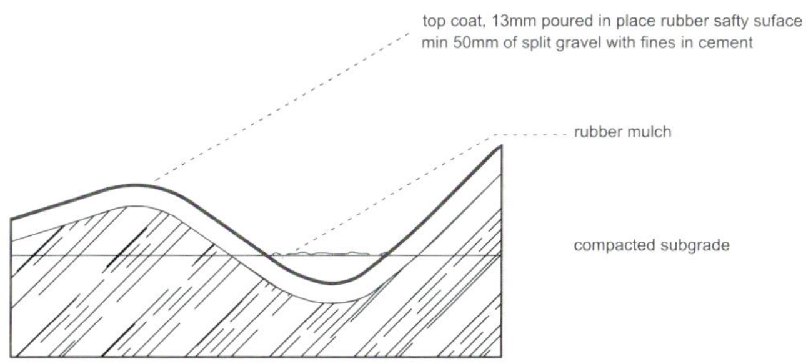

3.5 material utilization for surface run off

Problematic surface run off arose from existing waste recycling industries are resolved by analysing surface geometry and specifying the point where water would accumulate in order to introduce water collection ditches. The water collection ditches also provide axis for possible future access from surrounding area as well as connection between micro urban junctions.

Surface Area	18944m2
Permeable surface (soiled)	1990m2
Impermeable surface (concrete)	16954m2

Volume of surface runoff

Max: 132.24m3 monthly
Min: 86.5m3 monthly

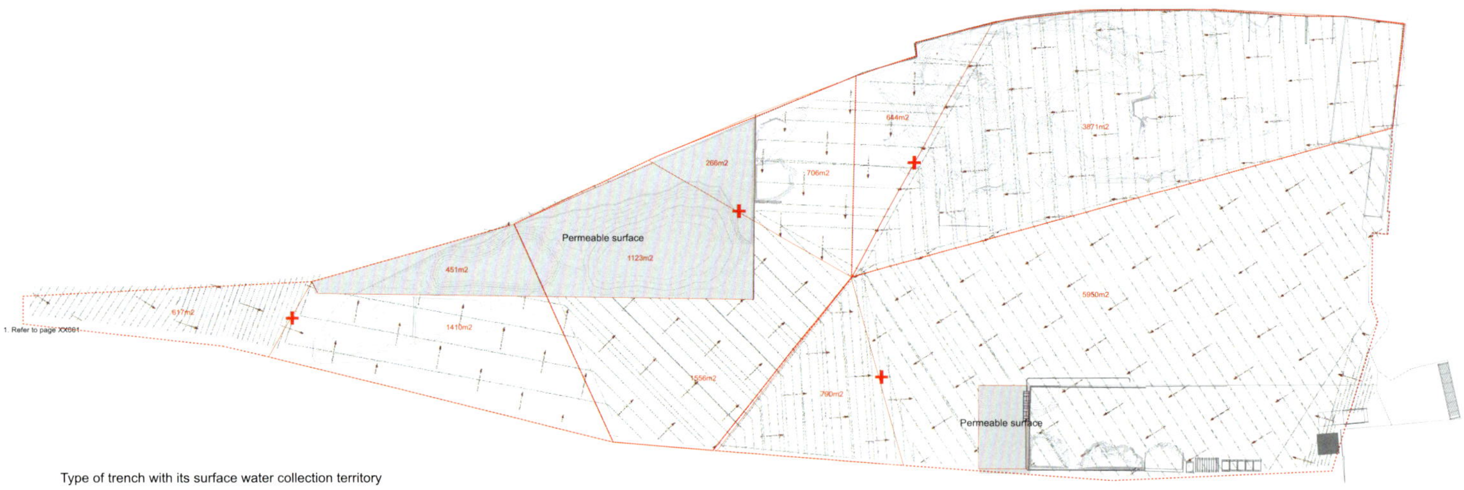

Type of trench with its surface water collection territory

future public access

captured surface area

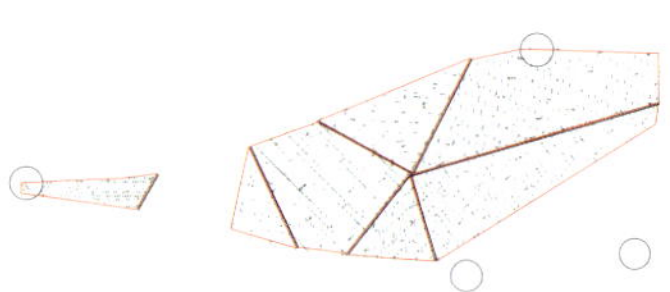

type A trench along the ridge of surface

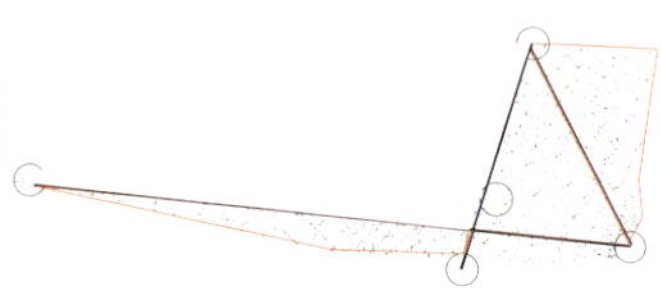

type B direct connection to public entry point

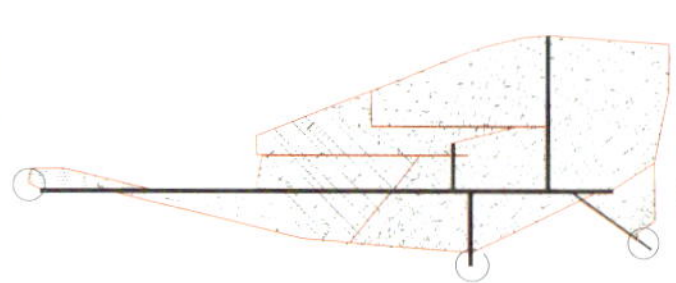

type C straight axis type

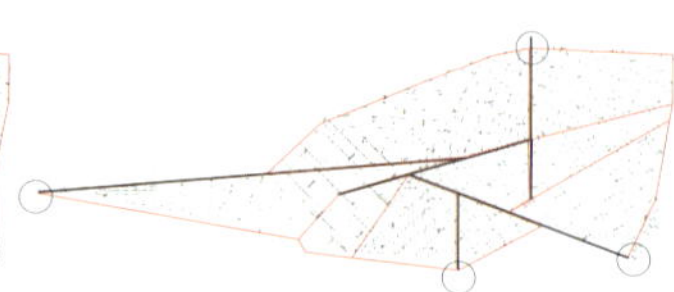

type D off set type

Utilaization of recycling materials for reed bed system

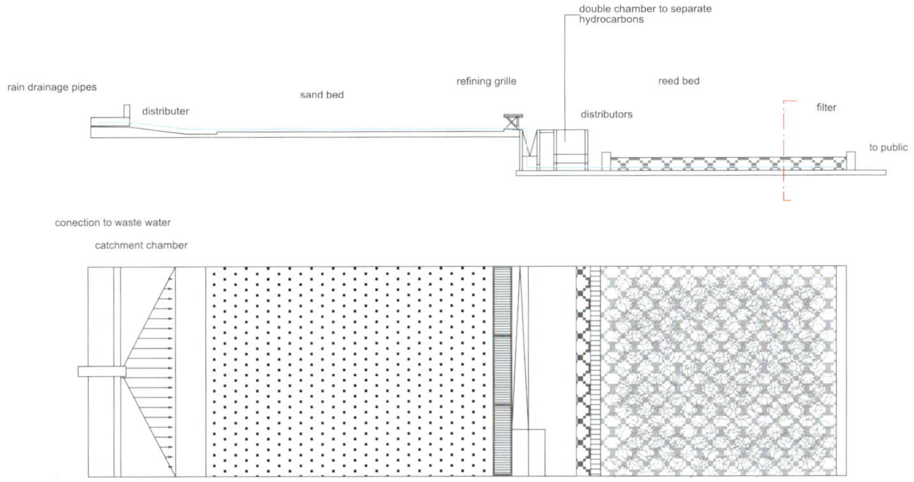

size of container should hold at least 5 days residense of water volume in order to ensure quality of the treated water.

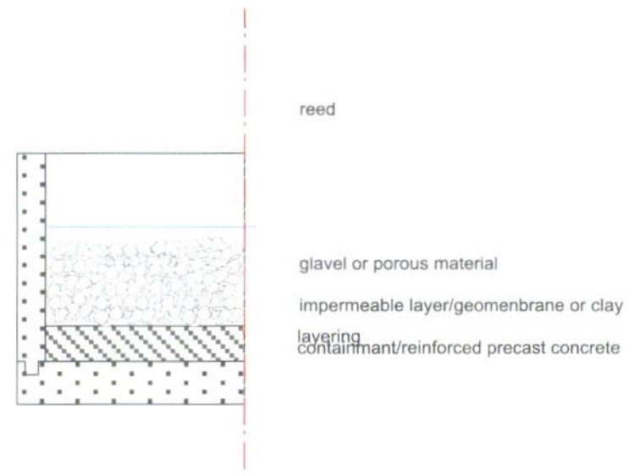

tipical reed bed installation

Layering technique and filtering material experiments

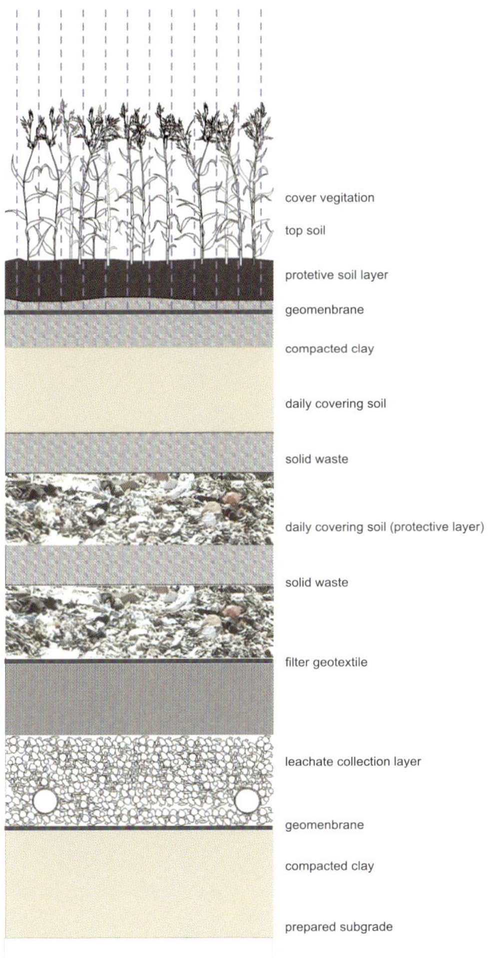

layerd materials
sawdust / fine metal
sample size
sawdust 300ml
fine metal 100ml
water input
300ml
duration
120 min
filtered water
90 ml
retained water
210 ml

layerd materials
gravels / fine metal
sample size
gravels 300ml
fine metal 100ml
water input
300ml
duration
2 min
filtered water
200ml
retained water
100 ml

Keita Tajima

3.6 Re-organizing the land of recycling industries

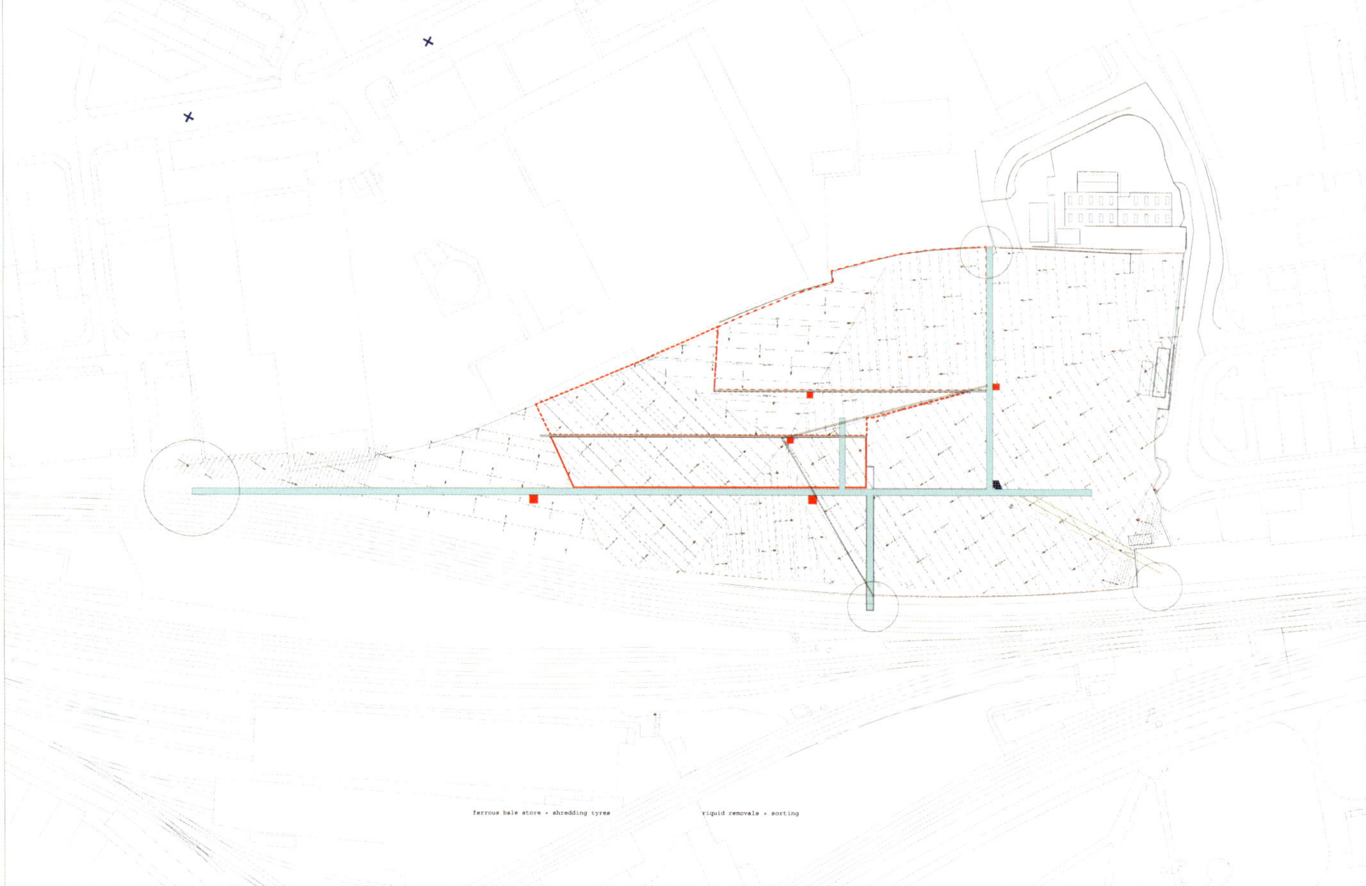

phase 1 - Insertion of drain trench for collecting surface run off from existing industries. the axis made by new trench also indicate future public access.

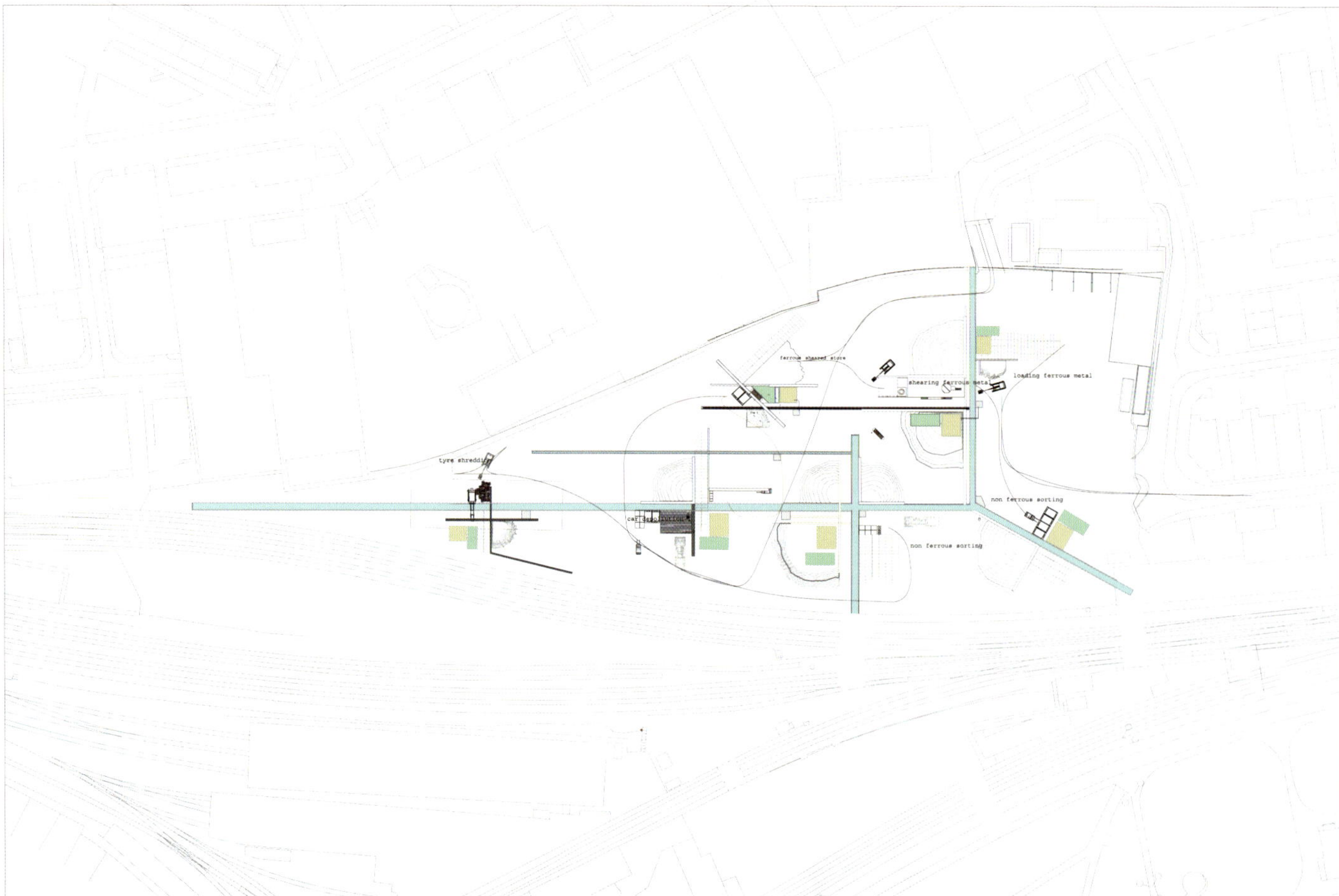

phase 3 - Implementing water purifying facilities (sand bed and reed bed) to each industrial unit for irrigations, and the water is utilized to provide public shared garden space.

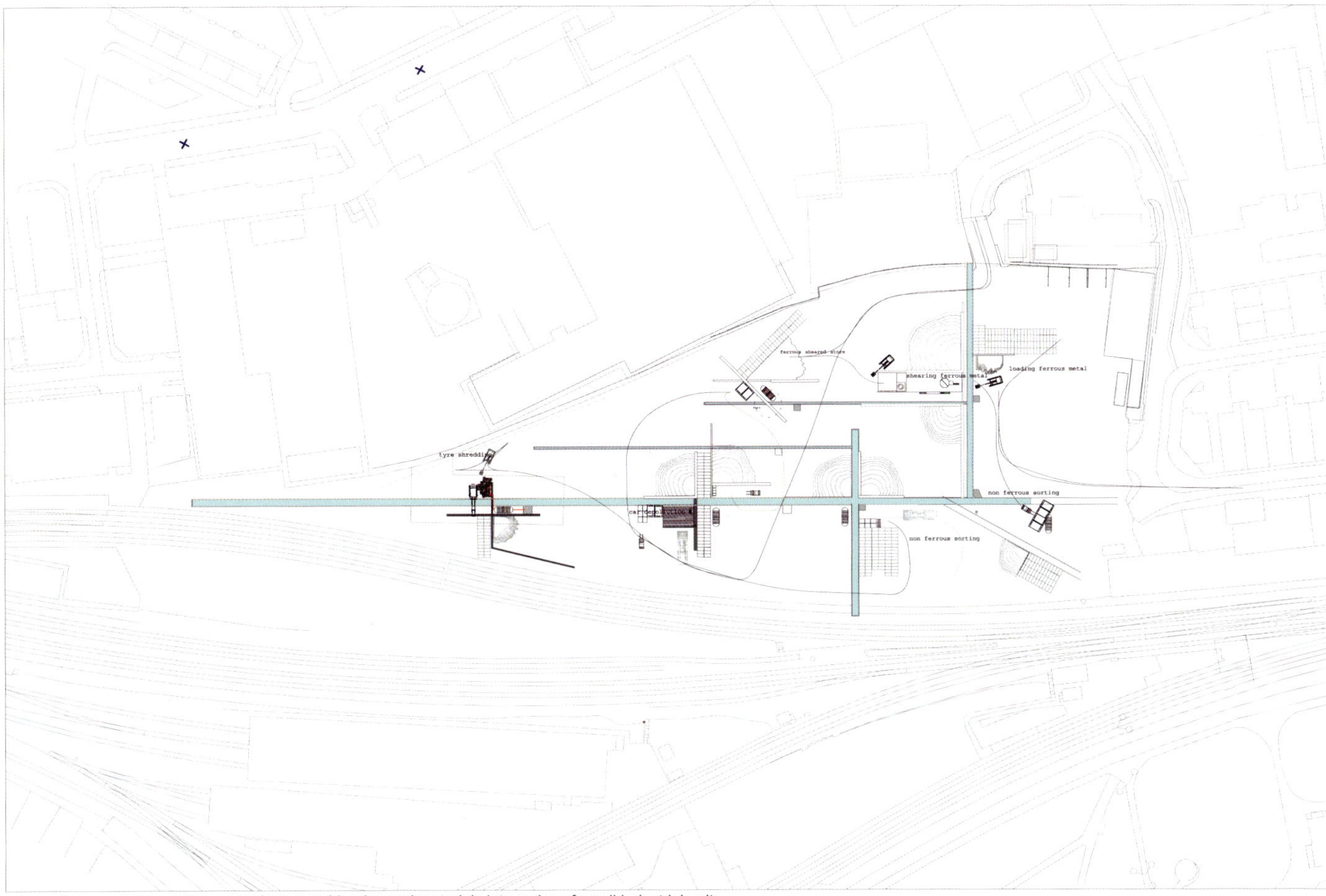

phase 2 - organizing existing operations, machineries and materials into series of small industrial units.

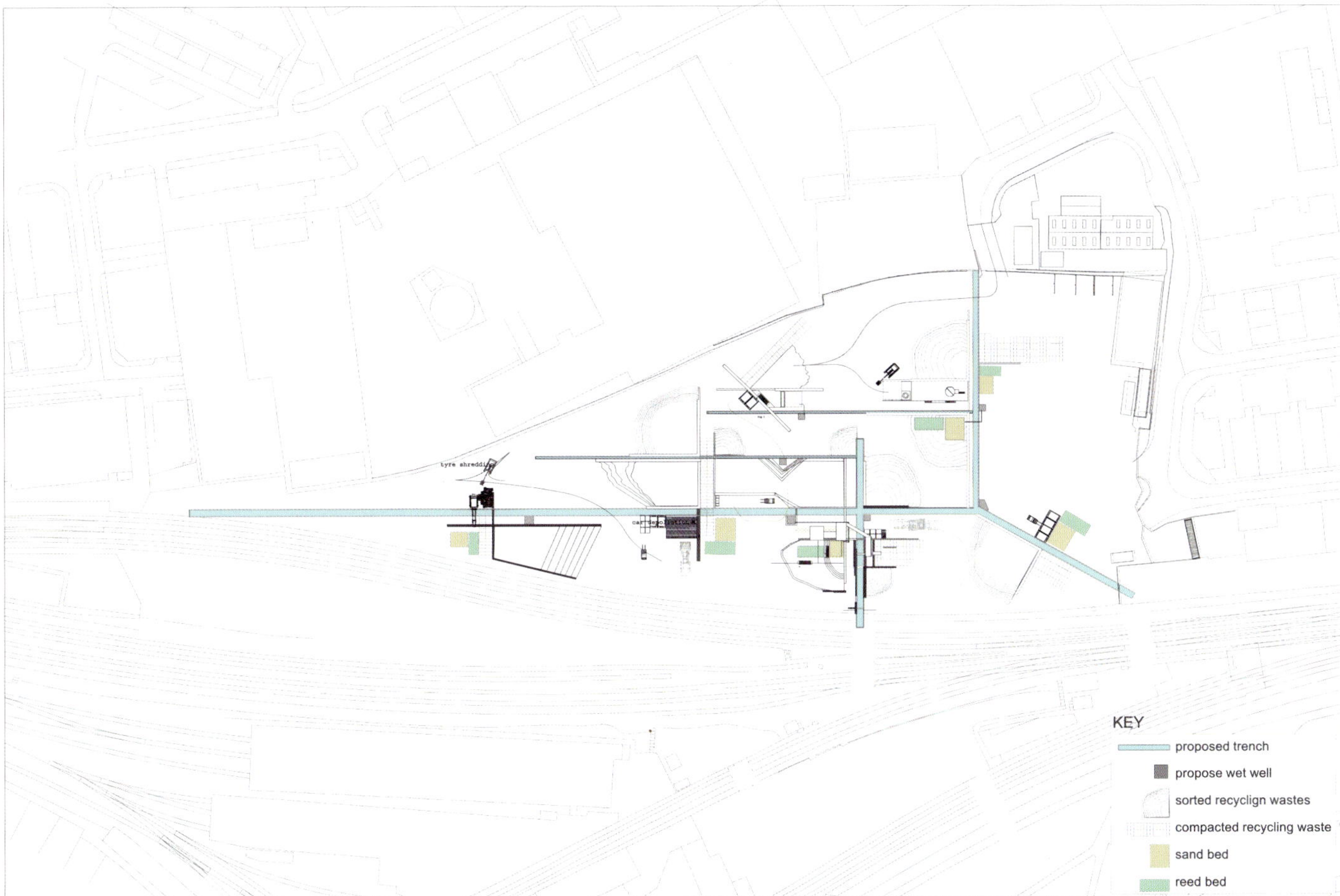

phase 4 - Territorial expansion of each units by its sedimentary process and continuous insertion and addition of bunker walls allow more public accessible spaces while industries are slowly declined.

3.7 Inverted space

The relationship between perceived volume and space is firstly depended on its volumetric growth but also the distance between volume and subject who perceives. Thus the distance between volume become closer, and the height increases, there is a point when inversion happens, and the volume is perceived as space. The photograph below (middle, bottom) shows the example of inverted space (Yaodon) in China. The court yard houses are built underground, separating dwelling space and work field on ground level.

The volume is no longer perceived as a mass, but the part of space.

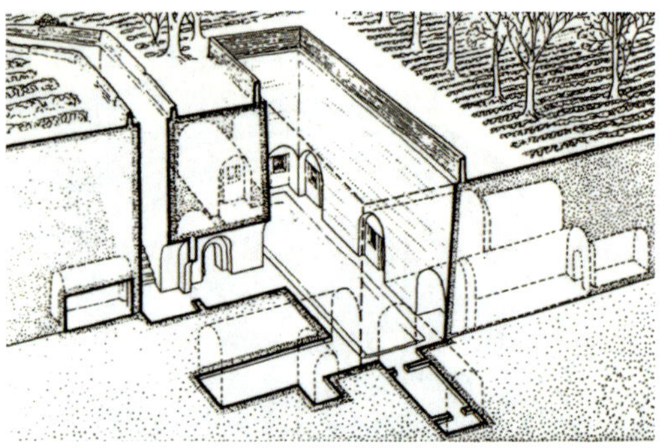

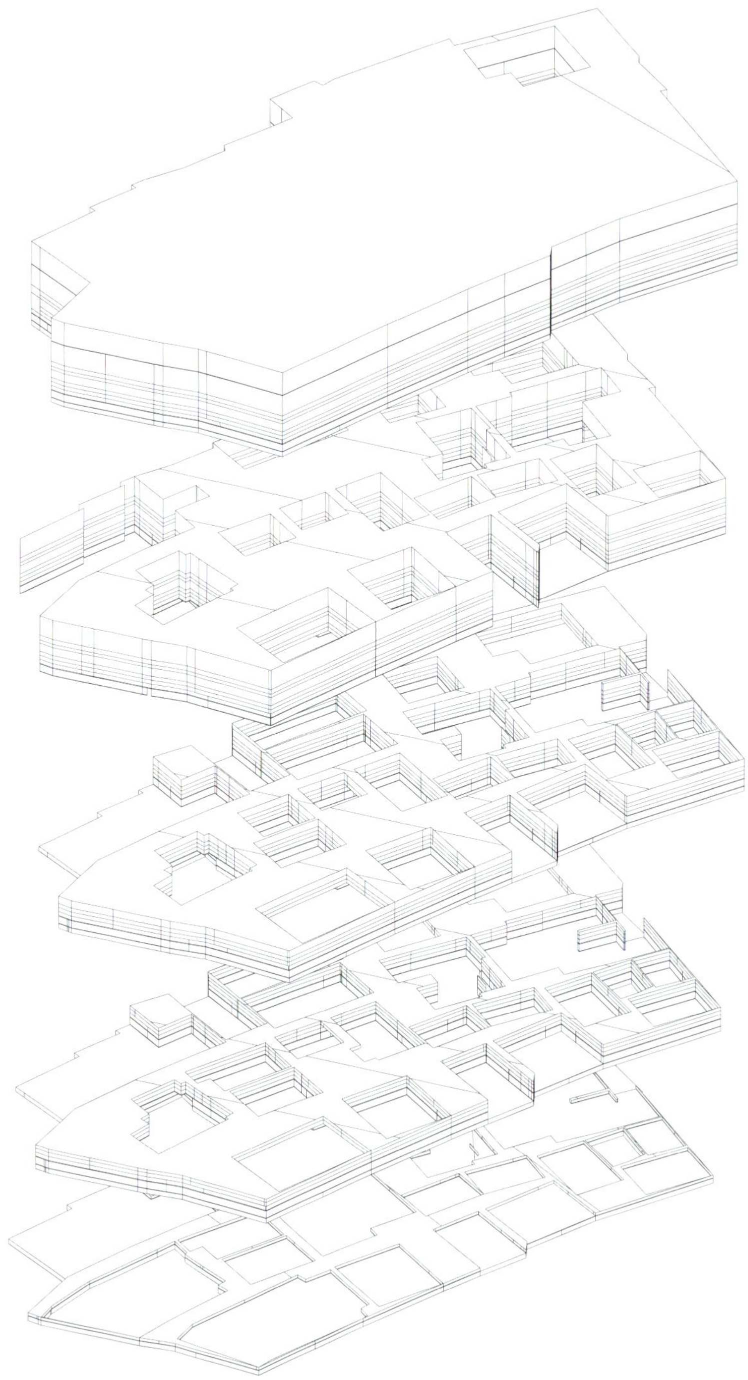

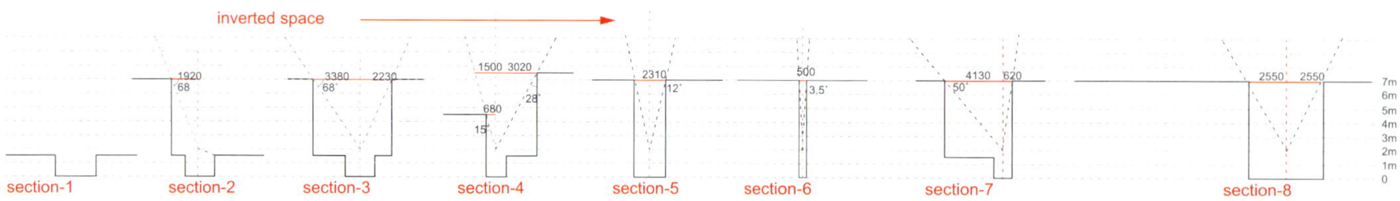

sedimentary process of waste materials as well as process of inversion

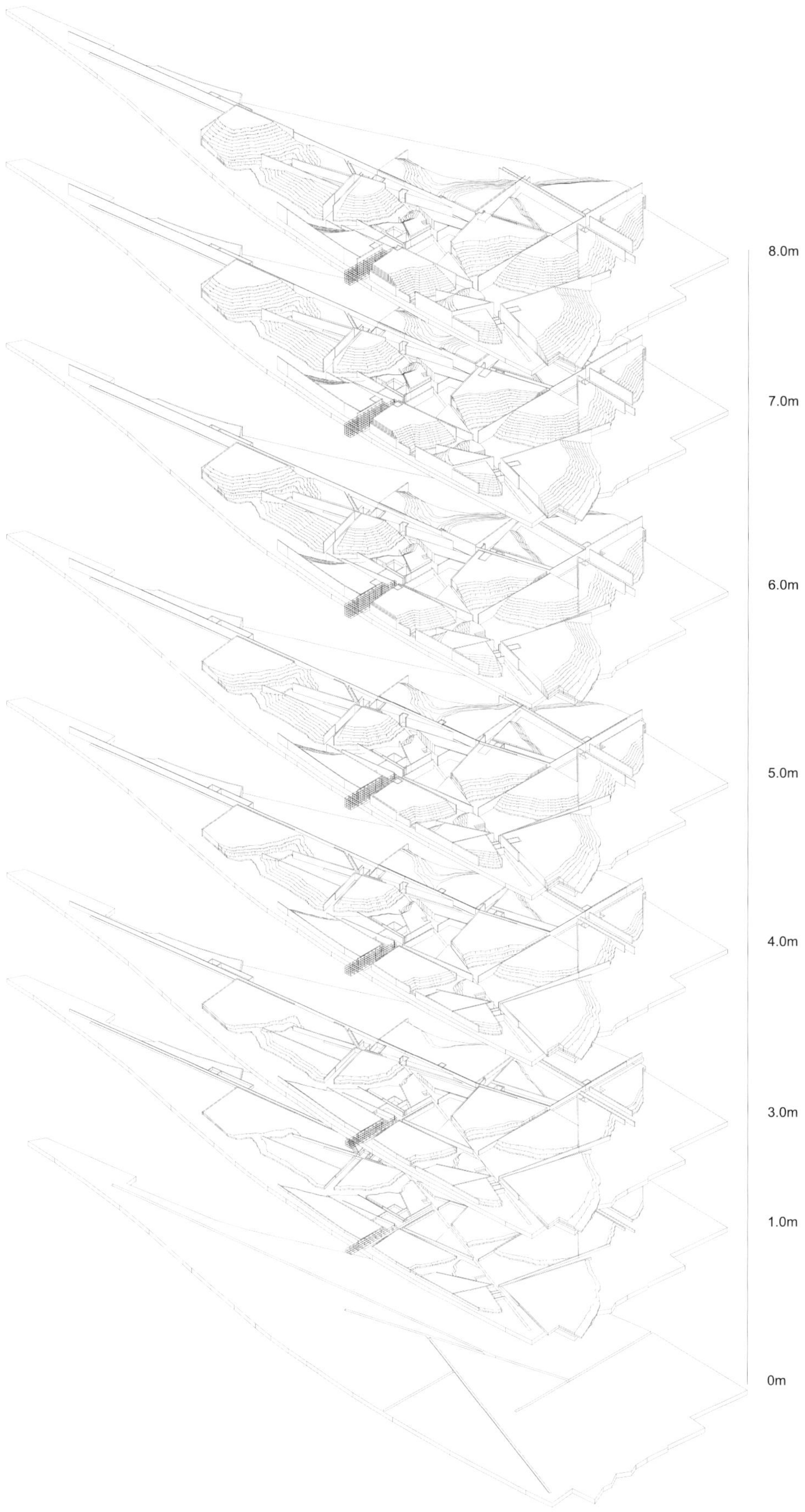

3.8 Material experience - Retaining wall study

Angle of Repose

The angle of repose is an engineering property of granular materials. The angle of repse is the maximum angle of a stable slope determined by friction, cohesion and shape of the particles. When bulk granular materials are poured onto a horizontal surface, a conical pile will form. The internal angle between the surface of the pile and the horizontal surface is known as the angle of repose and is related to the density, surface area, and coefficient of friction of the material. Material with a low angle of repose forms flatter piles than material with a high angle of repose.

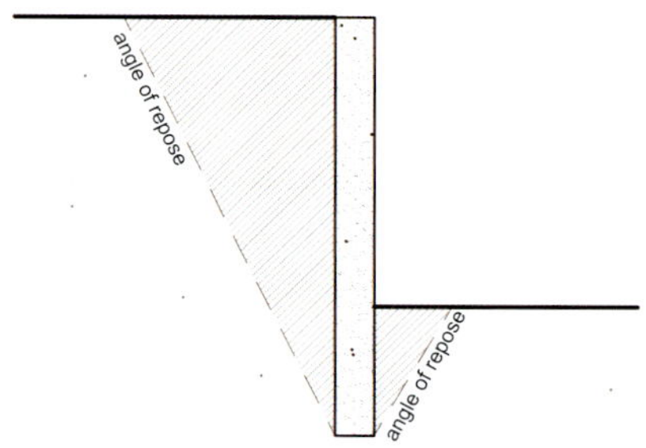

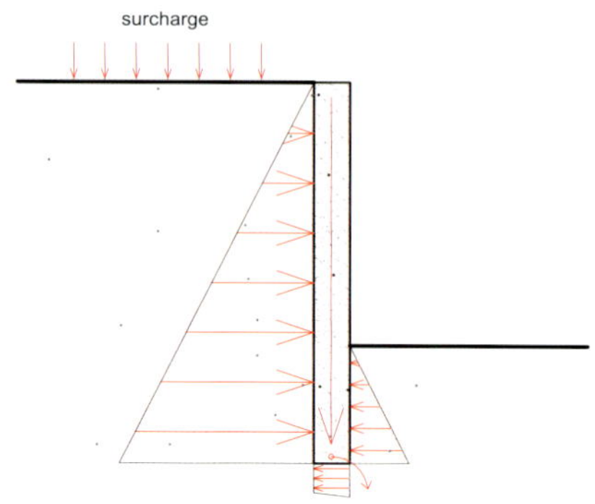

typical force diagram for retaining wall
(sheet pile used as a simplefied example)

Backfill

Backfill is a common term to describe the material used to refill an excavated area. In terms of retaining wall, backfill is used to refill area in front and behind of retaining wall. In general, the role of backfill is to work as a drainage medium behind the wall to prevent hydrostatic water pressure building up on the back of the structure. The selection of material therefore also works in combination with effective drainage system. In the case of crib walls, gabion walls and cantilever walls the mass of the back fill also become important because it contributes weight to the gravity wall principle.

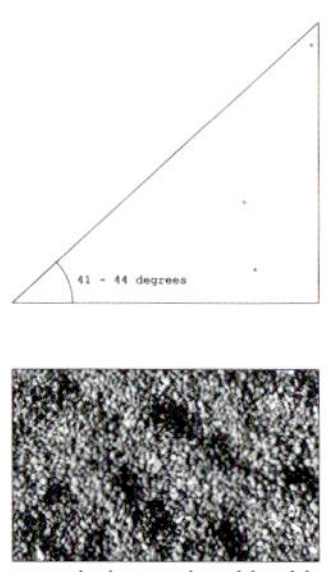
rounded gravels : 41 - 44

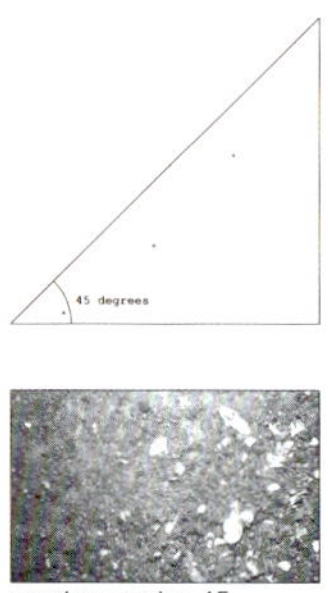
sandy gravels : 45

sandy clay (with 45% sand) : 29

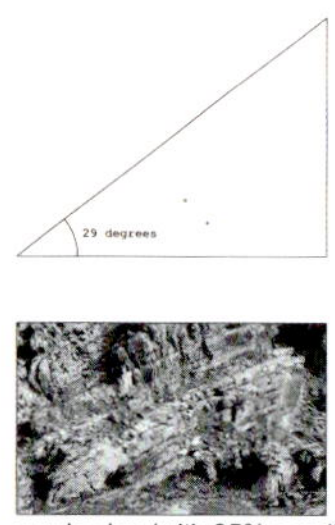
sandy clay (with 35% sand) : 38

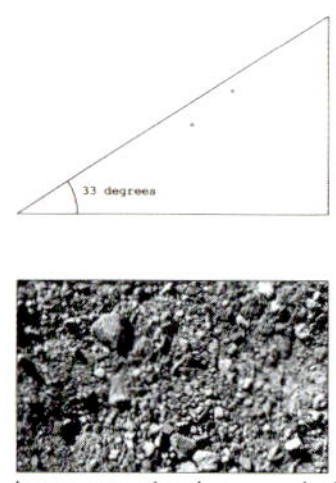
loose non circular gravel : 33

cement : 35

Backfill zones for drainage of sheet pile retaining wall (same principle for gravity walls)

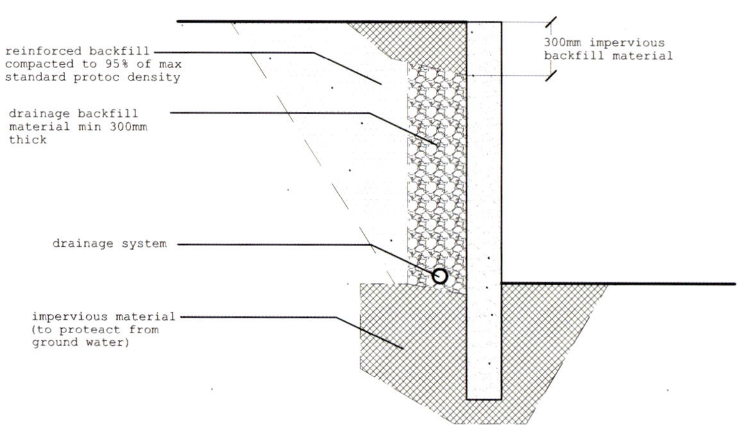

impervious material

clay

geomembrane

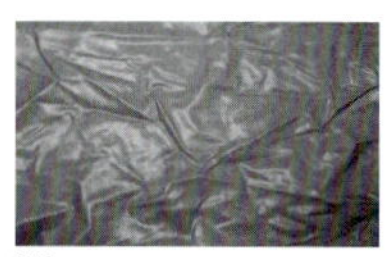
pvc

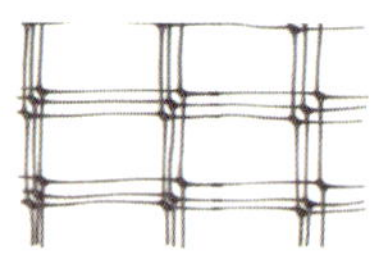
geonets

granular drainage material

aggregate

shredded car tires

reinforced compacted backfill

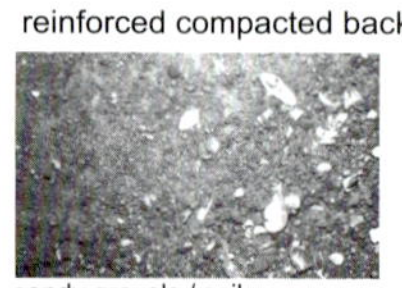
sandy gravels / soil

Specified retaining walls and types

gravity wall

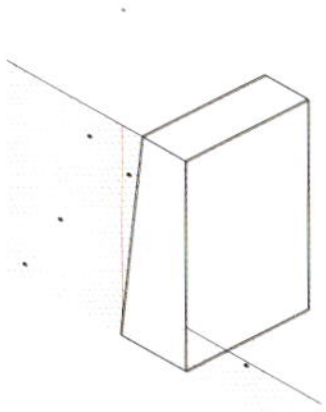

Gravity walls are most commonly constructed from concrete or stone but can be formed from almost any heavy materials.

cantilever wall

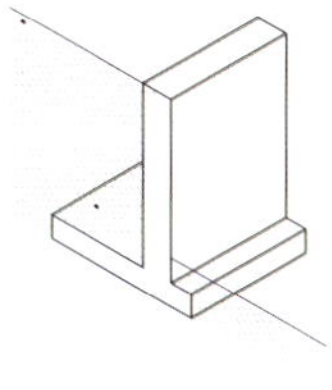

Cantilever walls are constructed from in - situ or precast reinforced consrete.

gabion wall

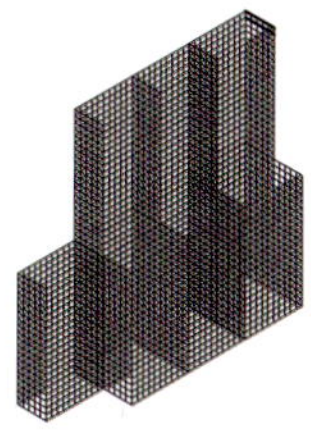

Gabion wall is constructed as both gabion baskets and fill material. The baskets are constructed of steel or aluminium mesh coated with PVC to increase the life of the material. The fill material is most commonly stone but can vary depending on exposure.

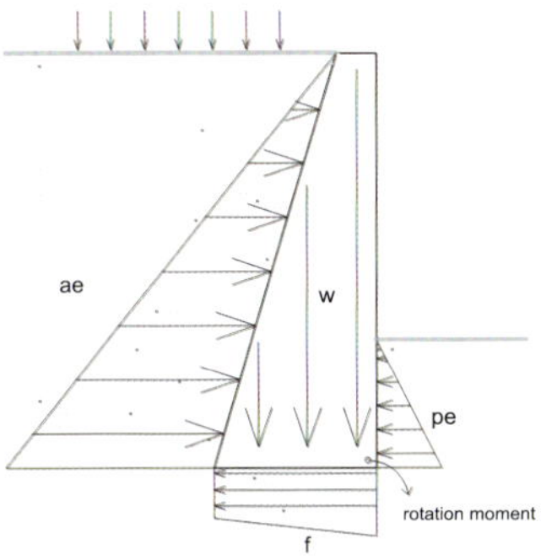

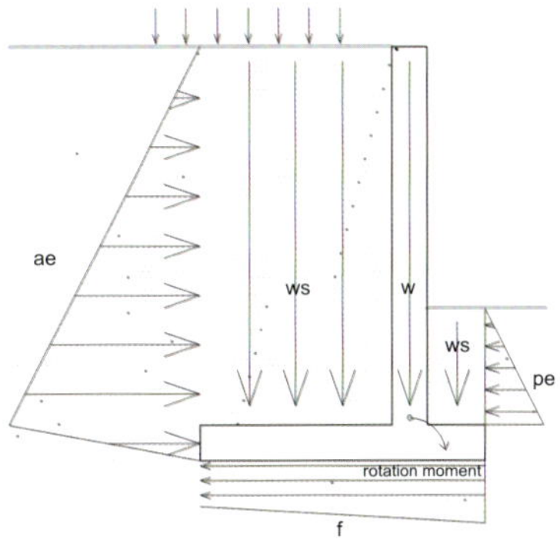

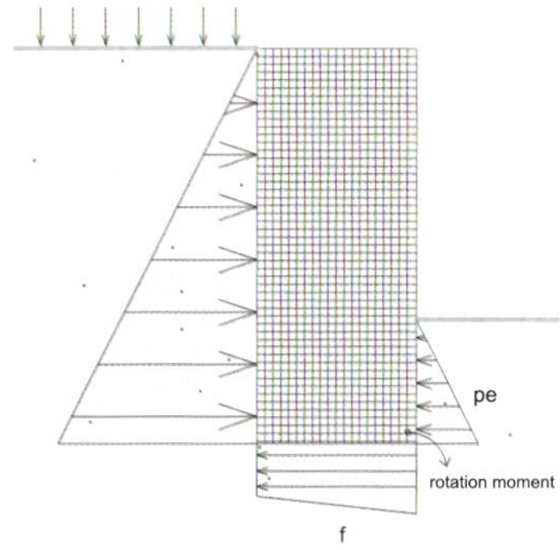

ae - active earth force
pe - passive earth force
f - friction
w - weight

Specified retaining wall thickness according to materials on site

material :

timber mulch

material density : kg/m3
220

angle of repose

30 degrees

height of wall

7m

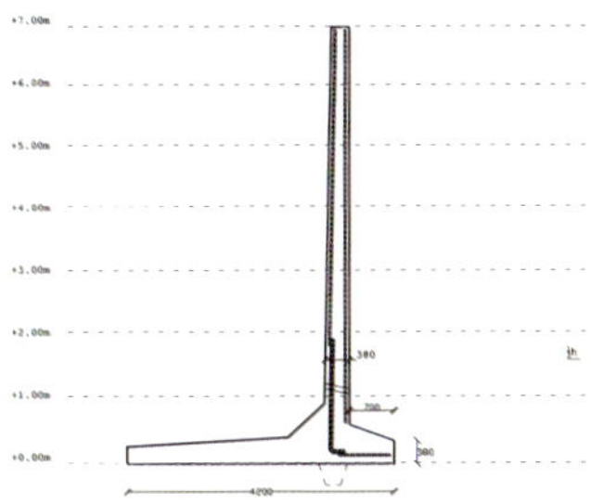

material :

crushed concrete

material density : kg/m3
2403

angle of repose

33 degrees

height of wall

7m

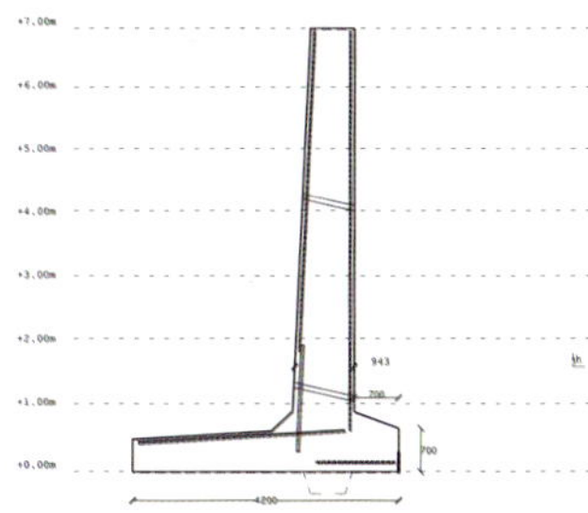

material :

dense earth

material density : kg/m3
2002

angle of repose

45 degrees

height of wall

7m

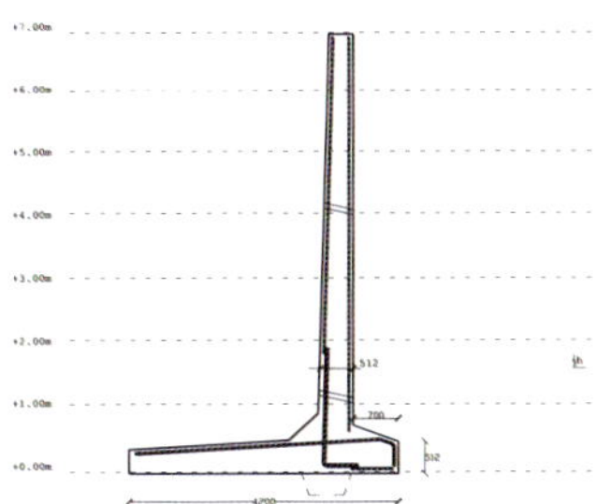

material :

shredded tires

material density : kg/m3
700

angle of repose

22 degrees

height of wall

7m

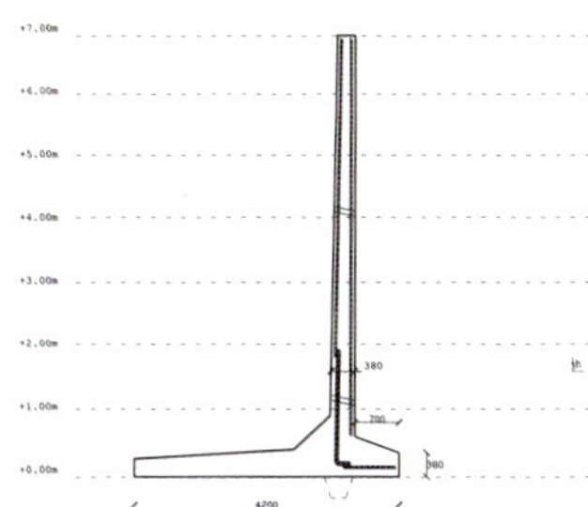

3.8 Material experiences - Urban bunker detail studies

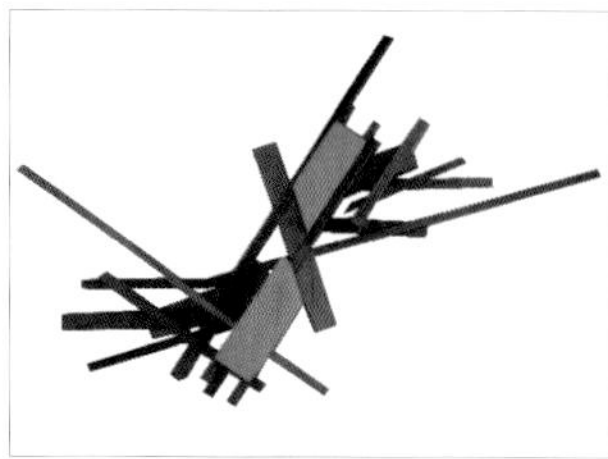

scale 1.100 model

scale 1.1000 model

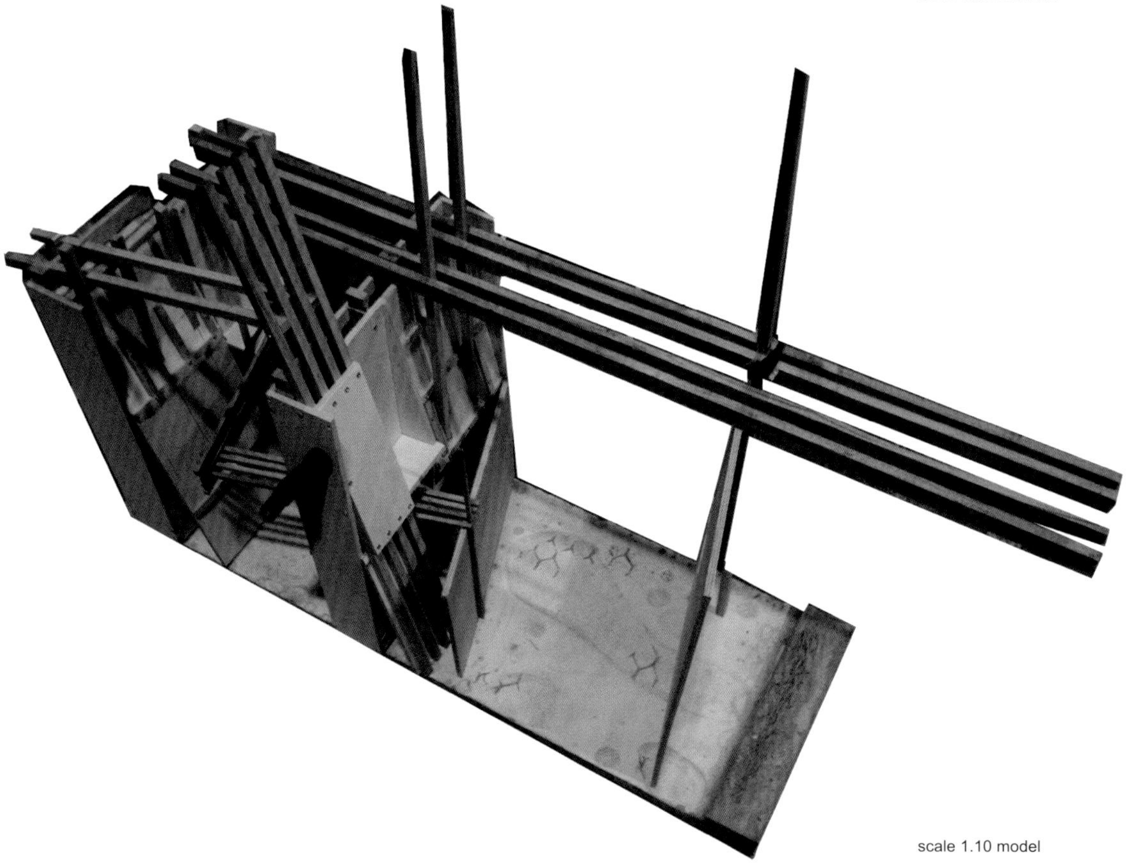

scale 1.10 model

Experimenting various combinations of timber structure and concrete wall in form work.

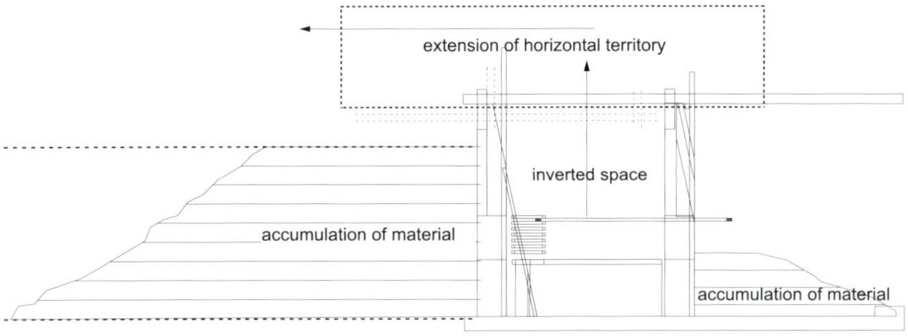

swimming pool at Leca de Palmeira

Architect : Alvaro Siza
construction period 1959-1973

structure
in situ gravity concrete retaining wall + timber frame structure

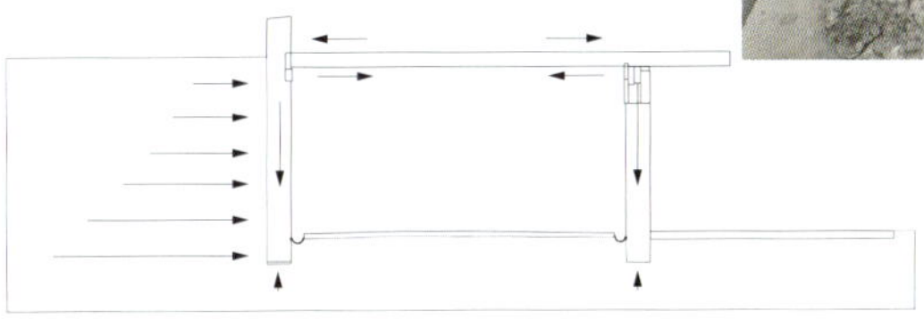

force diagram

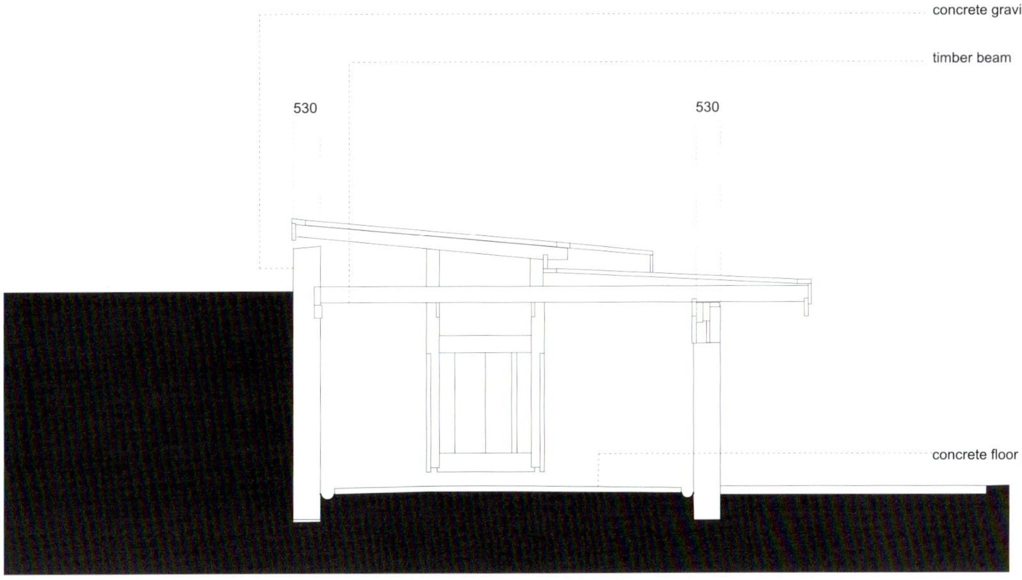

Keita Tajima

4. Proposal - micro urban landspace

4.1 Urban junction - composite drawing
4.2 Urban juncrtion - plan/section
4.3 Urban junction - axonometric
4.4 Erasure of industry - phase 1 - 4
4.5 micro urban landscape- axonometric
4.6 micro urban landscape- long section

This chapter introduces development of final proposal through series of drawings, each drawing is a complete and propositional, although each drawings are not mere representation of proposed architecture but tools for further exploration of architectural possibilities.
Final proposal intends to demonstrate densification process to facilitate and transform working landscape of recycle waste managements into micro city as a clustered junction spaces that arise from small architectural interventions of urban bunkers. Micro Urban junction emerges as a resultant of proposed strategic process, composes water infrastructure which compensates both existing operations and public activities, existing industrial machineries and re-inserted architectural artefacts.

The spatial quality of urban junction space is explored through negotiation and articulation of industrial infrastructure, sedimentary materials, public programs, and contextual details that make up de-collaged micro city within city. *1
Diagram shows densification process of new landscape by re-insertion of series of retaining wall components and additional timber structure in relation to sedimentary process of new ground height in time, and that corresponds to introduction of new public program into transitional post-industrial landscape. *1
Crushed concrete, shredded tires, saw dust metal fragments, all those recycle wastes that constitute new ground, create new spatial relationship to existing urban fabric and emerged inverted space.
The axonometric of the micro junction architecture show each urban component such as pumping station, reed bed system, water channel, existing industrial facility and complex of marginal public space (local studio, theatre, court yard etc..). *2
The phasing process drawing intends to explore erasure of waste management facilities over time, altered by gradual expansion of public program that compensate the rack of public facilities in adjacent area and re-connect them as a form of urban park. Finally, the relationship between several micro junctions reveals the internal quality of micro junctions to micro city as a whole. *3

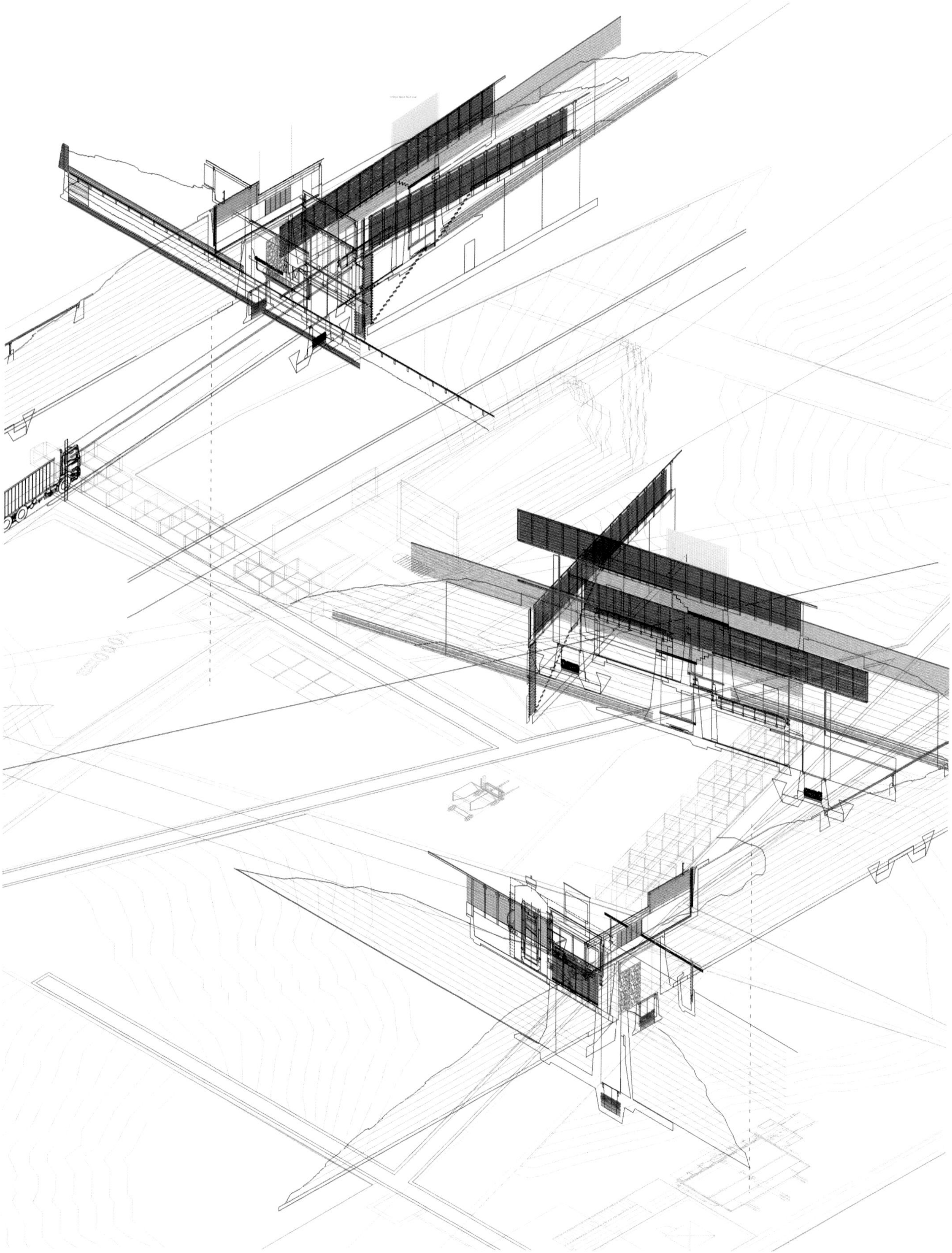

composit drawing

4.1 Micro urban junction - composite drawing

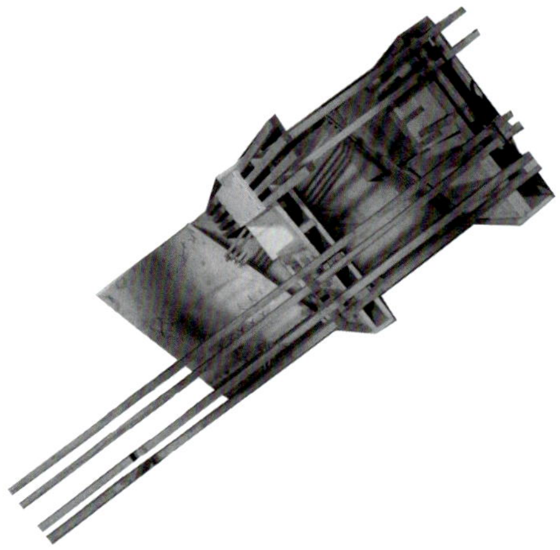

concrete/timber composite wall structure

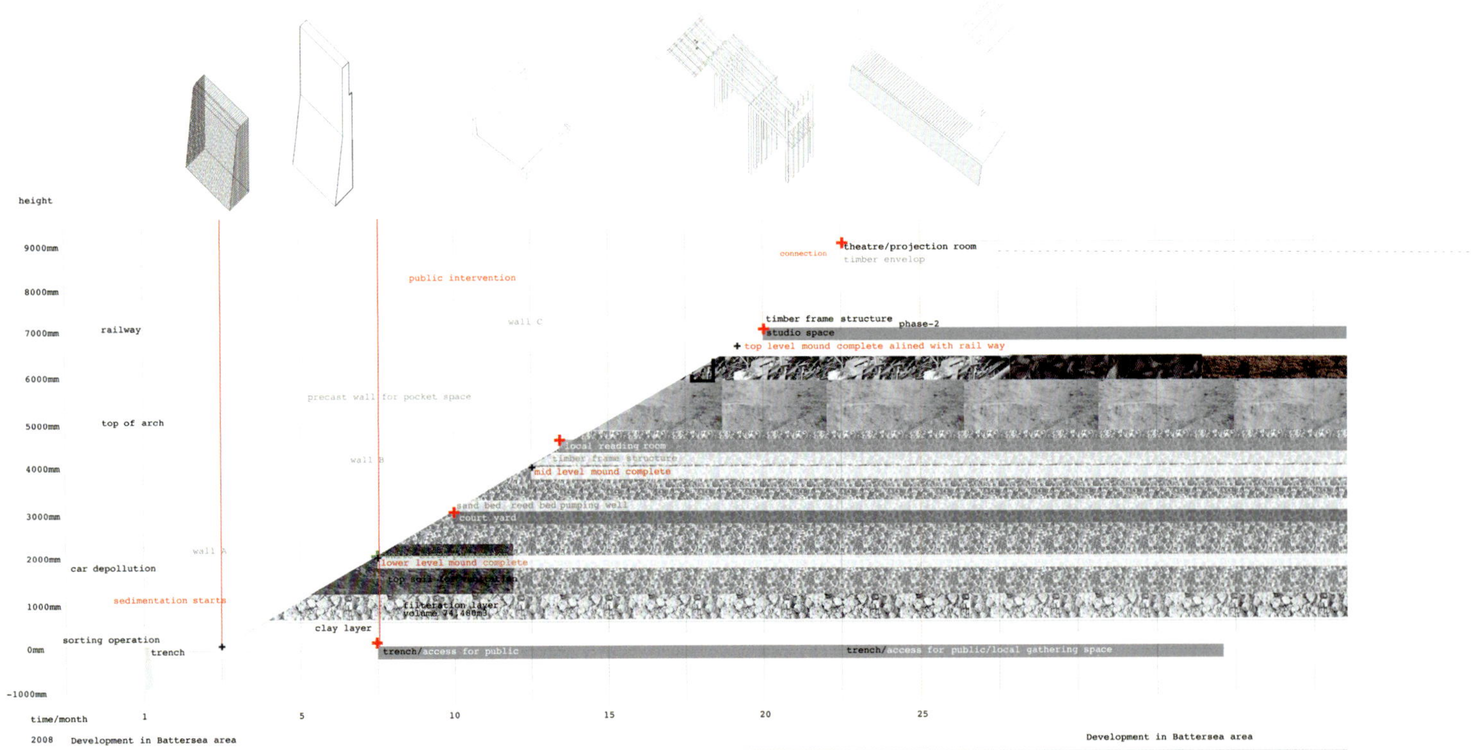

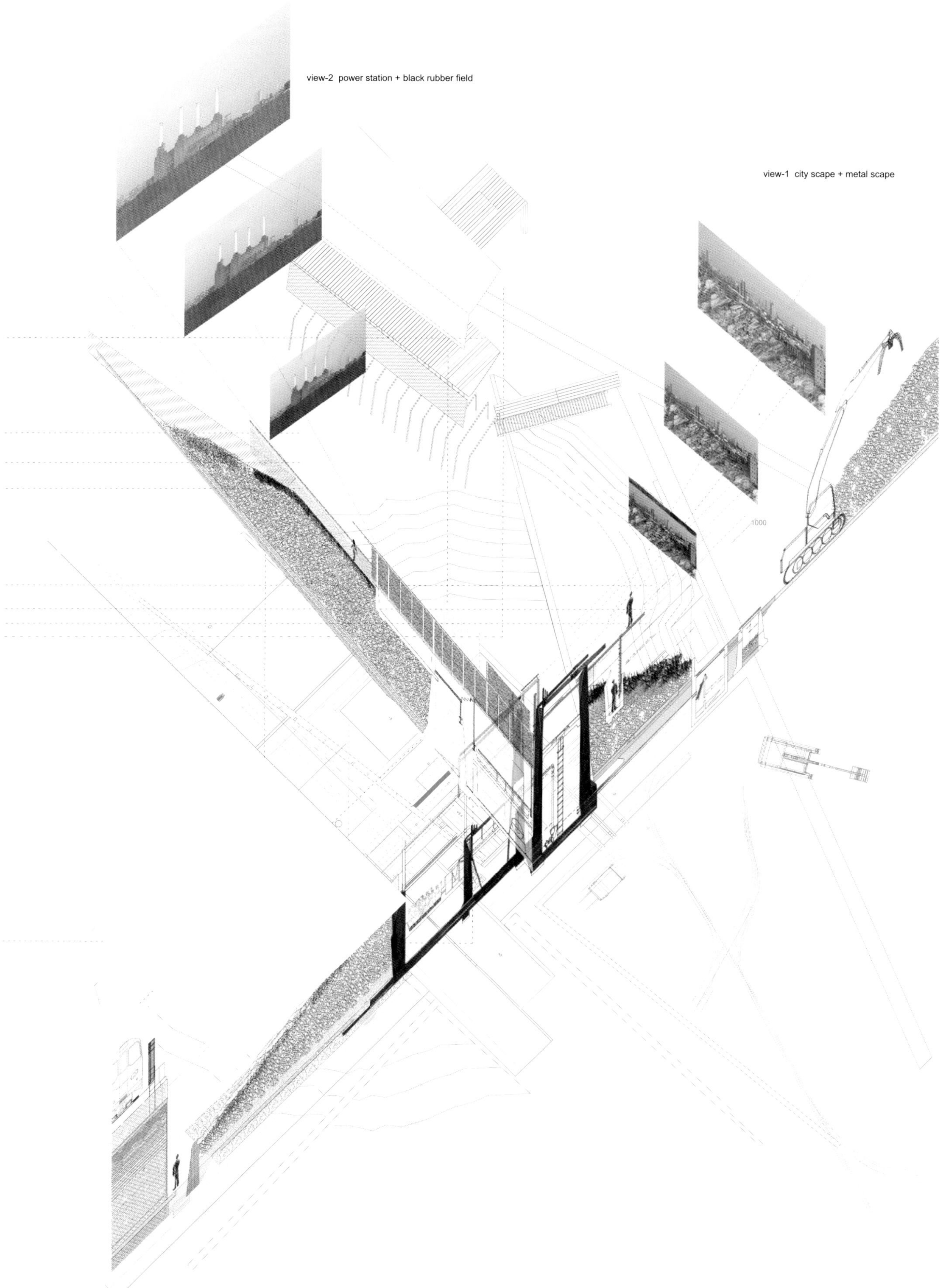

Keita Tajima

4.2 micro urban junction- section

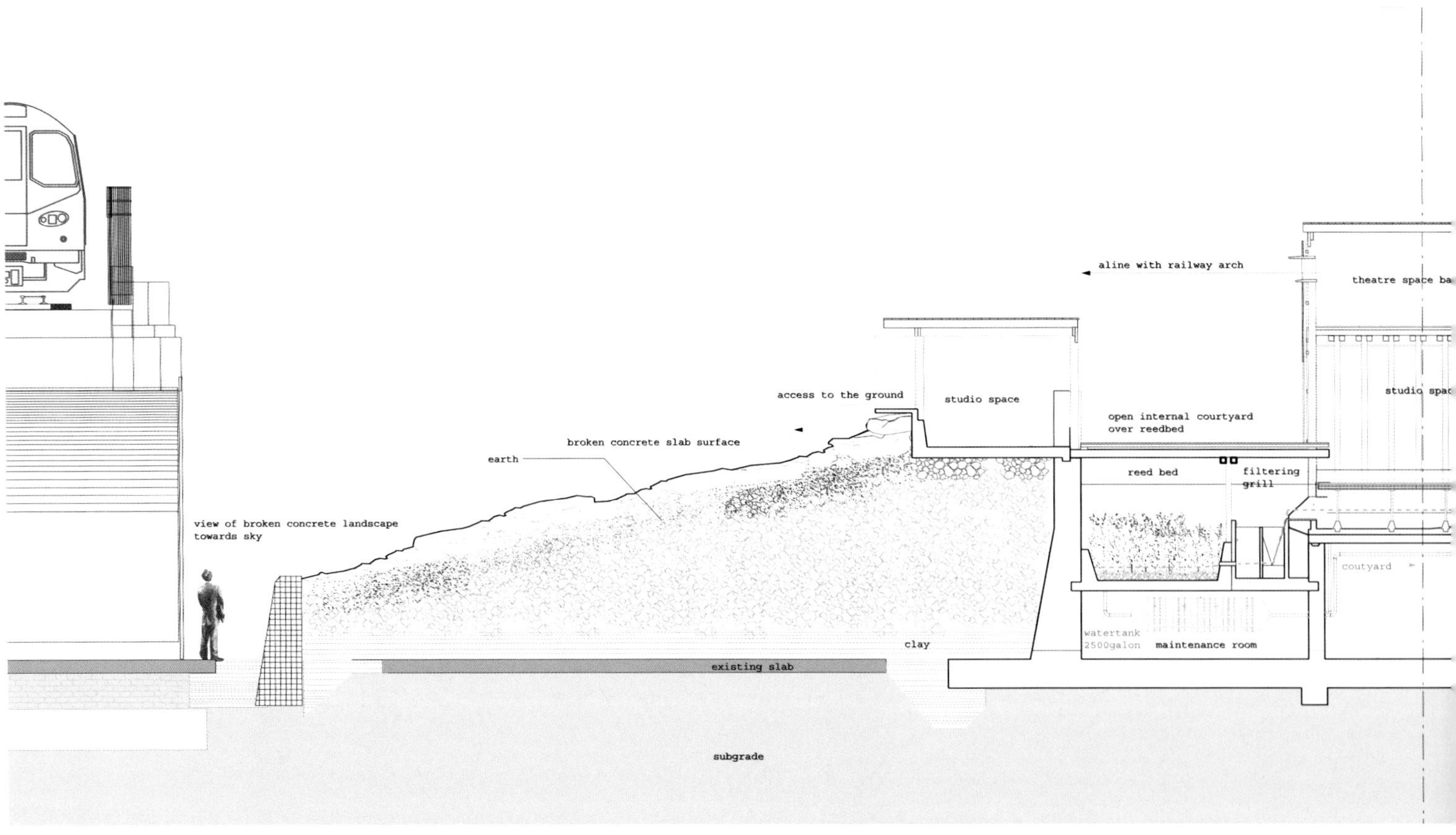

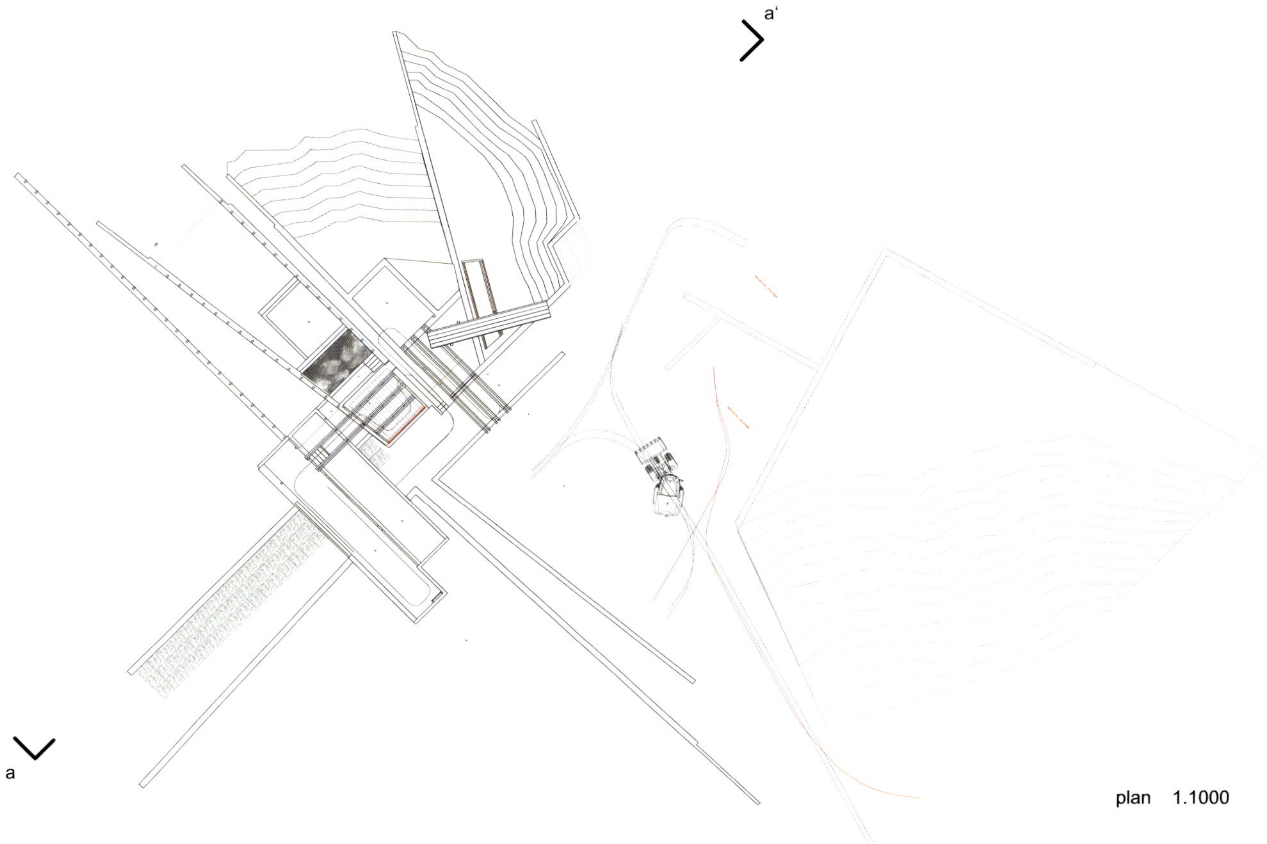

plan 1.1000

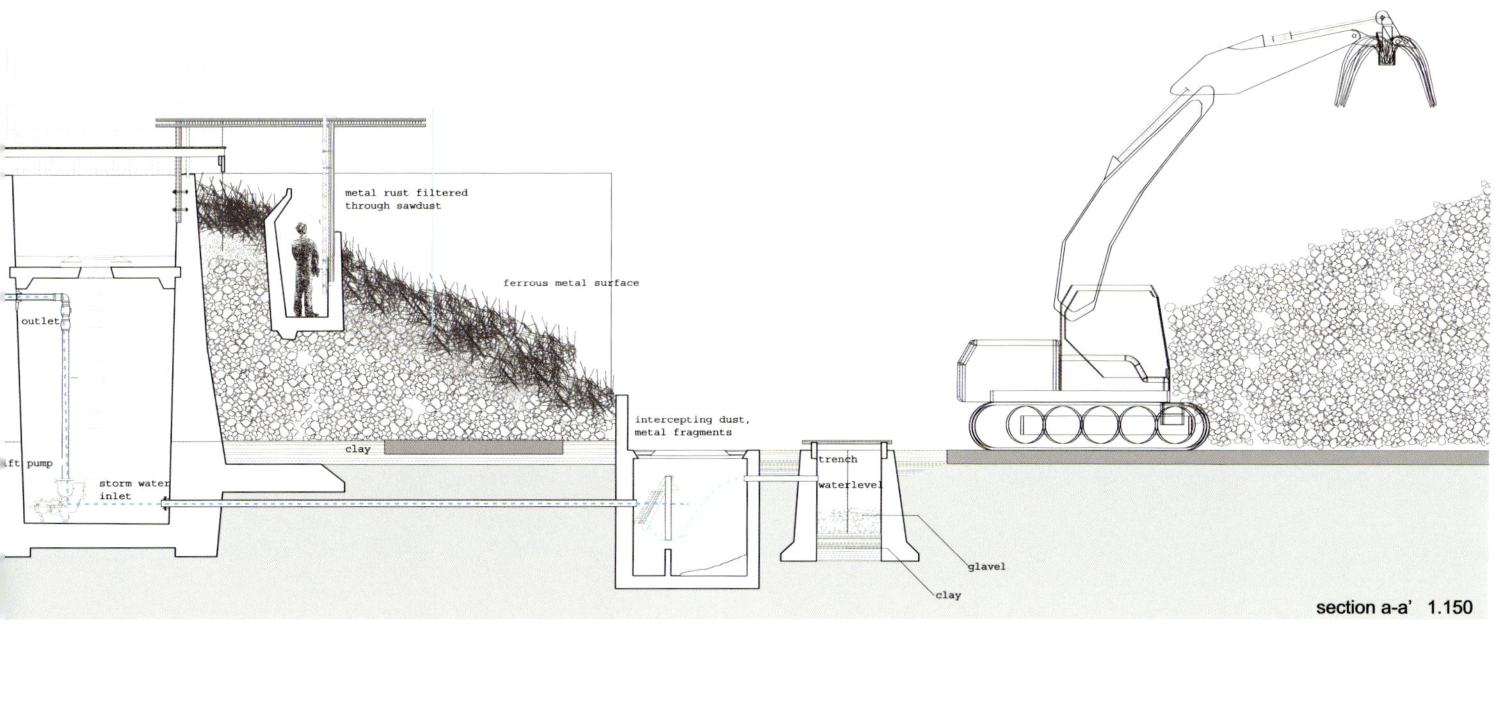

section a-a' 1.150

4.3 Urban micro junction - axonometric

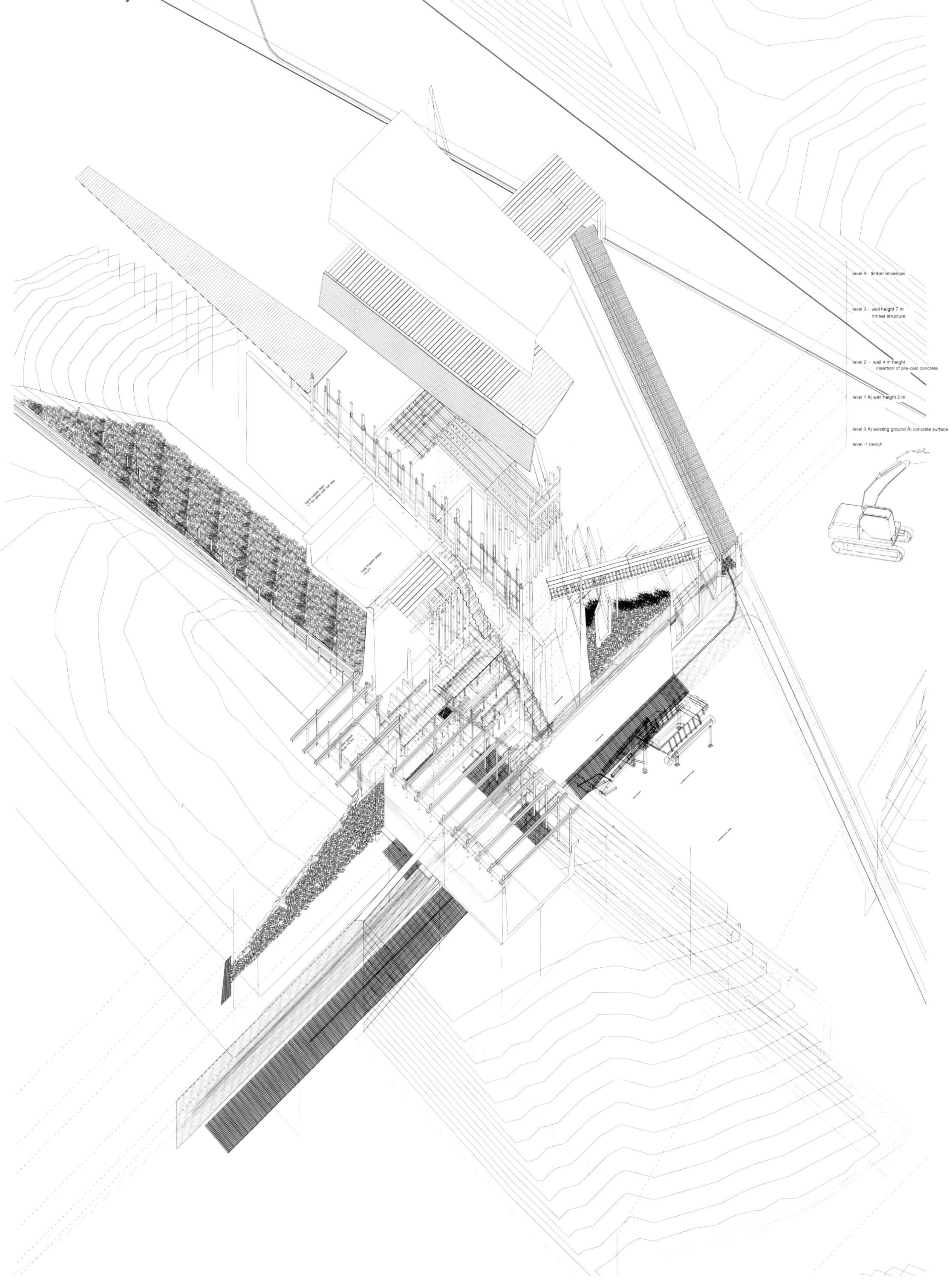

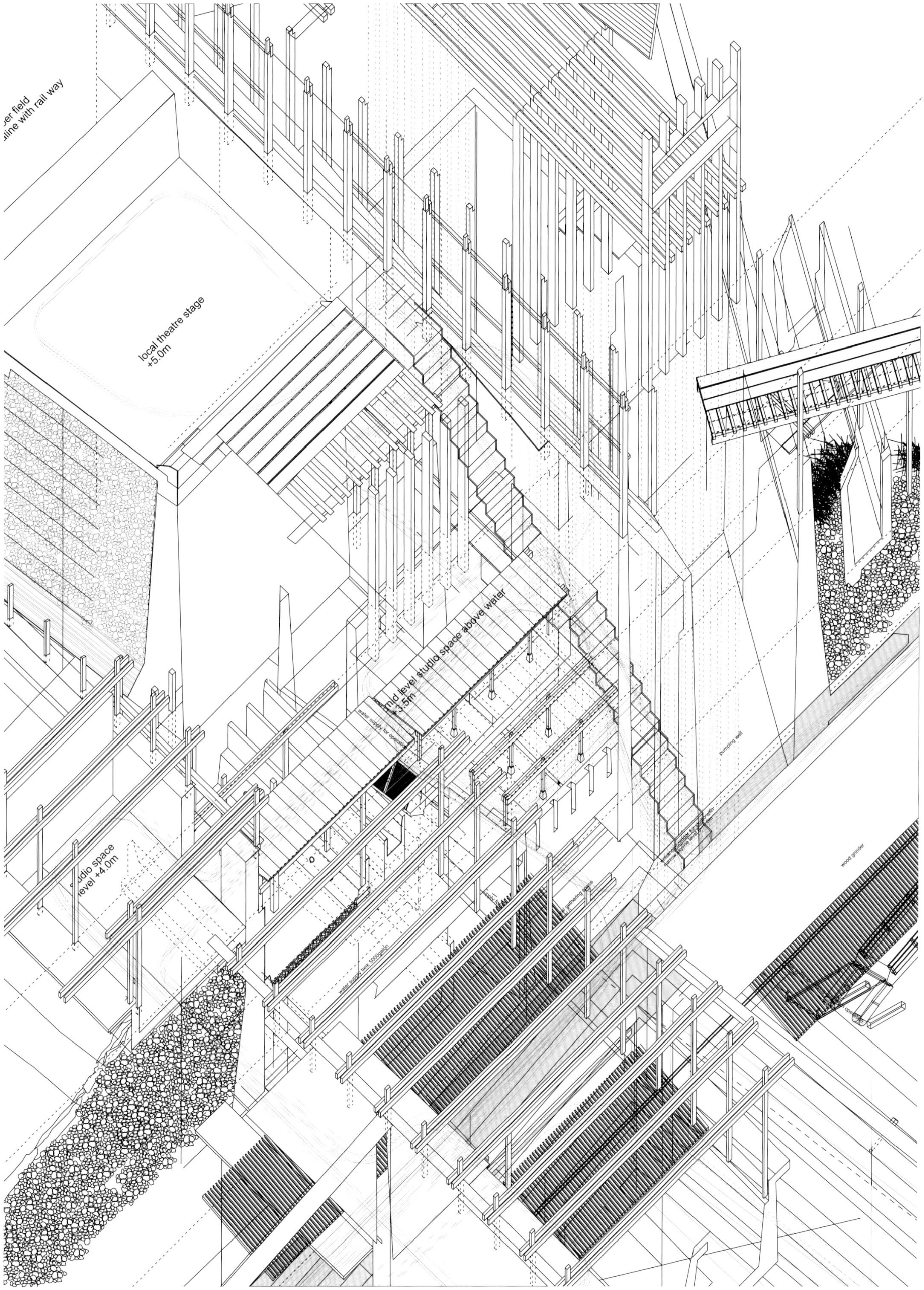
local theatre stage
+5.0m
mid level studio space above water
+3.5m
studio space
level +4.0m
pumping well
wood grinder

4.4 Erasure of industry

phase – 1

The alternation process of wastes recycling landscape is carried out in 4 phases in order to utilizing and organizing recycling materials as landscaping materials while maintaining industrial operations (sorting and trading) working. In phase-1, series of water courses are built within the site to ensure the collection of surface water from industries, and at the same time the water course indicates the axis for future public intervention and of micro urban junction.

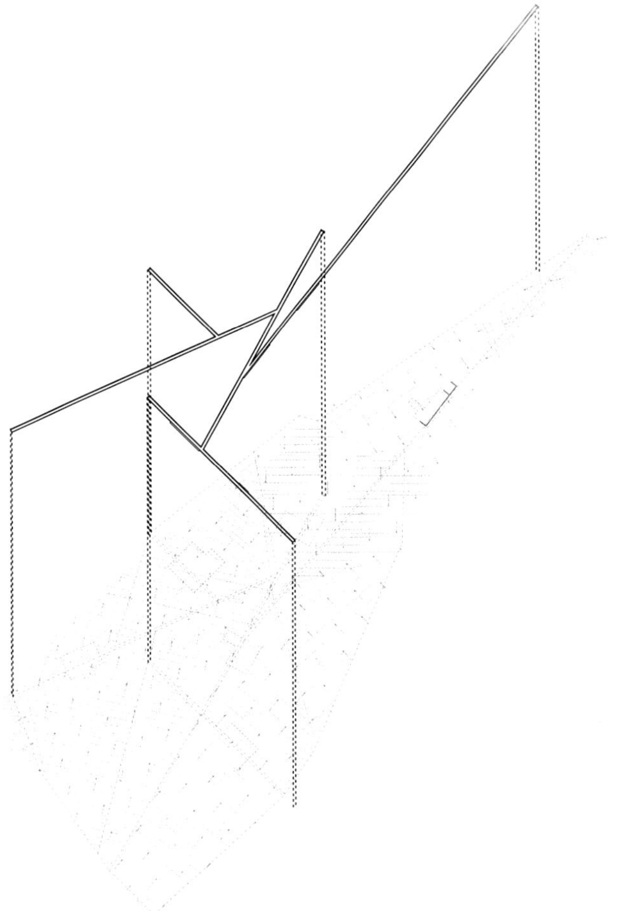

Phase – 2

The series of industrial operations and different type of recycling waste materials are organized and partly compacted, and sorted in relation to their characteristics as a landscaping material (differ from waste recycling sorting), create series of small units together with existing industrial operations and machineries. Each unit thus becomes junction point of coordination for new landscape to be built around.

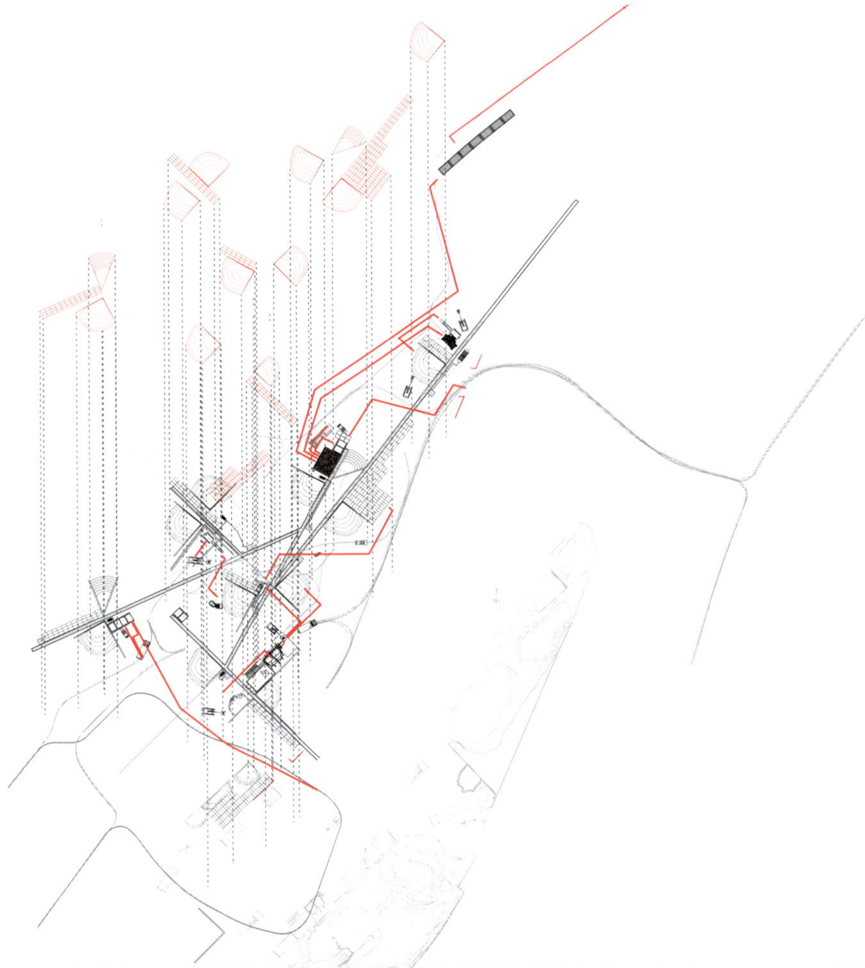

Keita Tajima

4.4 Erasure of industry

Phase – 3

While landscaping materials are gathered around each unit to facilitate new ground's stratification process, micro water treatment systems are implemented into each unit to utilize collected surface water for irrigation of newly built ground to be managed as series of small urban gardens that invite new public program such as shared studio / workshop spaces for locals. The territorial expansion of new ground thus allow more public accessible space within junction point by moderate recession of industrial activities.

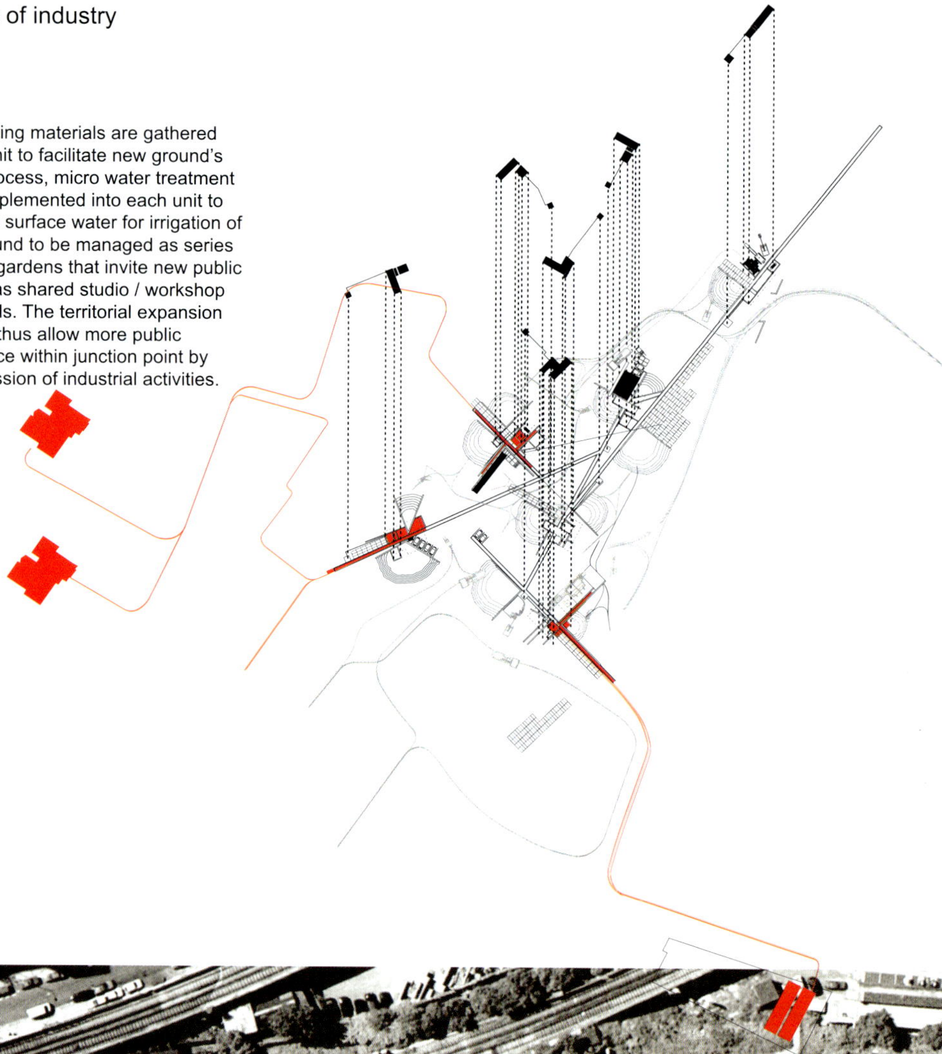

Phase – 4

The alternation process of recycling wastes landscape is completed when the stratification of new ground reach its plateau as same level as railway arch, to provide new perceptual connection between separated residential areas, and also to be served as public open spaces. Now, the retaining structure for accumulated recycling wastes materials becomes foundation to hold lightweight timber structures to be built upon as a part of stratification process of urban micro junction. This vertical projection (timber structure) of micro junction space brings internal quality of junction space up to new surface condition above ground, and provides various combinations of spatial effects and views over Battersea through textural quality of ground surface and the view from elevated ground. Finally, the densified junction point is seen as micro urban junction by layers of different building techniques, materials, infrastructure and various programs, while network of those micro urban junctions complete integration to the existing fabric of city and become part of post urban industrial collage within Battersea.

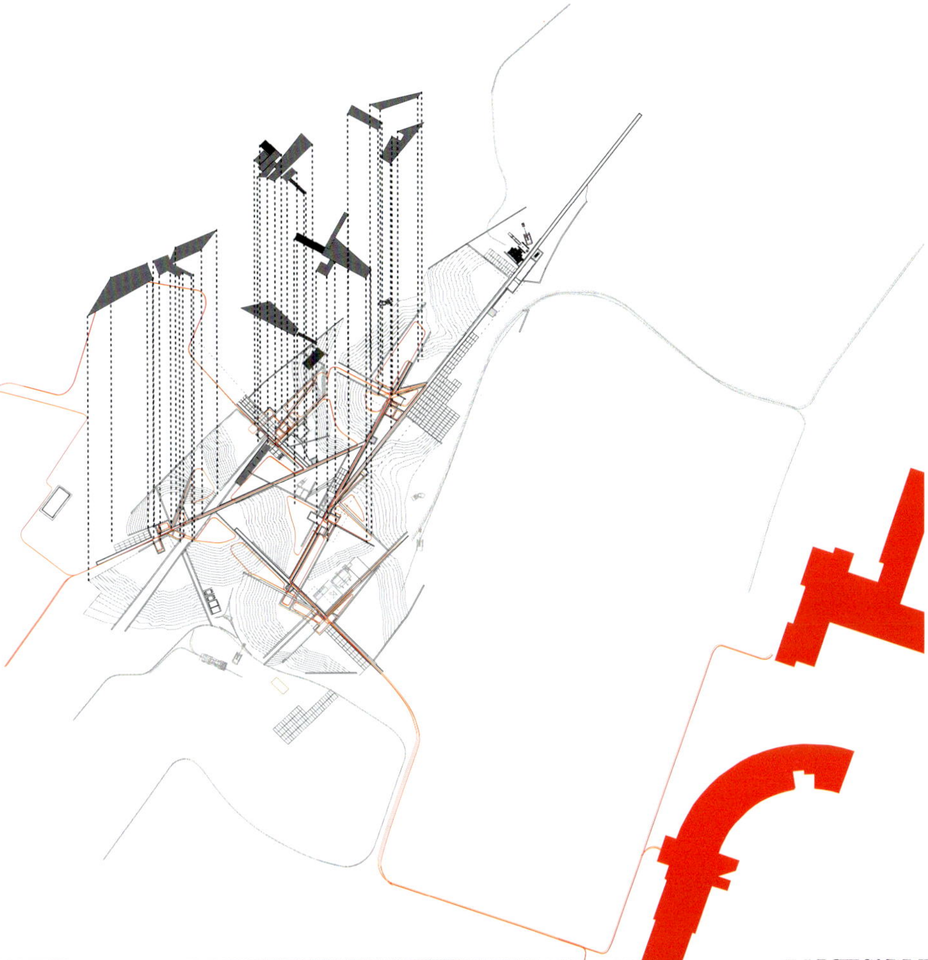

4.5 Micro urban landscape - axonometric

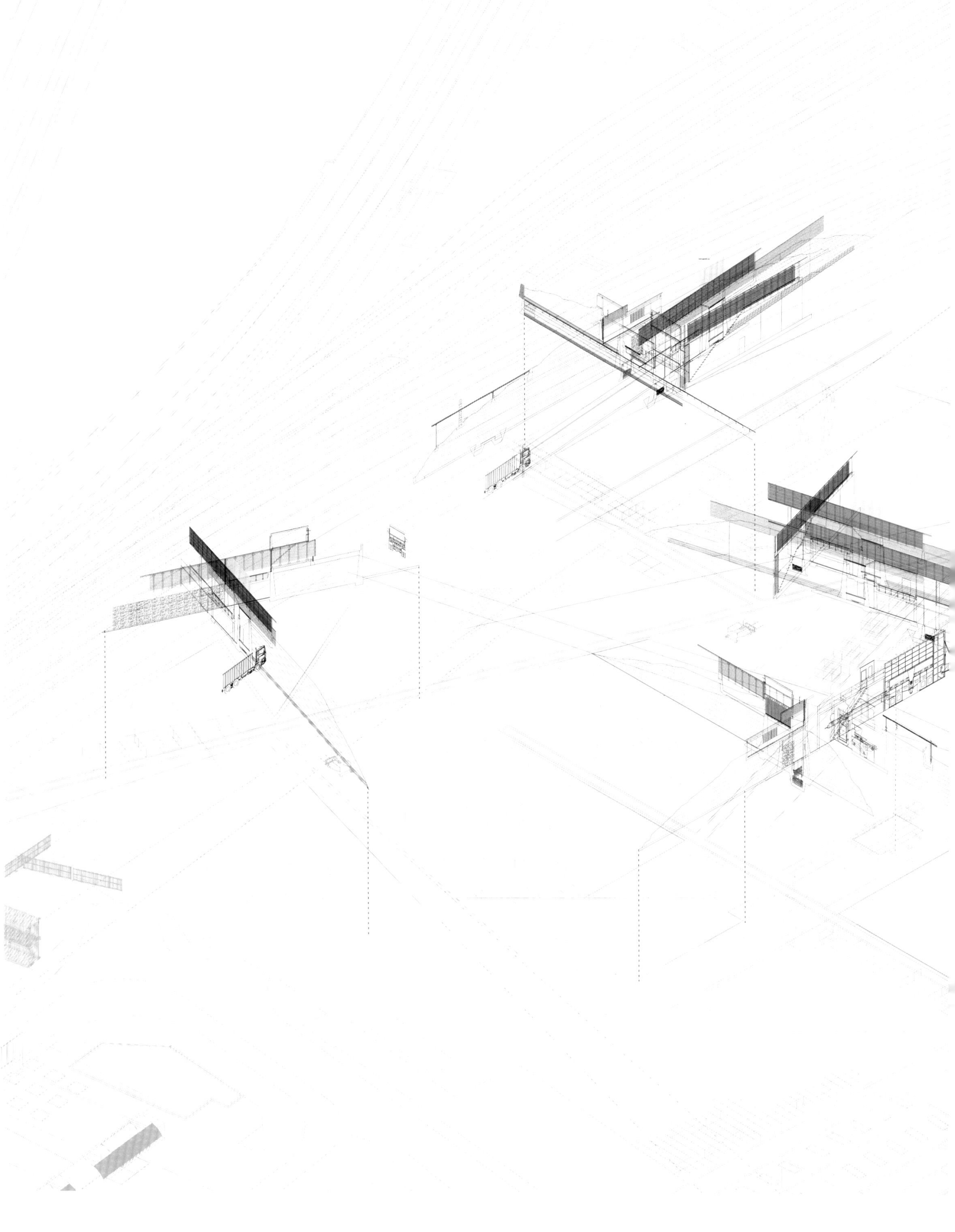

References

Micro city text - written by Shin Egashira
Micro city collage by all students from Diploma 11

Selected works , Alfred Jarry, Grove Pr (June 1980)
Merzbau, Elizabeth Burns Gamard,
Princeton Architectural Press
Living Systems, Liat Margolis, Alexander Robinson
Birkhäuser Basel
Same Landscapes, Gali-Izard, Teresa
http://www.emrltd.com/index.asp
http://www.wm.com/index.asp
http://www.niquecorp.com
http://en.wikipedia.org/wiki/Retaining_wall
http://en.wikipedia.org/wiki/Lateral_earth_pressure

I'd like to thank

Fernanda Arrillaga Sanches,
Joo - han Baek,
Katherine Batanova,
Viola Carnelutti,
Shin Egashira,
Emu Masuyama,
David Racz,
Nikolay Slavchev Shapazov,
Christopher Thorn,
Mike Weinstock
Mo Wong Woonyin,
Felix Xylander Swannell

and my family in Japan.

Siberian Photo(re)synthesis

Erlend Bakke-Eidsaa
Diploma unit 16

AA Diploma Honours presentation, 2008
The SOM award at RIBA Silver Medal awards, 2008
Shortlisted for RIBA Silver Medal, 2008
RIBA Part II High Pass, 2008
Technical Design Thesis High Pass, 2008
The Hollaway Trust award, 2008

SIBERIAN PHOTO(RE)SYNTHESIS

Pneumatic actuated large span structure to harvest methane and encourage agriculture along the thawing Siberian tundra

Erlend Bakke-Eidsaa AA Dip 16

The predictions on global warming might not on all levels be negative. Previous heat cycles of the earth has been argued to boost the growth of vegetation. While some species are threatened with extinction, others adopt through migrating with the moving favourable conditions. The Tundra, which so far has consisted of frozen ground, is as a result of global warming beginning to melt. Additionally the rain patterns which ominously seems to be disappearing from certain areas, are predicted to arrive at the tundra areas, making them possibly the Earths new main agricultural land..

One of the impending predictions of global warming is its contribution to accelerate the melting permafrost. In its frozen state it stores large amounts of methane and co2 which is released along its melting edge. Due to underdeveloped vegetation to absorb the gas, the methane is released straight into the atmosphere further accelerating global warming. If all the gas is released, this is predicted as one of the points of no return for global warming.

The project attempts to tap into the thawing edge as fuel for a new industry, utilizing the gas to actuate the structure and accelerate agricultural crop production through increased co2 levels. The proposal is an algae farm producing bio fuel, fertilizer and agriculture. As the structure moves with the melting edge to maintain its gas supply, it leaves a trace of agriculture and inhabitation in undeveloped land.

The structure is an attempt to challenge the use of double curvature in pneumatic structures through local form optimization. The formal language arrives from the behaviour of buckling strips, tested in physical models with up to 4000mm spans. Pneumatic cushions are cut to fit the form of the strip to actuate the span, and add support to the strips from failing, making the strip language achieve larger scales.

The proposals short term goal is to increase the interest into taking advantage of new land before the conditions are optimal, making an investment profitable even if the conditions do not change. The long term goal is to create new structures that better responds to man-made emissions, and if possible even could contribute to reverse some of our impact.

Photo Sue Barr

Photo Valerie Bennett

Erlend Bakke-Eidsaa

AA Dipl ARB/RIBA Part II MNAL

At the Architectural Association Erlend Bakke-Eidsaa developed chain effect mel scripts and category theory with Alistair Gill and Veronika Schmid. Then mathematical parametric design and Mies inspired translation from diagram to mass with George L. Legendre.

The graduate project Siberian Photo(re)synthesis was developed with Steve Hardy and Jonas Lundberg who over the five last years at the AA has focused on adaptation towards the threats of global warming, developing designs inspired by new technologies. The project was awarded the SOM award at the RIBA student medals 2008.

After graduating he joined Jonas Lundberg to set up a Norwegian branch of the world wide Urban Future Organization architects office, focusing on competitions in Scandinavia. www.no-urbanfuture.org

Table of Contents

1. Phenomenon and sketch proposal

1.1 Two climatic models
1.2 Phenomenon
1.3 Implemented vs. not implemented
1.4 Project ambition
1.5 Structural technologies
1.6.1-2 Algae photosynthesis, methane absorption
1.7 Ideas for sketch proposal_01. Module from array of algae tubes
1.8 Material behaviour model for sketch proposal_01
1.9.1-4 Sketch proposal_01

1.1 There has been a lot about global warming in the Medias lately, mainly negative predictions. There are however two scientific models on global warming. The Canadian model is the one predicting intensified storms, while the Hadley model refers to how increased co2 levels actually had a booming effect on the world's vegetation in the past.

1.2 In the book 'Six degrees' by Mark Lynas the author describes how the permafrost belt is beginning to melt as a result of global warming. In its frozen state it stores large amounts of methane, but at the melting edge, due to undeveloped vegetation to absorb the gas, its released straight into the atmosphere. If all the gas is released it's seen as one of the points of no return for global warming. Due to the scale it's feasibly impossible to catch all the methane, but what we could do is to tap into the thawing edge as fuel for a new industry.

1.3 On the top you see diagrams of the predicted outcome with no interference. And below you see a possible implementation, utilizing some of the gas. As the proposal would have to move with the thawing edge to maintain its gas supply, it would leave a trace of vegetation and inhabitation.

1.4 So my ambition this year was the opportunity to design a large span structure to harvest methane and encourage agriculture in an area of the world where the climate so far has made this very difficult.

1.5 As the project needs to be moved, I preferred a modular system.

1.6.1 One of the most efficient methods of dealing with the gas is an algae farm producing bio fuel and fertilizer.

1.6.2 This is a unit which absorbs co2 from the atmosphere by condensation on its surface. A similar method could be developed for methane.

1.7 The most efficient algae farm is the U.K. produced 'Biofence', due to smaller diameter and high sun penetration. To free up the ground and integrate all the components into one structure I began developing the module with a series of these algae tubes, forming a strip.

1.8 I investigated their behaviour, the way they boucle and encloses space simply from manipulating their edge conditions.

1.9.1 The ground releases methane gas, this is burnt into Co2 which expands by 2.74 creating pressure, this pressure I intend to use for the structure. The strips have been changed out with tubes, when their pressurized they contract, manipulating the stiffer edge frame, which again creates the behaviour of the strip model.

1.9.2 Several modules are attached together to cover an area. As gas is released locally, the structure responds by curling up like a piece of paper, like its feeding of the ground, but also showing the workers where to extend the structure for optimal gas absorption.

1.9.3 A single surface is attached at both sides to economically harvest larger areas. Gas is sucked towards the absorption units from a solar chimney effect, as they are placed higher and heated up by gas and sun. in this way it performs like a filter.

1.9.4 The gray circle indicates the city Khanty-Mansiysk in Siberia, containing 60 000 people, and here you can see the project measuring 100 x 1000 meters.

1.1

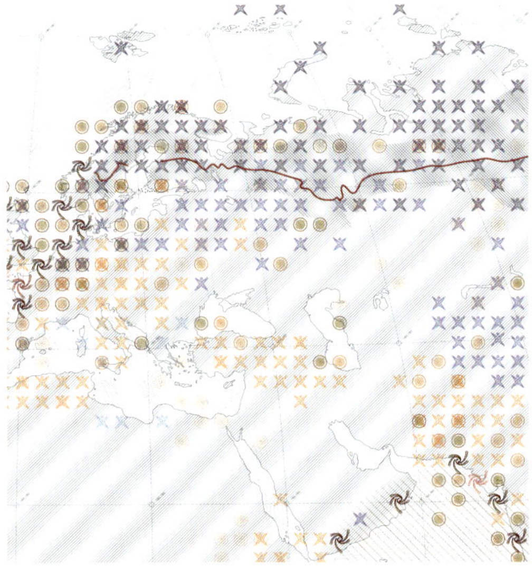
1.2

1.8

1.6.1

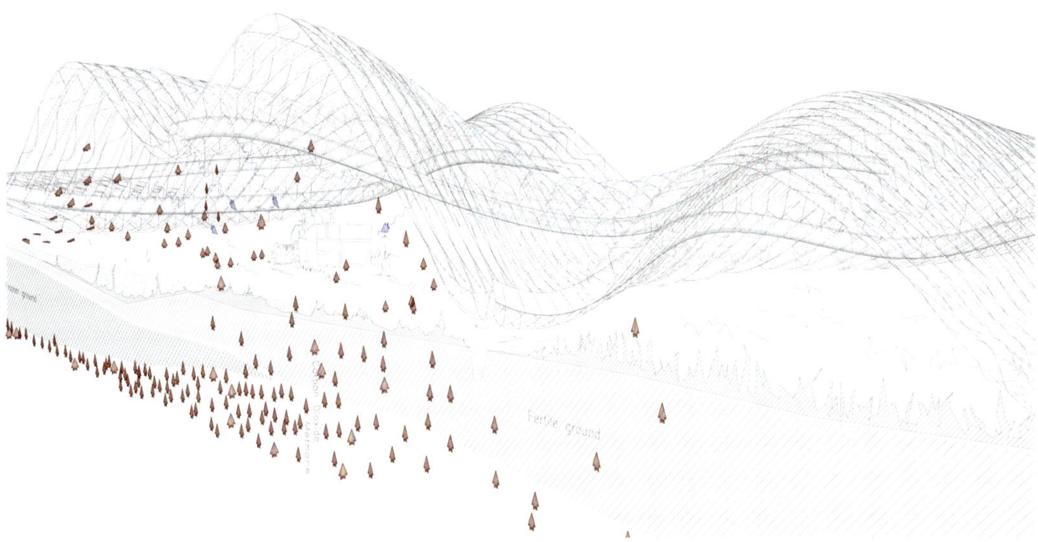

1.9.1

1.1 Two climatic models

Global concern

There has been a lot about global warming in the Medias lately, mainly negative predictions. There are however two scientific models on global warming. The Canadian model is the one predicting intensified storms, while the Hadley model refers to how increased co2 levels actually had a booming effect on the world's vegetation in the past.

From both models, at some areas the effects might actually be positive. As certain areas is getting to hot, others are going from frozen unused land to fertile and prosperous. I am interested in its positive effects and areas where opportunities might emerge.

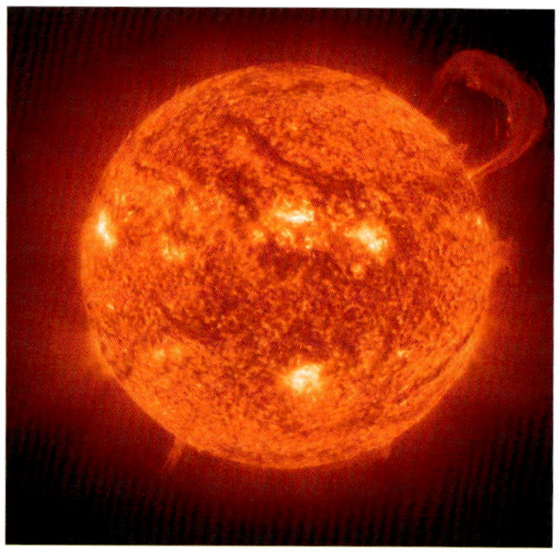

Left
Hurricane Katrina r01
No more gas r02
Melting Alaska r03

Right
Gun campaign photo r04
The Sun r05
Field in Artesia r06

Opposite
Cloud pattern r07

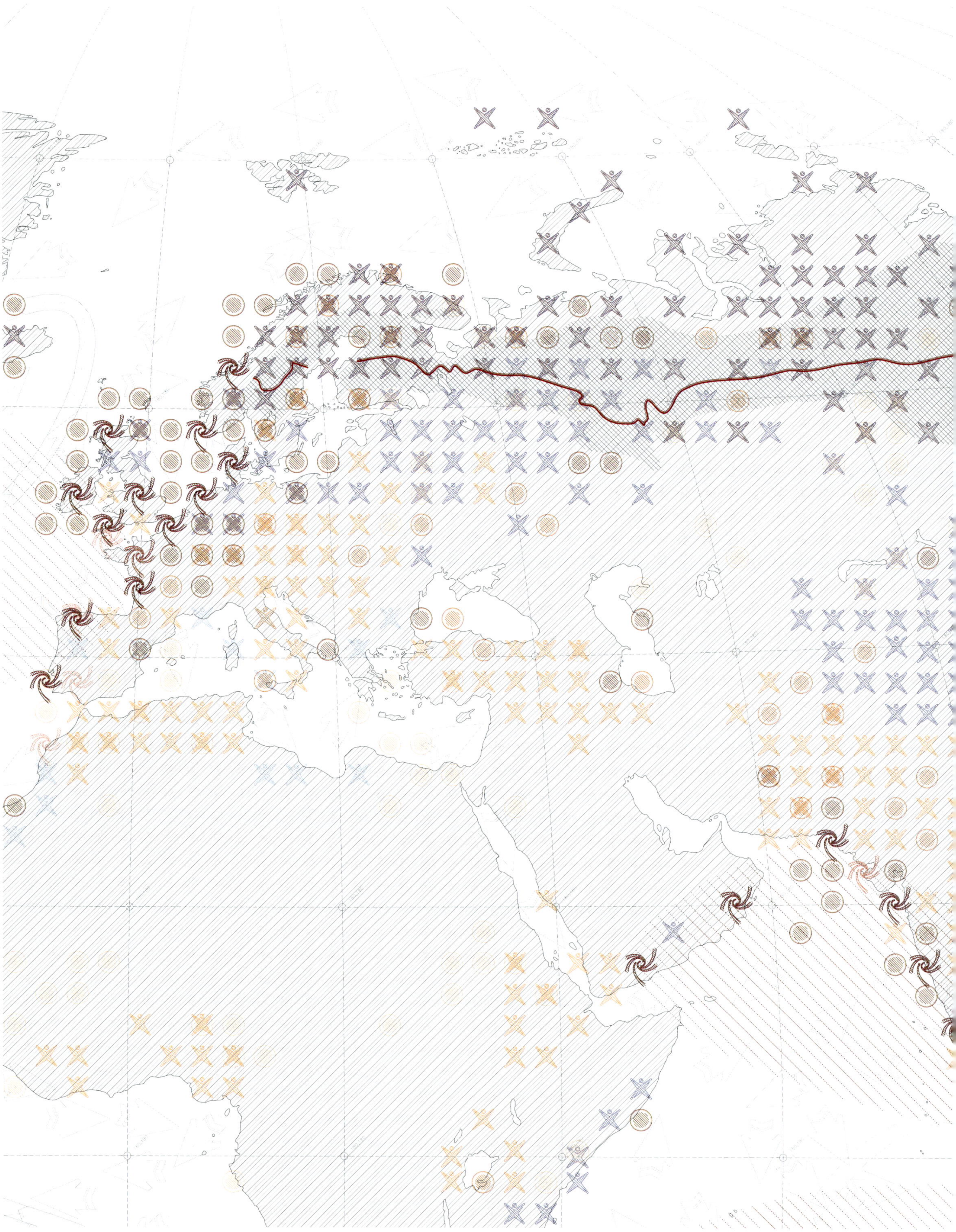

1.2 Phenomenon

Released methane and co2 along the melting edge of the tundra

In the book 'Six degrees' by Mark Lynas the author describes how the permafrost belt is beginning to melt as a result of global warming. In its frozen state it stores large amounts of methane, but at the melting edge, due to undeveloped vegetation to absorb the gas, its released straight into the atmosphere. If all the gas is released it's seen as one of the points of no return for global warming. Due to the scale it's feasibly impossible to catch all the methane, but what we could do is to tap into the thawing edge as fuel for a new industry.

Methane is 30-50 times more harmful than co2 in relation to build up greenhouse gasses and raise the temperature of the earth.

The proposal would be a closed loop zero energy emission industry, tapping into the methane to produce fuel for internal operation, and some for export in terms of bio-fuel production. In its trail it would leave developed land as it moves along the edge to maintain its gas supply. On a local level, it might also have an effect on the methane released to the atmosphere.

On this map all the predictions of six degrees are mapped out, storms, conveyor belt and migration, so if the predictions on global warming is true it seems like the Tundra area might be very attractive in the future.

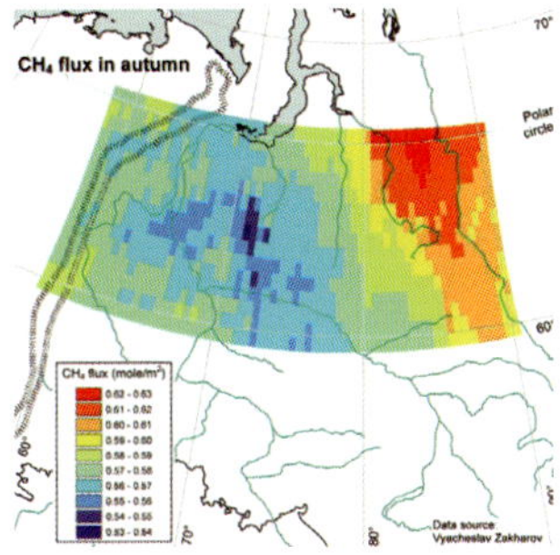

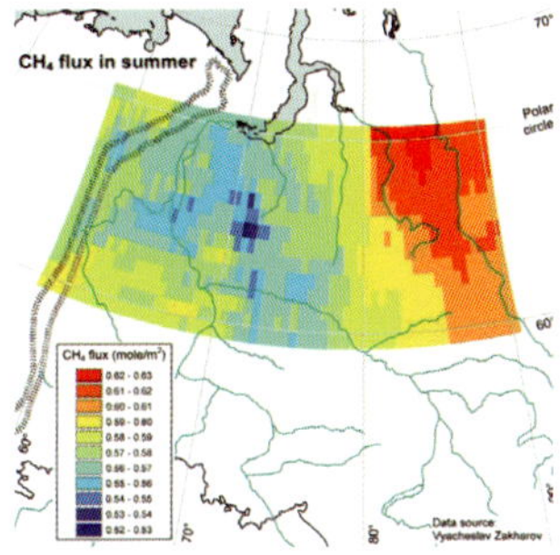

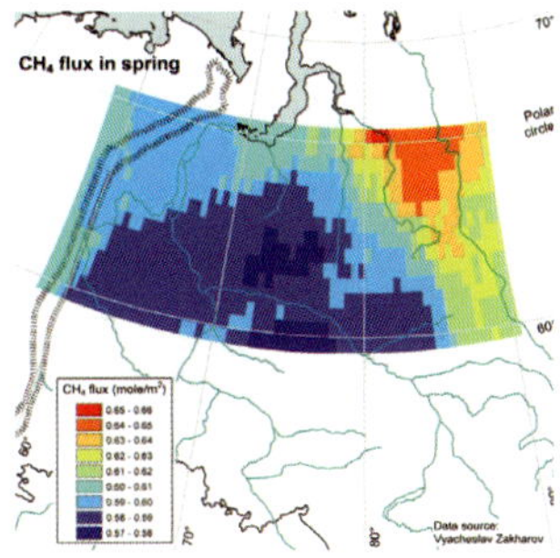

Melting Siberia r08
Methane release autumn r09
Methane release summer r10
Methane release spring r11

Erlend Bakke-Eidsaa

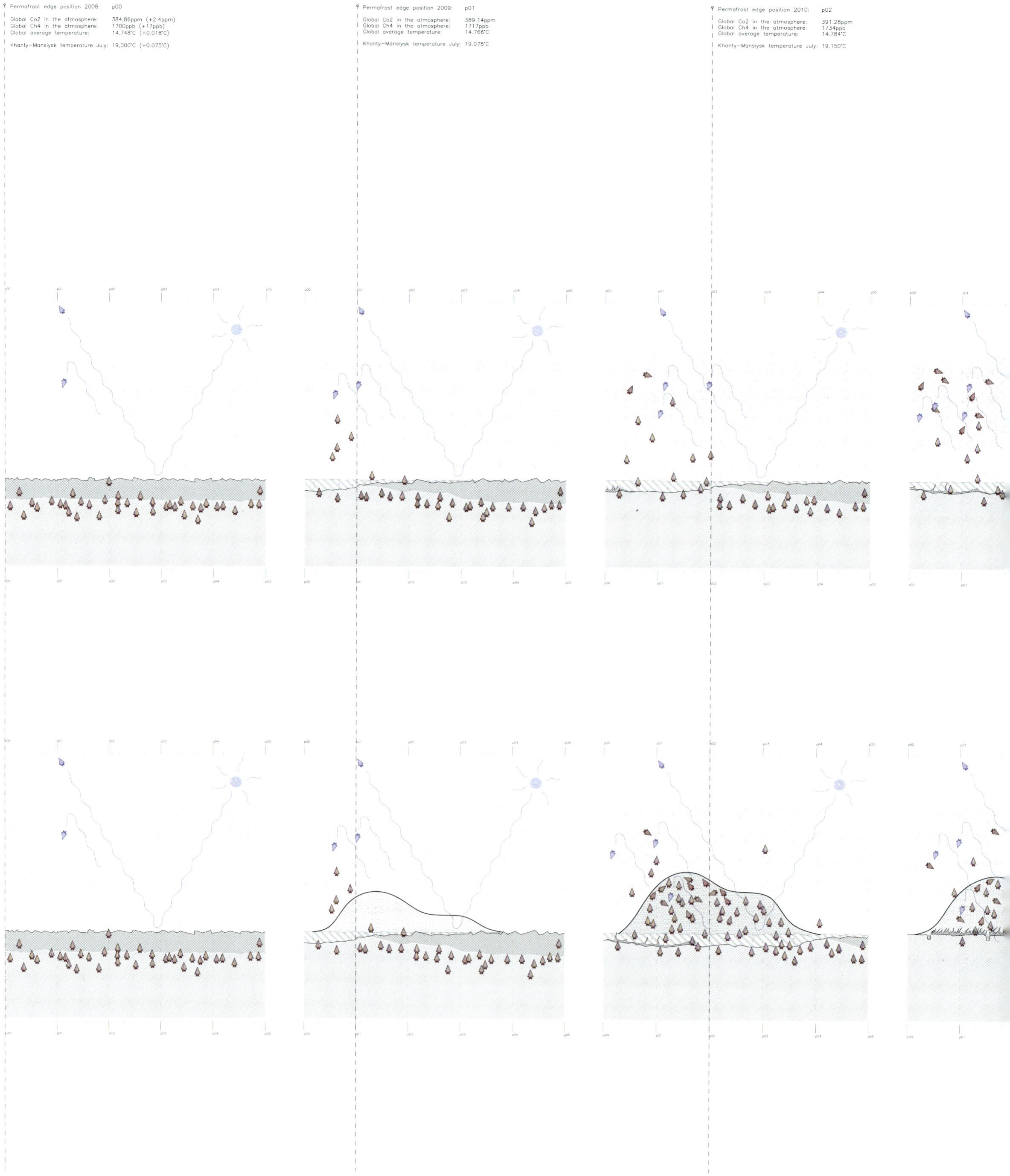

1.3 Implemented vs. not implemented
On the top you see diagrams of the predicted outcome with no interference. And below you see a possible implementation, utilizing some of the gas. As the proposal would have to move with the thawing edge to maintain its gas supply, it would leave a trace of vegetation and inhabitation.

afrost edge position 2011: p03

al Co2 in the atmosphere: 393.42ppm
al Ch4 in the atmosphere: 1751ppb
al average temperature: 14.802°C

ty–Mansiysk temperature July: 19.225°C

Permafrost edge position 2012: p04

Global Co2 in the atmosphere: 395.56ppm
Global Ch4 in the atmosphere: 1768ppb
Global average temperature: 14.820°C

Khanty–Mansiysk temperature July: 19.300°C

Permafrost edge position 2013: p05

Global Co2 in the atmosphere: 393.42ppm*
Global Ch4 in the atmosphere: 1751ppb*
Global average temperature: 14.838°C*

Khanty–Mansiysk temperature July: 19.375°C*

* Speculative project impact on local atmosphere

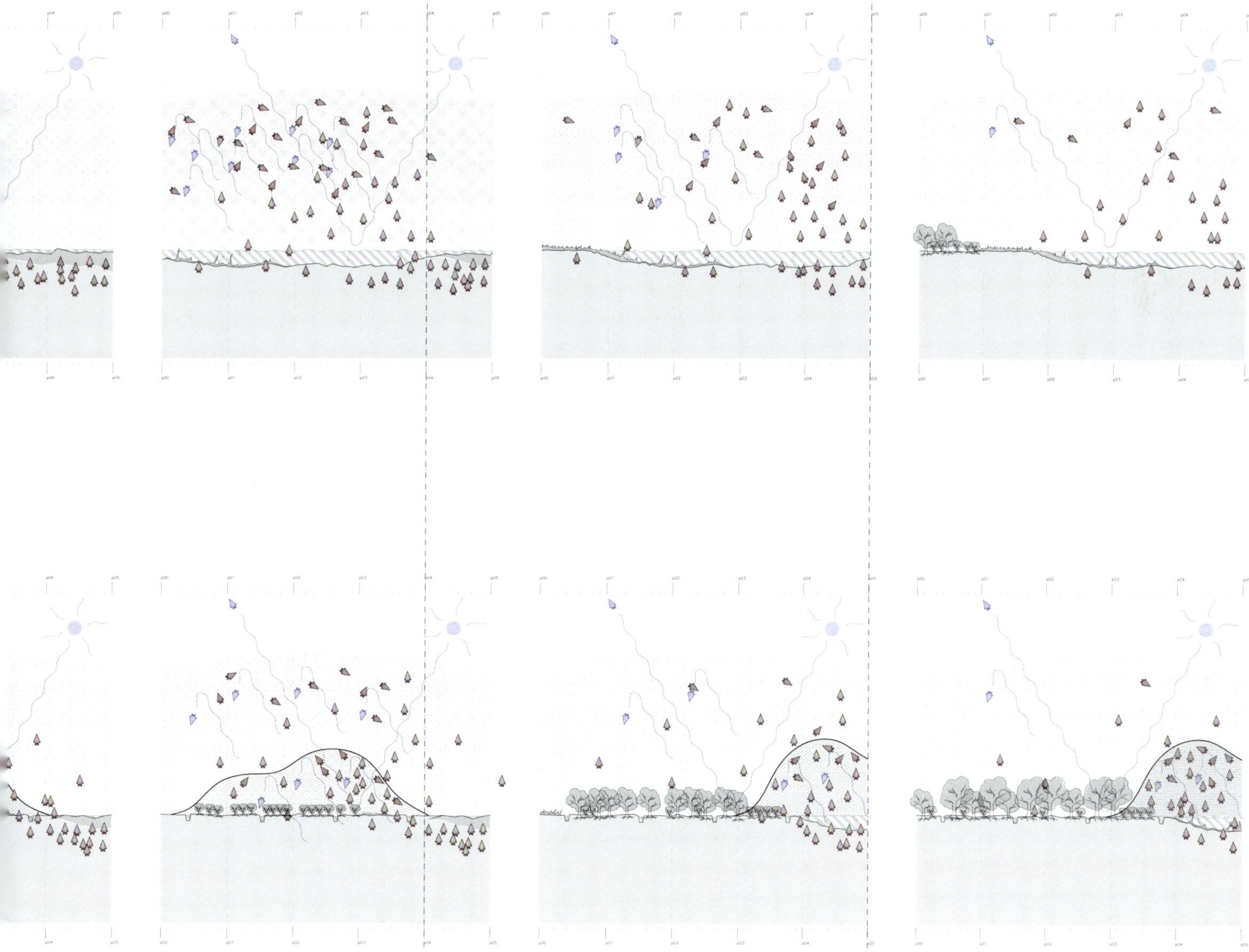

Global CO2 in atmosphere 384,86ppm, increase 2.4 CO2 Parts per million increase per year r12
Global CH4 in atmosphere 1700ppb, increase 17 ppb per year r13
Global temperature 14.748°C, increase +0.018 Celsius per year (average the 30 last years) r14
Siberia temperature july 19.00°C, increase +0.075 Celsius per year (average the 40 last years) r15

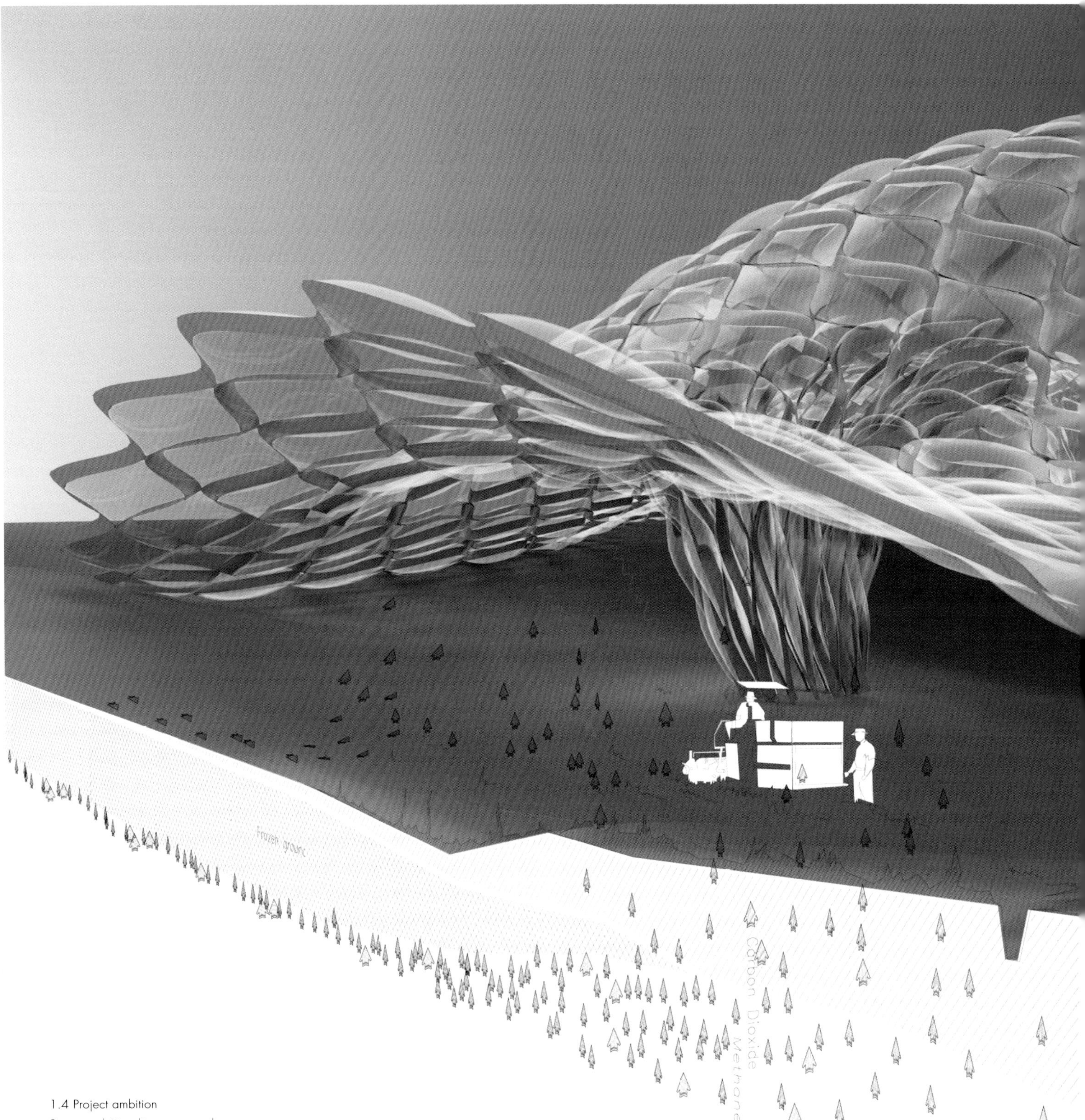

1.4 Project ambition

So my ambition this year was the opportunity to design a large span structure to harvest methane and encourage agriculture in an area of the world where the climate so far has made this very difficult.

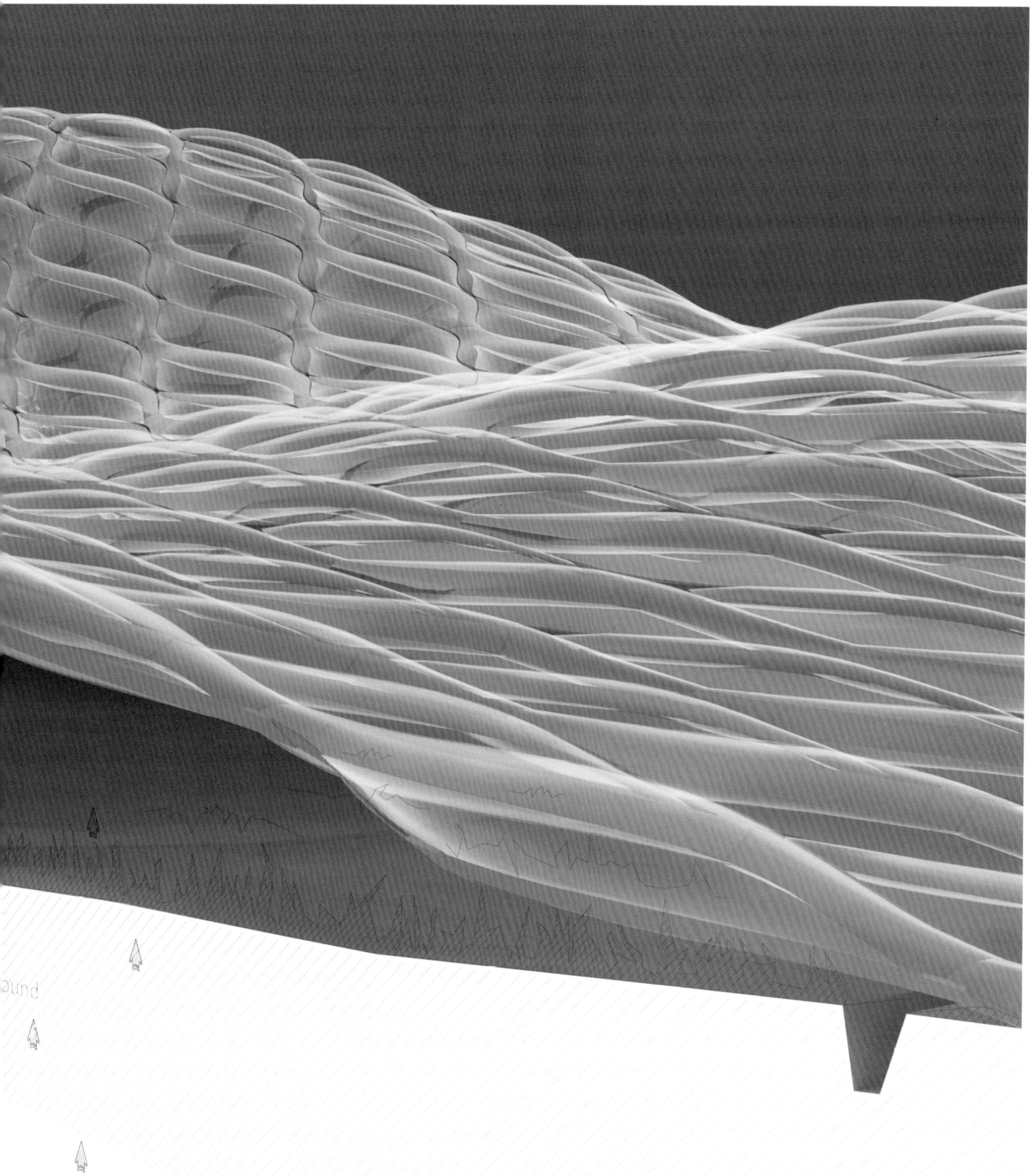

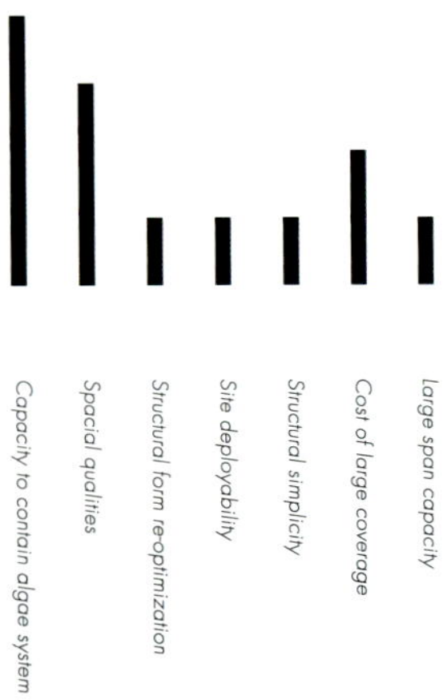

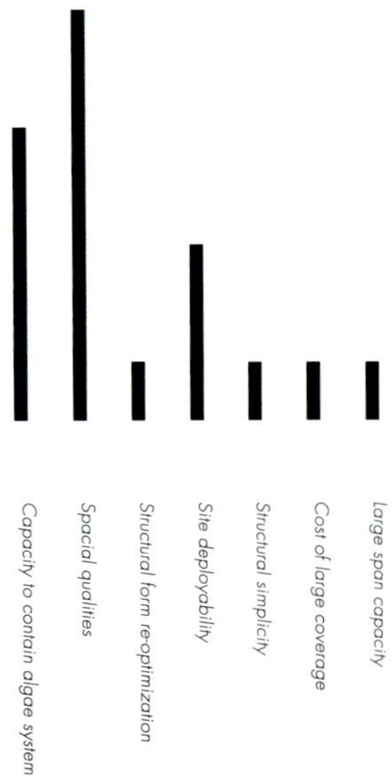

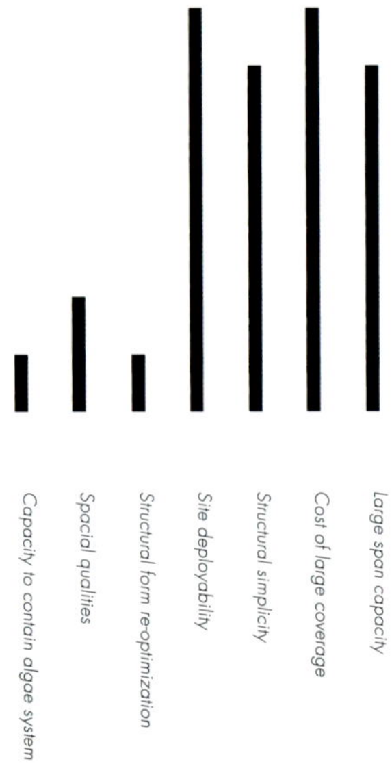

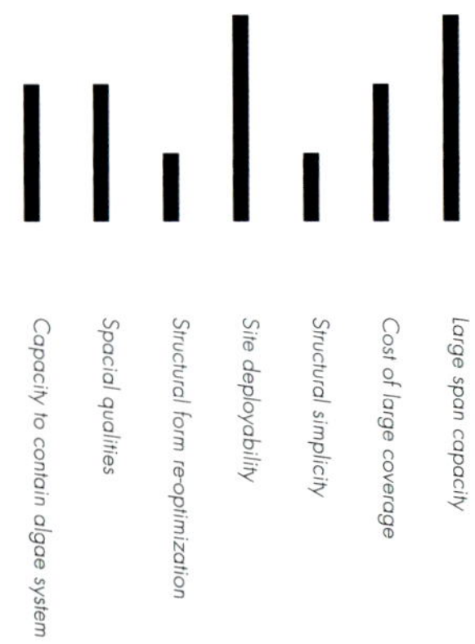

Compression brick

By utilizing refined optimal forms, larger domes can be achieved from mainly compression capacities of materials.

Triangulation

By triangulating a surface the form gains pure compression and tension lines making several forms beyond the dome achievable. The surface can be relatively thin as the structure is held up by the curvature of the larger assembly. But again is form optimization crucial to distribute the stress and avoid buckling.

Pole and skin

Refining the elements in to compression pole and tension skin might be the most economical method to achieve vast enclosures. As the skin is soft, it refines itself towards spreading the tension, and as such creates a optimization of the form.

Project evaluation of method:
Most economical large enclosure, but large pieces would demand more manpower, and it would not be able to respond to the ground conditions by refining its form, indicating gas release.

Pole and cable

The millennium dome is a cable structure that is clad with skin, As the cable can be bent, it refines itself towards spreading the tension, and as such creates a optimization of the form. The difference from the pole and skin is that the refinement of the form depends on the density of cables.

Project evaluation of method:
To many members, and complicated to erect

Optimal form

The Eden pro… ematical calc… similar to the … benefits from …

Project evaluc… not be easy to… does not fully… in place.

1.5 Structural technologies
From compression to tension

As the project needs to be moved, I preferred a modular system.

As the covered ground develops vegetation which absorbs the gas, the structure has to be moved to maintain its gas supply. So, I preferred a modular unit for low manpower mobility, structurally capable of a collective large span. On a more poetic level, I wanted the surface to respond to its ground conditions, utilizing gas locally, making the structure raise and span. Like it is feeding off the ground, but also showing the workers where to extend the structure.
I arrived at a proposed synthesis between high pressure modules used in Achim Menges designs, supported by the internal low pressure of Buckminster Fuller, while having ambitions on a more adoptable form optimisation.

Casa De La Madre Tierra r16
Fuksas glass triangulation r17
Circus tent r18
Millennium dome r19
The Eden project r20
Buckminster Fuller dome proposal r21
Pneumatic structure proposal r22
Pneumatic swimming hall r23
Single surface pneumatic skin r24

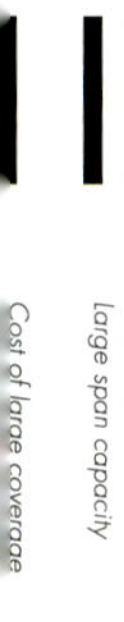

ecise math-
n that sense
pers, but it
les.

ucture would
ule, as it
nents are

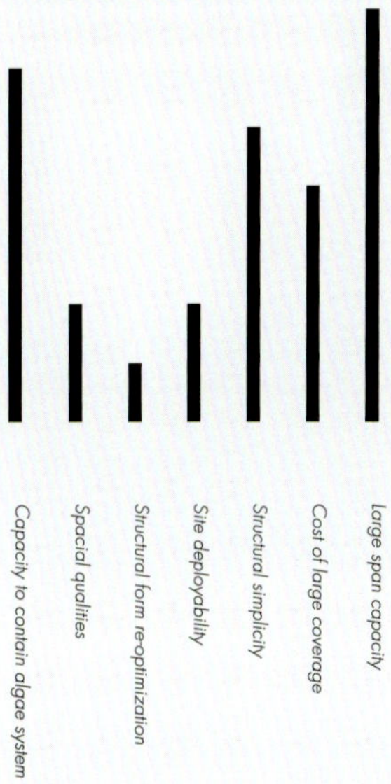

Structural dome and enclosure low pressure

The Buckminster Fuller dome proposals used modular units, and optimal dome form, but the most important factor to propose something at this scale was the claim that the internal low pressure would support the structure.

Project evaluation of method:
The internal pressure is something the project should take advantage of to avoid collapse, but the rigid form of the dome would not allow the structure to respond to ground conditions by refining its form, indicating gas release.

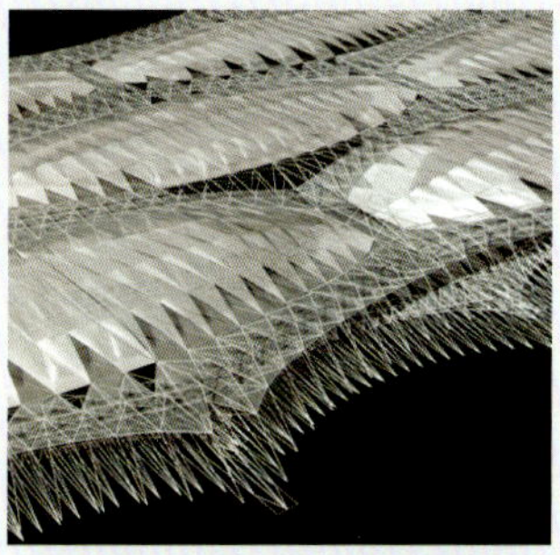

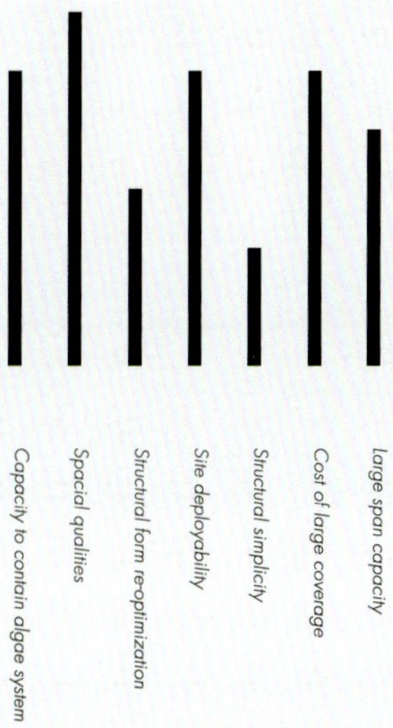

Pneumatic modular high pressure and cable

Achim Menges high pressure modules takes advantage of the modules geometry to make it arch. The modules overlap each other, averaging out the stress as an collective air beam.

Project evaluation of method:
The modules utilize high internal pressure to generate the form. This is something my project could take advantage from. But the rigid form of uniformed module would not allow the structure to respond to ground conditions by refining its form, indicating gas release. The structure can not rely on internal high pressure alone, but it can use the pressure as form triggering.

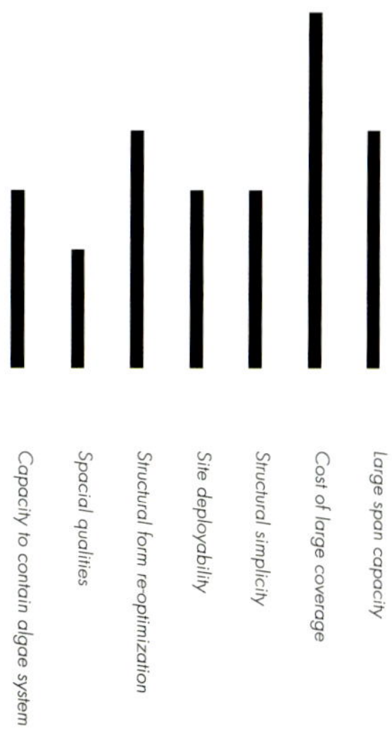

Pneumatic high and low pressure

Inflated high pressure tubes, with internal low pressure support could.

Project evaluation of method:
Uses both internal low pressure, and module high pressure, but the form is again fixed to the module design. Which would not allow the structure to respond to ground conditions by refining its form, indicating gas release

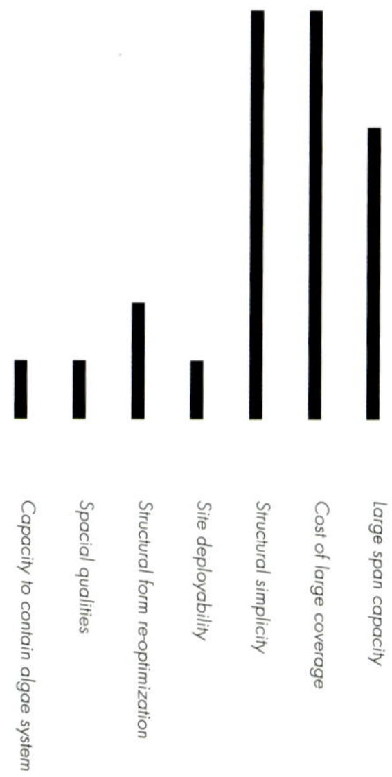

Single surface and low pressure

The single surface inflated skin might be one of the most material subtractive methods. It relies only on internal low pressure.

Project evaluation of method:
One large skin, without modules would have to be moved as a whole, which explains the reduced the scales it so far has been applied to. A new strategy could be used by adding strips to the surface in sections, but it would still be a very monotonous structure. Subductive simplicity with this method, but as the covered ground develops vegetation which absorbs the gas, the structure has to be moved to maintain its gas supply. So, I preferred a modular unit for low manpower mobility, structurally capable of a collective large span.

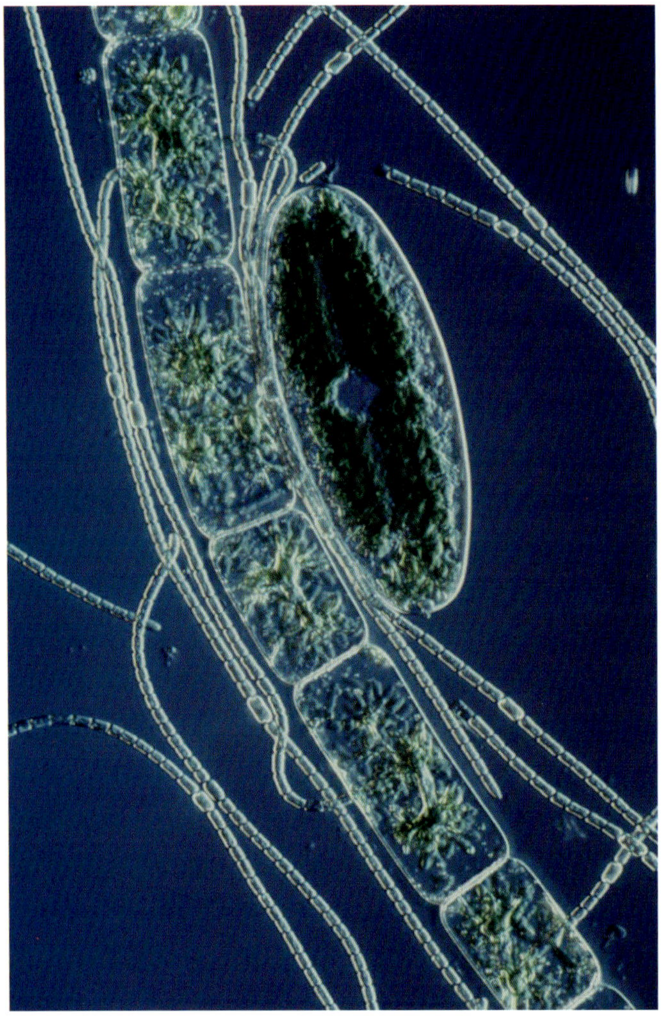

1.6.1 Algae photosynthesis

Conditions and production

One of the most efficient methods of dealing with the gas is an algae farm producing bio fuel and fertilizer.

"Scenedesmus Dimorphus - Categorized as a heavy bacterium, scenedesmus has a lipid content of 16-40%. According to Dr. Orcutt [8], scenedesmus is a very promising strain that should be further researched. The strain must be constantly agitated while grown because of the ease of sediment buildup which hinders growth. The optimal growth temperature falls between 30-35 degrees Celsius (86-95 degrees Fahrenheit) which may prove difficult considering a potential growth facility and the ambient temperature of the Michigan area year round. Scenedesmus will use any and all light it is given and should be further researched for use in mass production. Unlike botryococcus, scenedesmus can be acquired from a variety of public sources. Another strain of scenedesmus obliquus is the strain that the group has currently been working with in the lab.

Euglena Gracilis - Euglena has a lipid content of 14-20% by dry weight. The optimal temperature requirement is 27-31 degrees Celsius (80.6-87.8 degrees Fahrenheit). Euglena was one of the few strains that had information on the nutrient medium content for optimal photosynthesis. The strain enjoys a carbon dioxide concentration within the medium of 4% and an oxygen concentration of 20%. The lighting requirement for the strain is a photosynthetic photon flux of 100 micromoles m-2 s-1. Euglena can be acquired easily from a variety of public sources.

Prymnesium Parvum - Known as a golden algae, prymnesium has a lipid content on average of 22-38%. The difficulty that comes with prymnesium is that it is considered toxic algae which could prove problematic when dealing with it in large quantities. If it were to be mass produced for our processes, various safety hazards would have to be taken into consideration when working with it. The strain optimally grows in salinity from 4% to three times the salinity of regular water which is a constraint that should be further analyzed to narrow its window. Most documentation dictates that prymnesium parvum is a nuisance and is usually only mentioned in reference to how it should be destroyed [5]." r27

"IN THE PRESENCE OF LIGHT, THE SINGLE-CELLED ALGAE TAKE UP CO2 TO PRODUCE THE ENERGY THAT FUELS PLANT LIFE – WITH A GENERAL RULE OF THUMB BEING THAT TWO TONS OF ALGAE REMOVE ONE TON OF CO2. ONCE THE ALGAE ARE HARVESTED, THEY CAN BE CONVERTED TO GENERATE COMMERCIALLY VIABLE BYPRODUCTS SUCH AS ETHANOL OR BIODIESEL." r25

Anabaena sp. (cyanobacterium) and Netrium and Zygnema sp. (green algae) (100x) r29
BioFence Algae farm r30
Green Research Technologies co2 capture r31
Green Research Technologies Technician r31

1.6.2 Methane absorption
Existing techniques for capturing methane and co2

This is a unit which absorbs co2 from the atmosphere by condensation on its surface. A similar method could be developed for methane.

"How to capture co2? - Artificial trees LLC (GRT), a technology research and development company, and Klaus Lackner from Columbia University have achieved the successful demonstration of a new technology to capture carbon from the air.
The machine is a chemical system, and it attracts and absorbs carbon dioxide from the atmosphere," he says. "In that sense, it is stuck to the surface of a panel which we then wash off with a liquid chemistry, and that chemistry is then pumped over to a separator which splits that liquid into its carbon dioxide gas and the liquid then gets sent back to the collector for another round of air capture."

A device with an opening of one square meter can extract about 10 tons of carbon dioxide from the atmosphere each year. If a single device were to measure 10 meters by 10 meters it could extract 1,000 tons each year. On this scale, one million devices would be required to remove one billion tons of carbon dioxide from the atmosphere. According to the U.K. Treasury's Stern Review on climate change, the world will need to reduce carbon emissions by 11 billion tons by 2025 in order to maintain a concentration of carbon dioxide at twice pre-industrial levels." r28

TREES ARE NATURAL AIR FILTERS, AND PLANTING A BILLION NEW ONES, AS A UN CAMPAIGN HAS CHALLENGED PEOPLE AROUND THE WORLD TO DO, WOULD REMOVE 13 MILLION TONS OF CARBON DIOXIDE FROM THE ATMOSPHERE EACH YEAR. BUT FOSSIL FUEL BURNING CURRENTLY ADDS 26 BILLION TONS OF CO2 A YEAR. r26

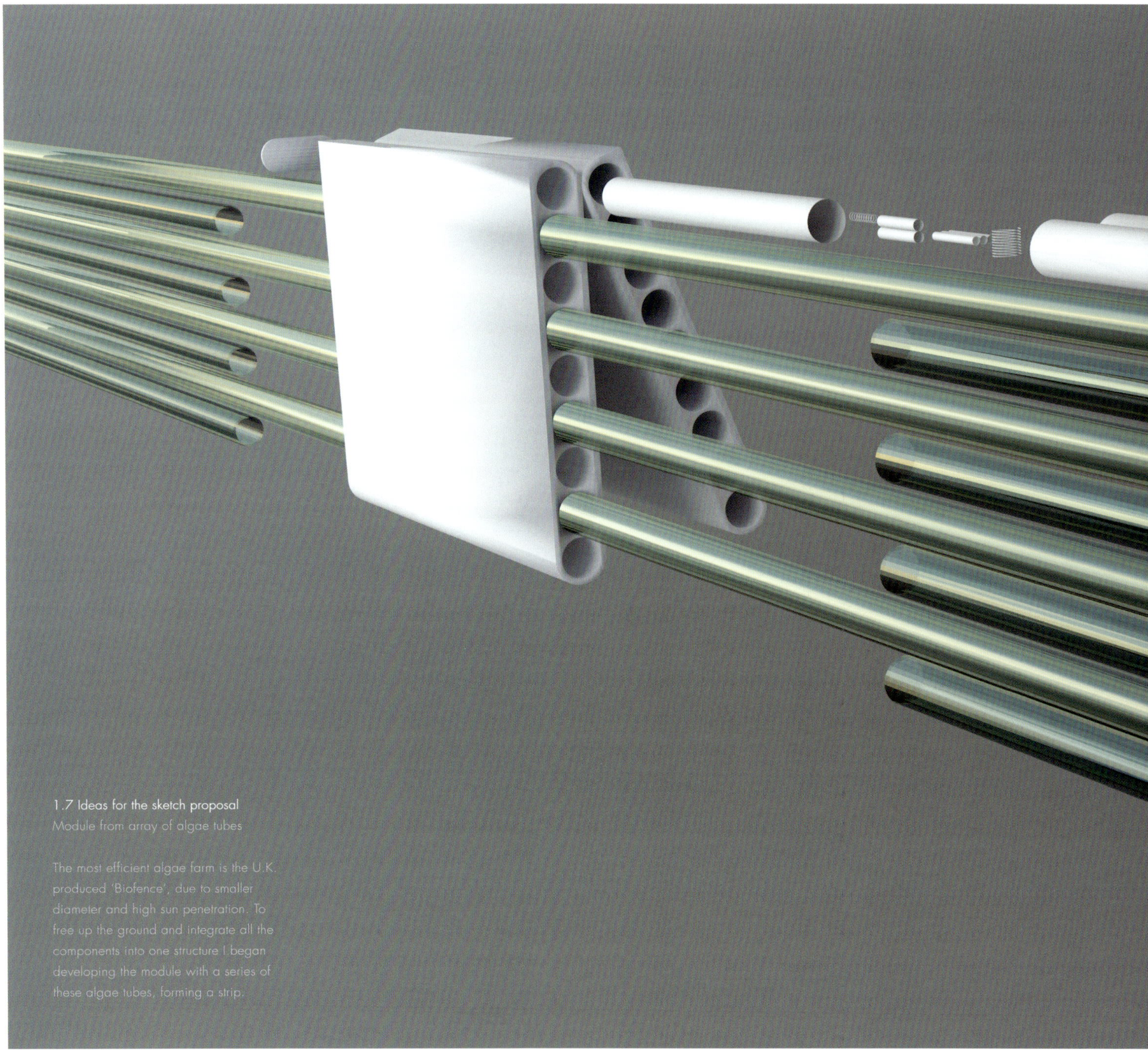

1.7 Ideas for the sketch proposal
Module from array of algae tubes

The most efficient algae farm is the U.K. produced 'Biofence', due to smaller diameter and high sun penetration. To free up the ground and integrate all the components into one structure I began developing the module with a series of these algae tubes, forming a strip.

Traditional laboratory production
Could be produced in soda bottles

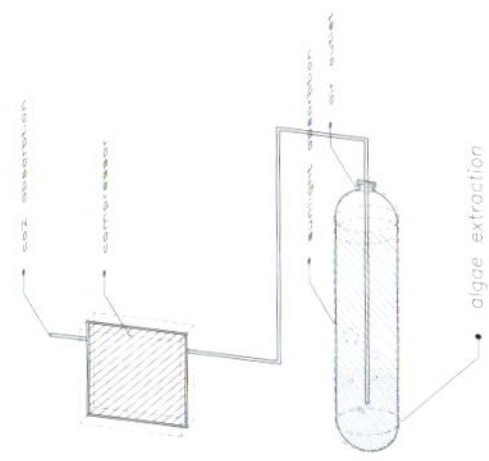

Green fuels technologies us
Already implemented at several power plants in the U.S., with good feedback of efficiency and economic gain

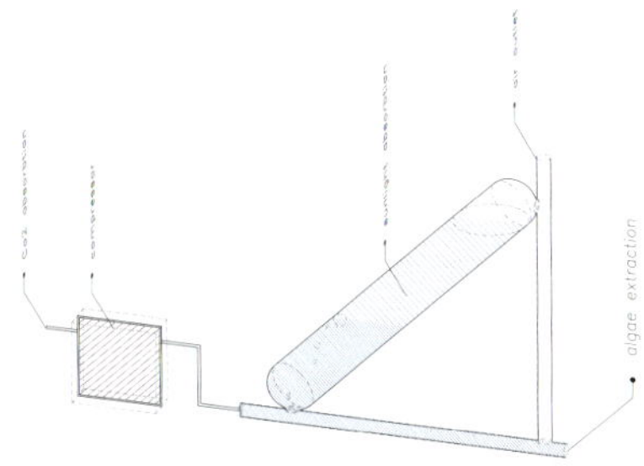

Bio fence uk
Claims to be the currently most efficient algae production system as they decrease the diameter of the tube to allow more sunlight to reach the algaes

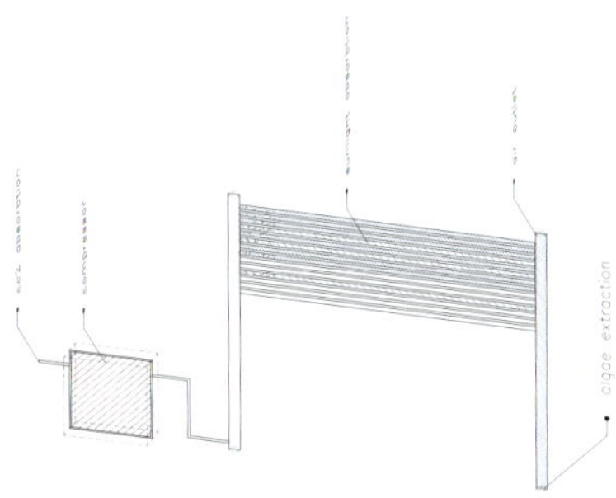

Open air algae farms
Far less production rates than the closed systems. It requires vast areas, and attempts to make this method economically beneficial has so far been less successful. (The Image is an artistic rendered visualisation)

Laboratory algae production r32
Green Fuel Technologies r33
BioFence Algae farmr30
Artistic visualisation of Open air algae farm r34

1.8 Material behaviour model for sketch proposal_01 I began investigating their behaviour, the way they boucle and encloses space simply from manipulating their edge conditions.

Melting tundra in Siberia r35

Field in Artesia r36

1.9.1 Sketch proposal_01

First sketch as an attempt to assemble all the components of the situation.

The ground releases methane gas, this is burnt into Co2 which expands by 2.74 creating pressure, this pressure I intend to use for the structure. The strips have been changed out with tubes, when their pressurized they contract, manipulating the stiffer edge frame, which again creates the behaviour of the strip model.

Methane is absorbed in the lower tubes, it's then locally burnt, producing heat and pressurized co2 which is sent into the muscle tube, the tube contracts, storing the pressure, before the co2 is distributed to the algae tubes on top, receiving maximum sun light.

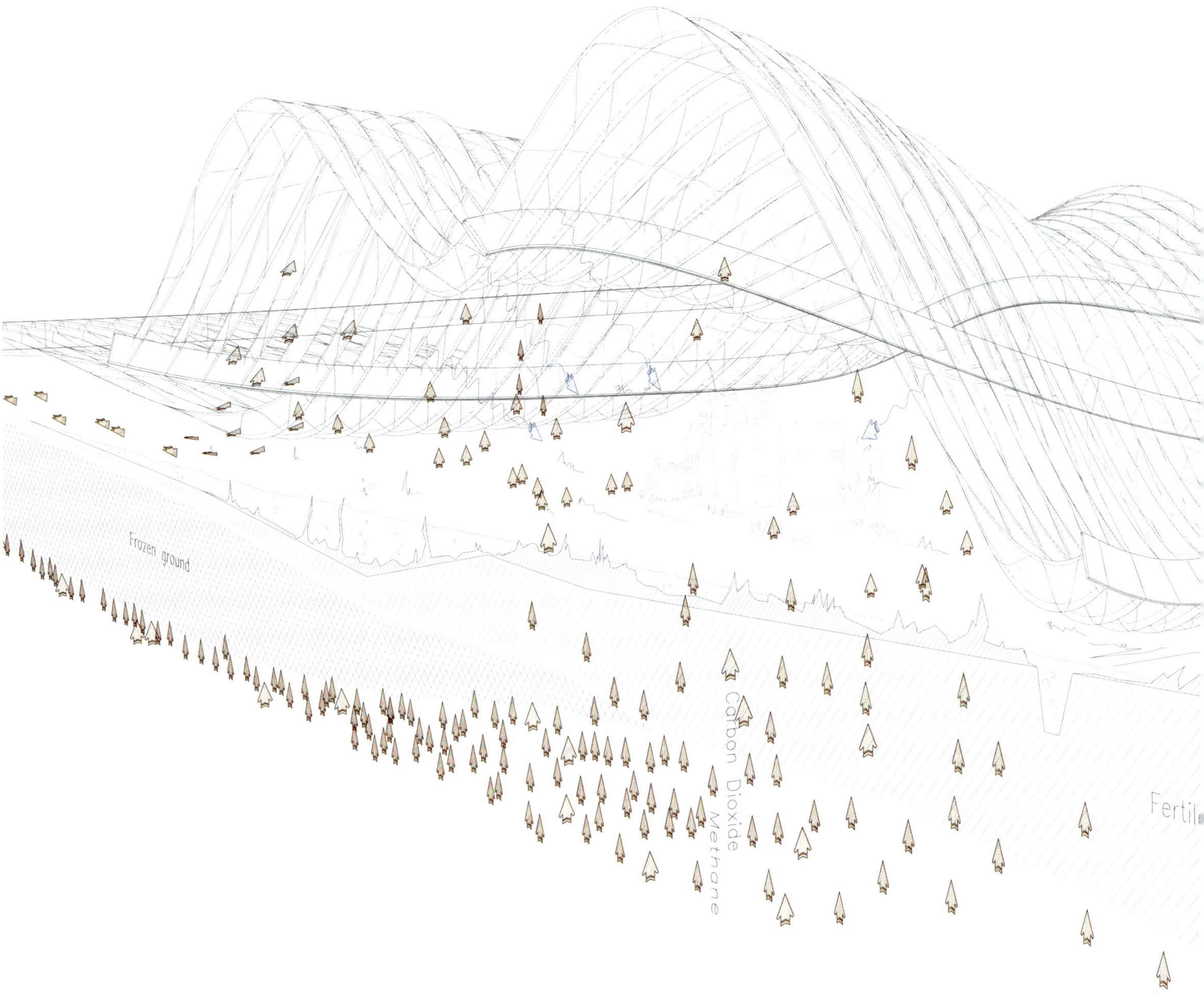

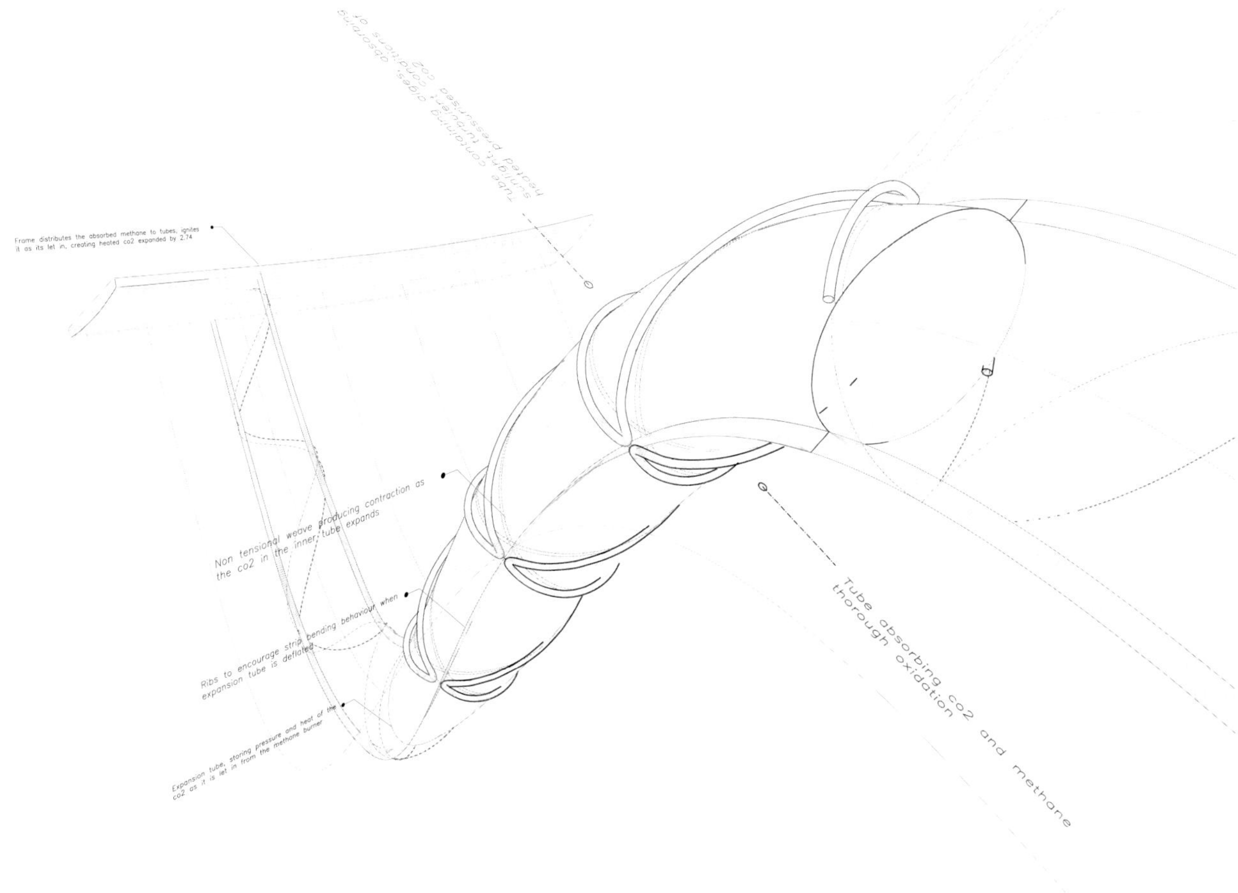

Erlend Bakke-Eidsaa

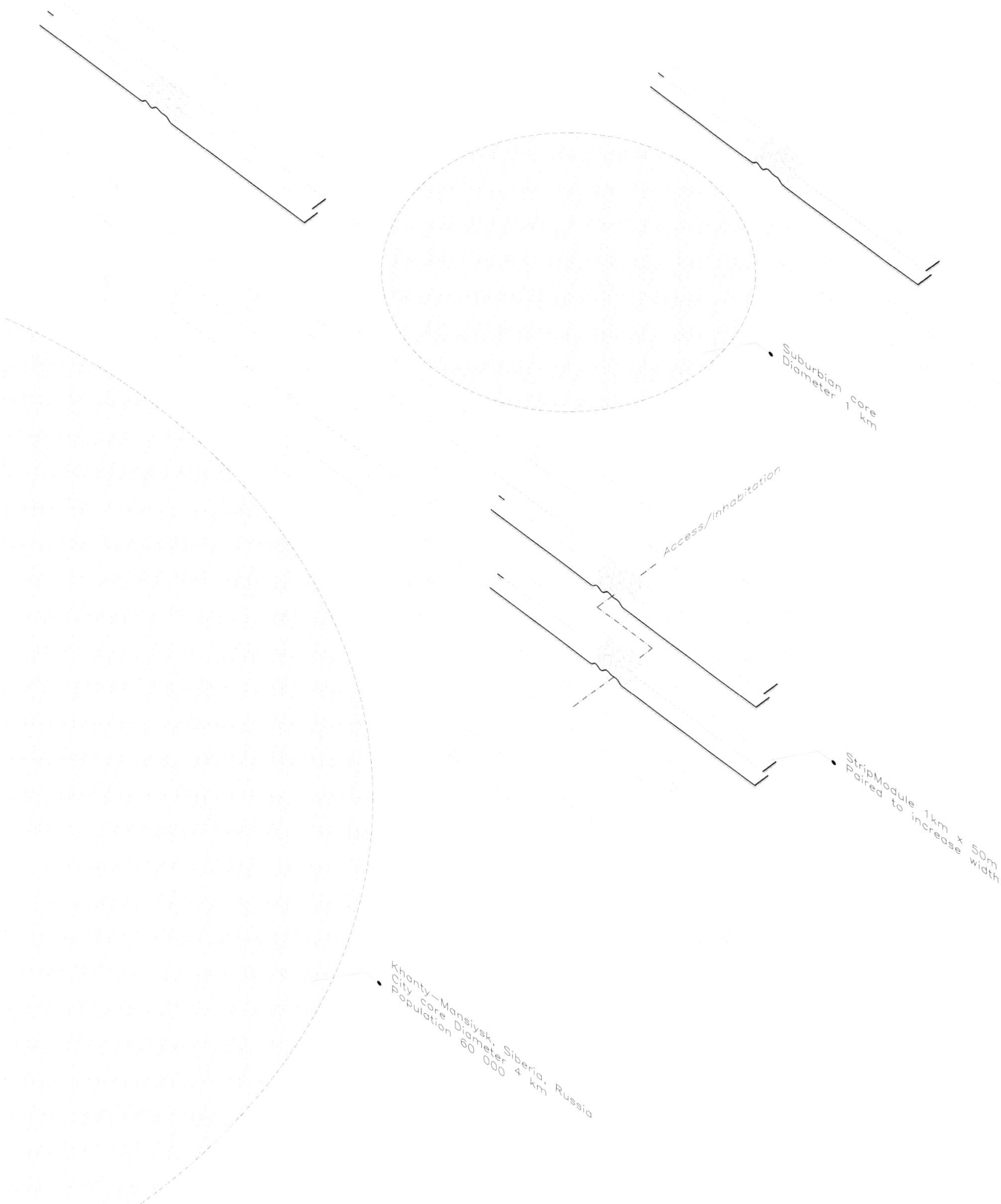

1.9.3 A single surface is attached at both sides to economically harvest larger areas. Gas is sucked towards the absorption units from a solar chimney effect, as they are placed higher and heated up by gas and sun. In this way it performs like a filter.

1.9.4 The gray circle indicates the city Khanty-Mansiysk in Siberia, containing 60 000 people, and here you can see the project measuring 100 x 1000 meters.

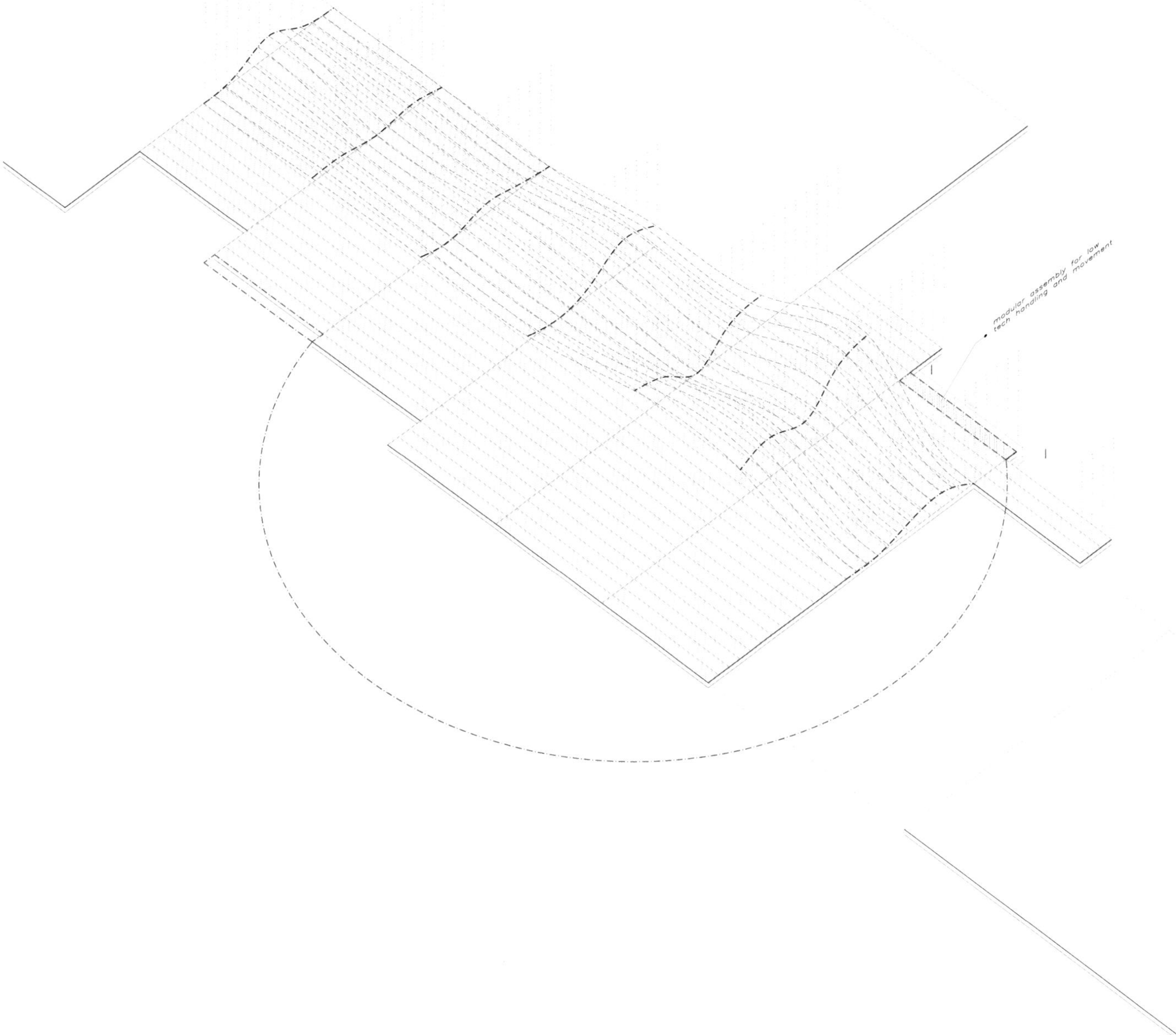

1.9.2 Several modules are attached together to cover an area. As gas is released locally, the structure responds by curling up like a piece of paper, like its feeding of the ground, but also showing the workers where to extend the structure for optimal gas absorption.

2. Site visit to Siberia and redeveloped proposal

2.1 Oil industry workforce
2.2 Russian Dubai
2.3 Ideas for proposal_02. Sealed enclosure pressure
2.4 Ideas for proposal_03. Connected module behaviour
2.5 Ideas for proposal_03. Pneumatic supported structure
2.6 Redeveloped proposal_03
2.7.1-2 Site excursion with the Siberian geologist team, soil fertility
2.8 Drainage and water distribution

2.1 Khanty-Mansiysk is Russia's main oil region, they actually contain 230 of Russia total 300 billion tons oil reserves.

2.2 Like Dubai they realize that the oil is limited, and want to become more than just an oil state, they even invite foreign architects like this Norman fosters project to try to put them on the world map.

2.3 After visiting the site I realized the area is too cold and windy for the open separated strips, so I moved the pressure tube between the strips sealing of an internal space. This also let me tap into the sealed single surface structure used in Buckminster Fuller or Frei Otto's dome proposals. The form would still be controlled and articulated by the high pressure in the modules, but not able to collapse due to the low pressure in the larger enclosement. It would still move and indicate gas, but it would be more subtle movement.

2.4 The other input comes from this model. If you expand the pockets it arches up and encloses space. This behaves fine at this scale, but if it's scaled up further it would eventually boucle and fail.

2.5 However, if the pressure tube is placed within the gap, it could perform the expansion, but more importantly ad support to the strips from buckling. It would still be strip language, not pneumatic, but the pneumatic support would make the strip language achievable on much larger scales.

2.6 Mushroom inspired columns where introduced to lift the edge and link enclosures to further increase the span. The space could be used as access to the sky and a breathing space for the workers due to increased co2 levels inside the structure, it would also be a maintenance point and gather rain water from the surface like the mushroom did.

2.7.1 I visited the site with the C.A.S.U.S geologist team which is monitoring the gas release.

2.7.2 What they showed me was this stream marked in pink, where the pete-lands, marked in blue, drains out, resulting in 100 meters of high vegetation on both sides. For the project this means that the land is fertile, all it needs is added fertilizer and a developed drainage pattern, both which is by-products of an algae farm.

2.8 Drainage control would be important to allow any agriculture. The current location of the melting edge in an immense flood plane makes it even more urgent. I would continue with Dutch bouldering techniques, reinforcing the natural sediment patterns, to create pockets of controlled space for the project.

2.1

2.7.1

2.7.2

2.3

2.4

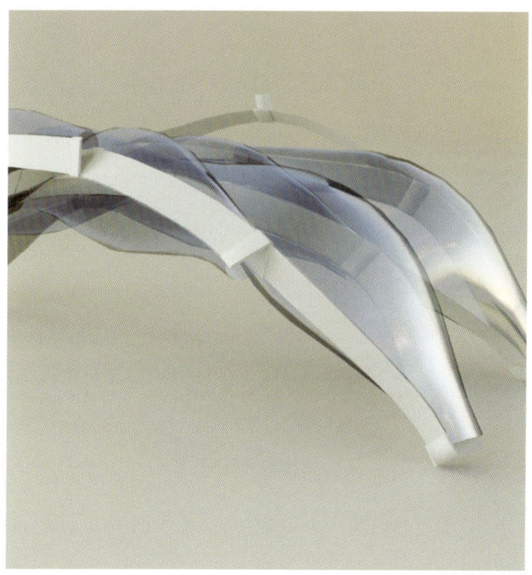

2.5

ЛЕНИН
ЛЕНИН
ПРАВД
Орган Центрального Комитета
Газета основана
5 мая 1912 года
В. И. ЛЕНИНЫМ
№ 219 (23015)
Пятница, 7 августа 1981 года
В ЦЕНТ
КОМИТ

2.1 Oil industry workforce
Brief history of site

Khanty-Mansiysk is Russia's main oil region, they actually contain 230 of Russia total 300 billion tons oil reserves.

Khanty-Mansiysk which is the capital of Ugra county was originally a fishing village, relying on fish and timber. However, 20 years ago oil was discovered, It actually contains 230 of the 300 units of Russia's total oil reserves, and as a result it has tripled its population and gained large influence in the Russian government. It is more or less allowed do what it wants with its county.

UGRA COUNTY CONTAINS 230 OF RUSSIA'S 300 BILLION TONS TOTAL OIL RESERVES

Left
Ugra becomes of great importance to Russia r37
Workforce adopted from fishing to oil r37

Right
Meeting with the head the university which seemed to have a great influence on politics and future development of Ugra, and Elena D Lapshina who is head of CASUS research team monitoring the gas release.
Ugra University main building
Poster which strengthens the claim of the Principal's involvement in politics.
Oil refinery. r37

Opposite
Oil discovered in Ugra county r37

2.2 Russian Dubai

City to become more than an oil state.

Like Dubai they realize that the oil is limited, and want to become more than just an oil state, they even invite foreign architects like this Norman Foster project to try to put them on the world map.

What it wants is to invest its current wealth into the future of the state. I would call it a Russian version of Dubai. Like Dubai they realize that the oil is limited, and want to become more than an oil state. They have built hotels, business centers, a TV station, and a university with 65 professors. They even invite foreign architects like this Norman Foster project to try to put them on the world map.

Previously the oil workers went to the county for two week work shifts, much like the oil rigs in the North sea. When they finished their work shift, they went back home. Khanty-Mansiysk is trying to make them stay.

Fishing and Timber was the main livelihoods before the oil, and it still is practiced. The loft house construction represents a strong contrast to the oil industry's steel and glass being built side by side.

Left
Frozen harbour in Khanty-Mansiysk

Right
Norman Foster proposal for Khanty-Mansiysk r38

Below
Fishing and Timber was the main livelihoods before the oil, and it still is practiced. r37
Older but reliable technology prevails.
Neon light decoration used to lift the spirit through the longer dark periods of the year.
Politics, Sports and Education seems to be the future of Khanty-Mansiysk, while oil is the engine that brings it forward. r37, r39, r37

2.3 Ideas for proposal_02

Sealed enclosure pressure

After visiting the site I realized the area is too cold and windy for the open separated strips, so I moved the pressure tube between the strips sealing of an internal space. This also let me tap into the sealed single surface structure used in Buckminster Fuller or Frei Otto's dome proposals. The form would still be controlled and articulated by the high pressure in the modules, but not able to collapse due to the low pressure in the larger enclosement. It would still move and indicate gas, but it would be more subtle movement.

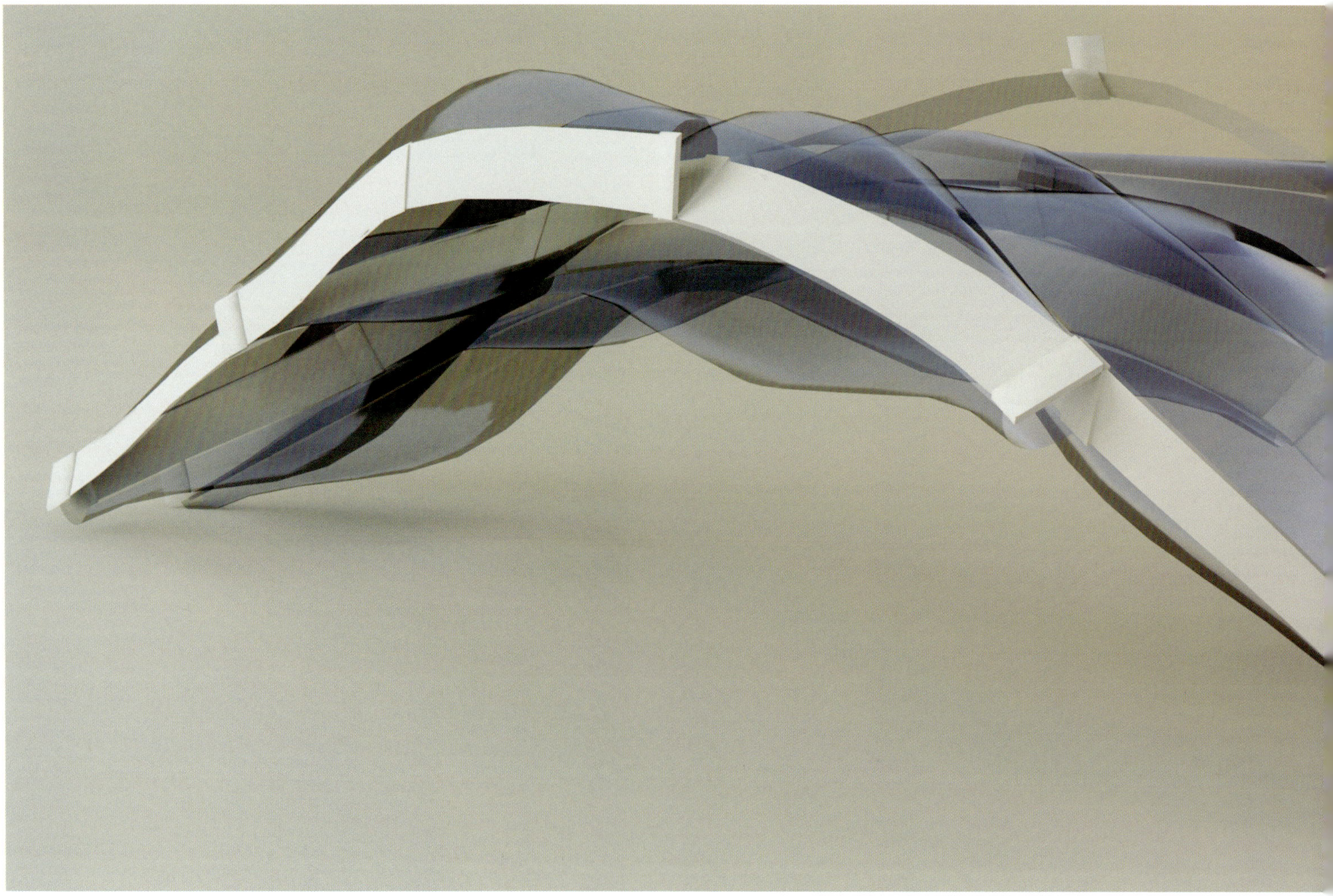

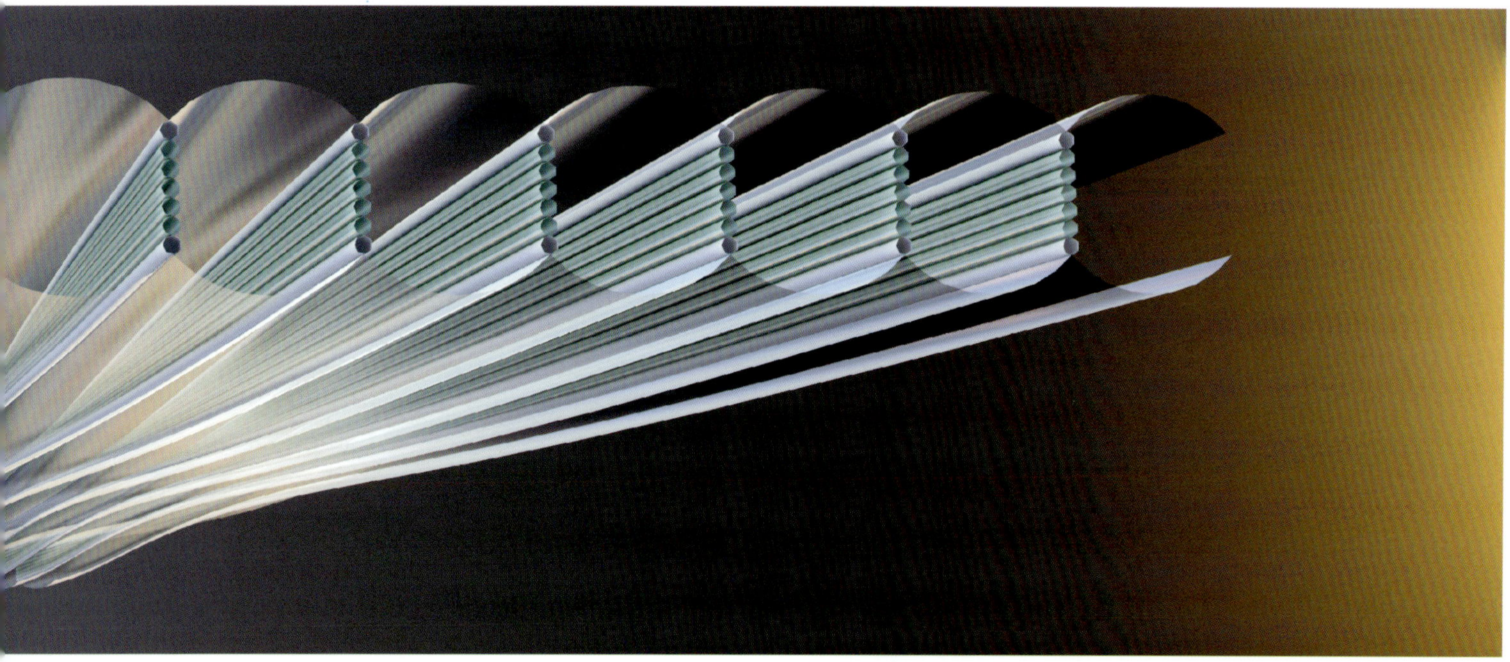

2.4 Ideas for proposal_03
Connected module behaviour

The other input comes from this model. If you expand the pockets it arches up and encloses space. This behaves fine at this scale, but if it's scaled up further it would eventually boucle and fail.

2.5 Ideas for proposal_03
Pneumatic supported structure

However, if the pressure tube is placed within the gap, it could perform the expansion, but more importantly ad support to the strips from buckling. It would still be strip language, not pneumatic, but the pneumatic support would make the strip language achievable on much larger scales.

Stratospheric weather balloon r40
Buckminster Fuller's dome proposals r41

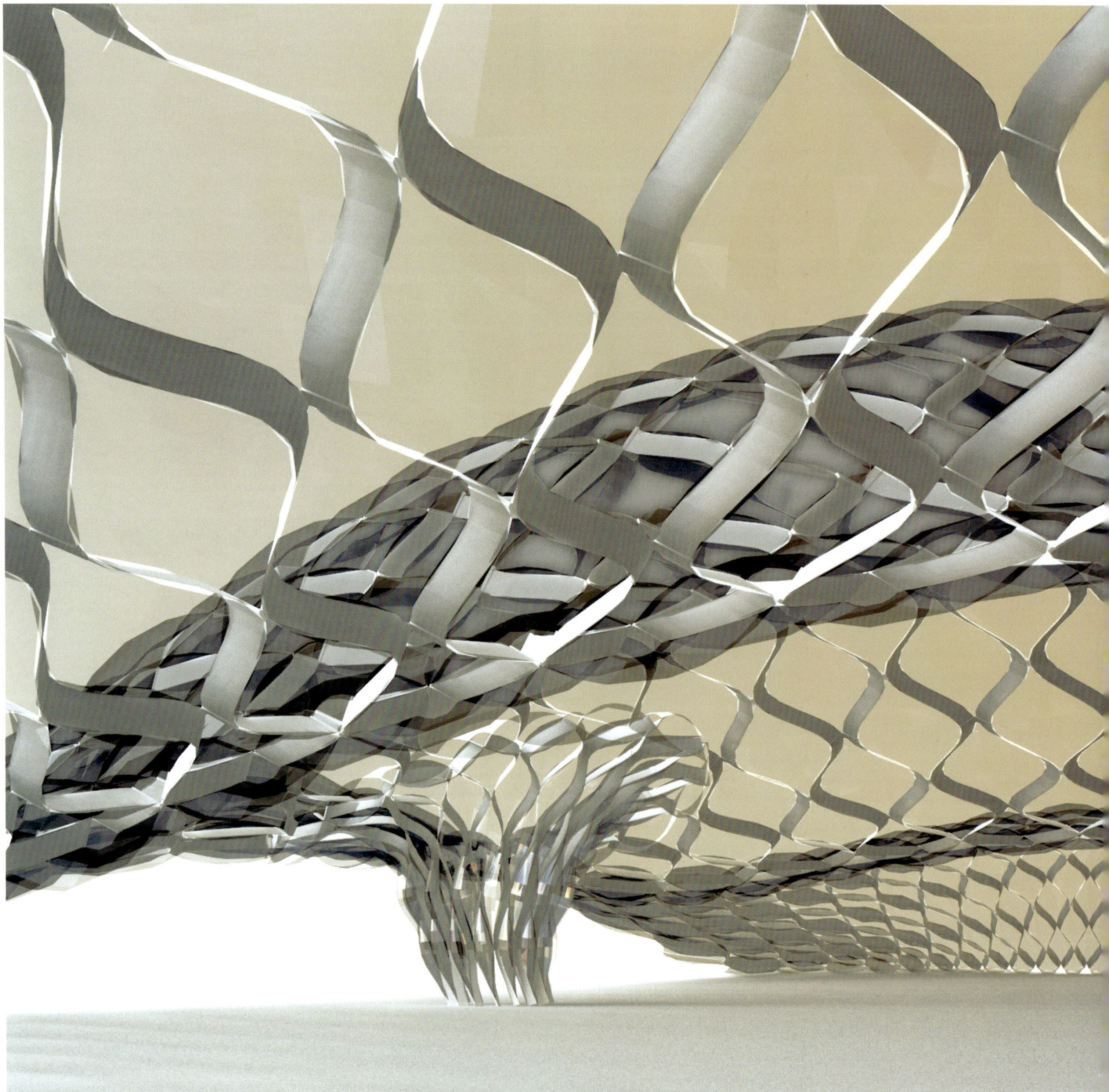

2.6 Redeveloped proposal_03
Columns and flexible triangulation

Mushroom inspired columns where introduced to lift the edge and link enclosures to further increase the span. The space could be used as access to the sky and a breathing space for the workers due to increased co2 levels inside the structure, it would also be a maintenance point and gather rain water from the surface like the mushroom did.

Local mushroom that inspired the columns. The surface arches and spans locally from high gas pressure in the module, while it's stabilized from low pressure in the enclosed space, giving it subtle movement.

Dr. Eugeniy Glukhmanchuk

Practical geologist
Specialist on mathematical simulation of oil-bearing strata
Worked in the oil industry since oil was discovered in Yugra region, now teaches geology at the university

2.7.1 Site excursion with the Siberian geologist team

The frozen Siberian Pete-lands

I visited the site with the C.A.S.U.S geologist team which is monitoring the gas release.

At the site I was taken out to the tundra by snow mobile, and discussed my project with the C.A.S.U.S. geologist and biologist team monitoring the gas release. Professor Elena D. Lapshina who is the team leader, informed the project on areas of gas release, ground conditions, and quantum of gas release. Dr. Eugeniy Glukhmanchuk worked in the oil industry since oil was discovered in Ugra county, he now teaches geology at the university. He informed the project on oil work force and finances. Alexander Gamov was also part of this excursion, he is the Head of International Field Station of Yugra State University and research assistant for the CASUS ecologist and biologist team. In 2009 the Field Station are opening UNESCO Chair Environmental Dynamics and Climate Change. Oleg Deyev of the International Cooperation Department Yugra State University was my guide and interpreter, and an invaluable resource for deeper understanding of the culture and way of life in this harsh climate.

Elena D. Lapshina

alist on ecology of peat land ecosystems leader of the CASUS ecologist and biologeam, monitoring the gas release. Informed oject on areas of biodiversity and types of and, ground conditions, and quantum of gas e.

2.7.2 Soil fertility

100 meters of intensified vegetation from increased mineral content

What they showed me was this stream marked in pink, where the pete-lands, marked in blue, drains out, resulting in 100 meters of high vegetation on both sides. For the project this means that the land is fertile, all it needs is added fertilizer and a developed drainage pattern, both which is by-products of an algae farm.

This stream even releases methane at wintertime as you can see on the photo above.
Map shown by Professor Elena D. Lapshina

Meltwater exposing optimal flow patterns.
The meltwater from most of central Siberia drains out in the enormous rivers Yen and Ob. Khanty-Mansiysk is situated at the rivers meeting point. Every spring these rivers overflow by 5 to 10 meters and create an enormous flood plane. In terms of the project, the flooding exposes optimal flow patterns in the ground sediments, which the project could reinforce and utilise to create protected pockets for its operation.

Dutch bouldering systems.
Even when the edge moves beyond the current flood location, the melting permafrost produces large amounts of water which needs to be drained to allow agriculture. The Dutch has since the 1400 been successful with windmill powered drainage techniques to extend their agricultural land and inhabited land into the Rhine river delta. Afsluitdijk completed in 1933 runs 32 km and separates freshwater from the sea. Also in England has pete-lands been drained for agriculture. The projects is currently situated in an enormous flood plane which builds up optimal flood line sand banks. The project could through Dutch bouldering techniques, attempt to reinforce and extend these patterns to create pockets of protected space.

2.8 Drainage and water distribution
Optimal flow patterns and Dutch bouldering systems

Drainage control would be important to allow any agriculture. The current location of the melting edge in an immense flood plane makes it even more urgent. I would continue with Dutch bouldering techniques, reinforcing the natural sediment patterns, to create pockets of controlled space for the project.

Opposite
Khanty-Mansiysk natural sediment patterns r42

Right
Dutch drainage wind mills r43
Dutch bouldering diagrams r44
Haarlemmermeer drained agricultural land r45

3. Developing the structural theories through physical models

3.1 Flexible double curvature behaviour and span capacities
3.2.1-2 Physical strip model animation
3.3 Analysing the strip model. Load deformation experiments
3.4 Strut expansion vs. pneumatic cushion actuators
3.5 Physical pneumatic model animation
3.6 Analysing the pneumatic model

3.1 So were back to the structure, I needed to reach the scale where the strips boucle and fail to find where to add the support of the cushions. This is a 3.5 meter long model. Here the expansion is done by struts, but in the final model it would be pressurized bags.

3.2.1 The modules are connected into a surface and begin to react to local gas release. It expands, raises and forms pockets of enclosed space.

3.2.2 Here you can see how the model is reacting to gas release by inflation and deflation. It does not collapse due to internal low pressure, but has local subtle movement. Like its feeding of the ground while indicating where to expand the structure for optimal gas absorption.

3.3.1-3 To find the forms and scales achievable with this structure, I went through a series of load deformation experiments. These are the three most important ones. You can see that it is actually the expansion angle of the end joint which is making the structure arch. Then its the shift and rotation which manipulates the form.

3.4 I expected some deviation between strut expansion and cushion expansion, the theory of the cushions supporting the strips without interfering with its formal language also needed to be tested.

3.5 So this is a prototype with all the components in place. Compared to the struts, the pneumatic cushions are now performing the expansion and acts as seamless actuators.

3.6 The cushions are cut from the form of the buckled strip, and seems to support the strip behaviour without too much influence. I would in that sense claim this experiment a success. This strengthens the claim of using pneumatics to support a structure, making it achieve larger scales.
In the animation you can see how the structure is sort of breathing from reacting to gas release.

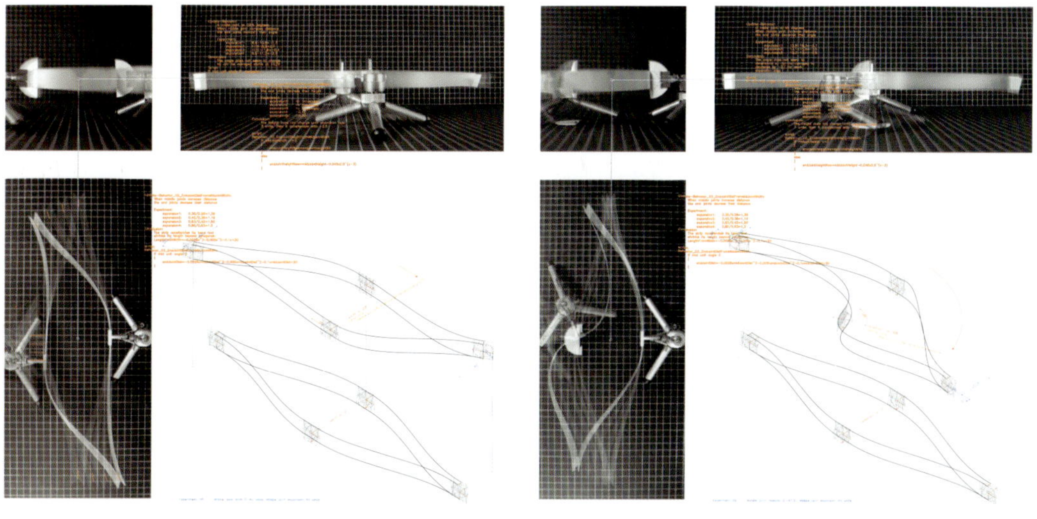

3.3

3.2.2

3.1

3.5

3.6

3.1 Flexible double curvature behaviour and span capacities

Large scale to reach buckling failure

So were back to the structure, I needed to reach the scale where the strips boucle and fail to find where to add the support of the cushions. This is a 3.5 meter long model. Here the expansion is done by struts, but in the final model it would be pressurized bags.

Detail Photos: Valerie Bennett

3.2.1 Physical strip model animation
Short side view

The modules are connected into a surface and begin to react to local gas release. It expands, raises and forms pockets of enclosed space.

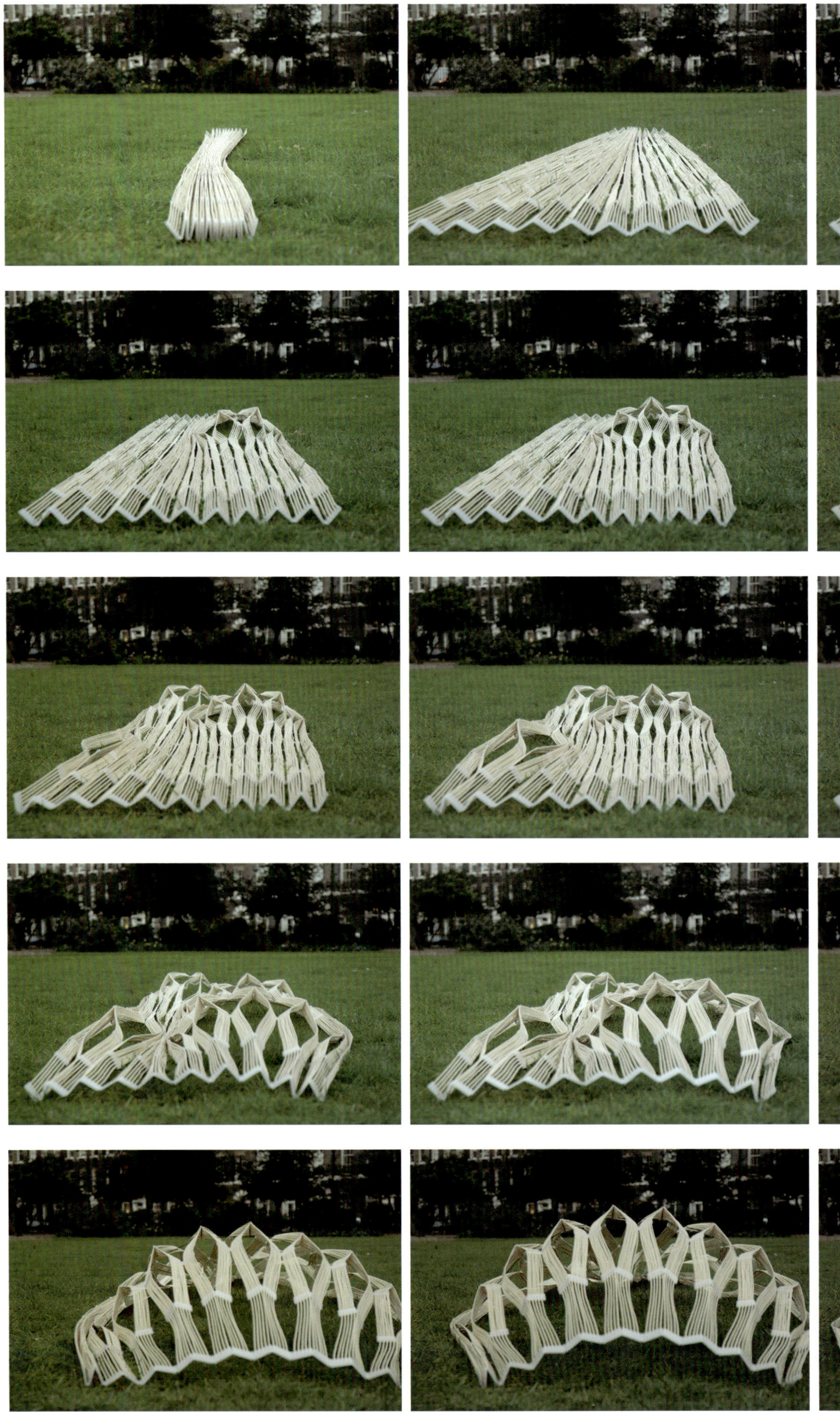

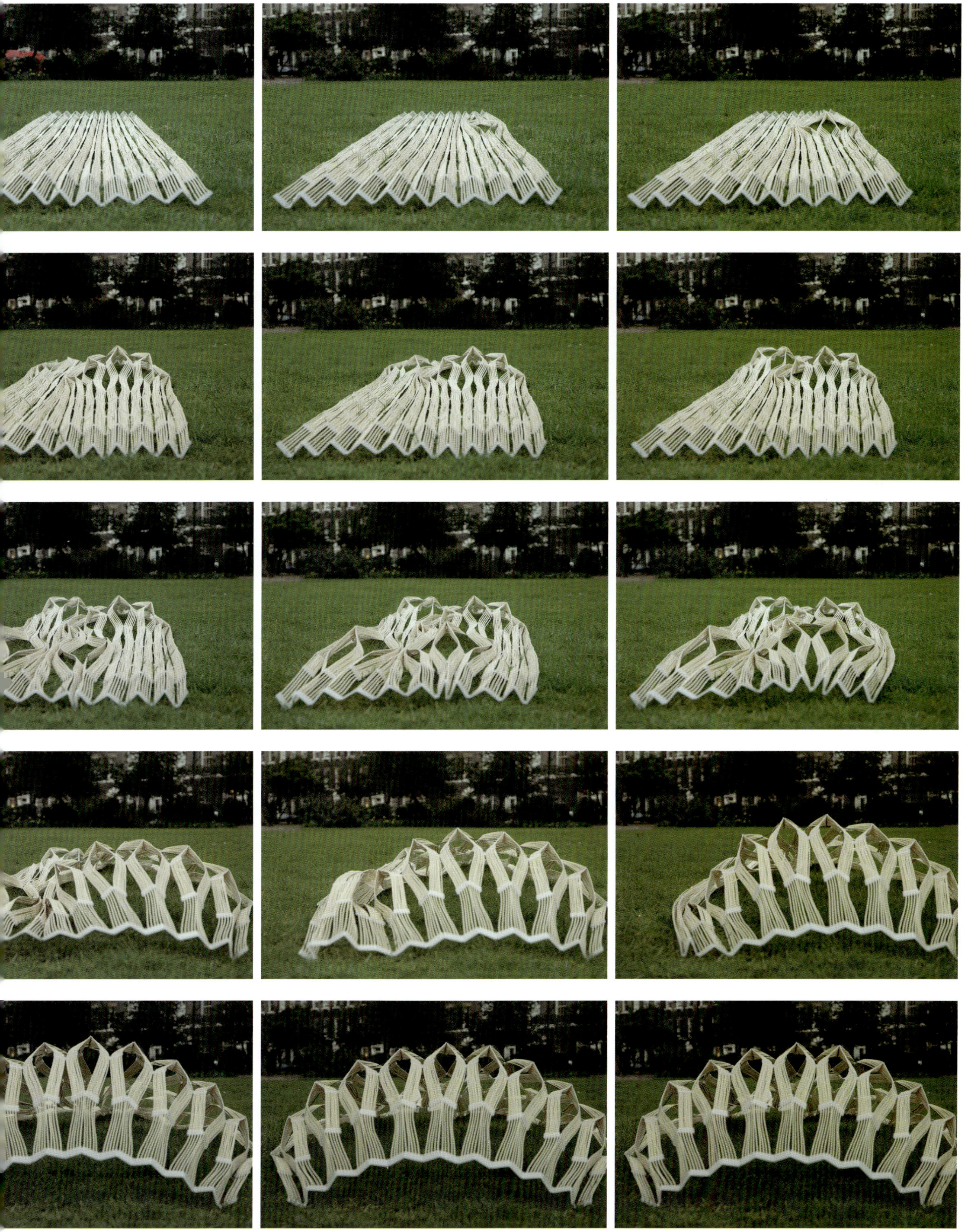

3.2.2 Physical strip model animation
'Bird' view

Here you can see how the model is reacting to gas release by inflation and deflation. It does not collapse due to internal low pressure, but has local subtle movement. Like its feeding of the ground while indicating where to expand the structure for optimal gas absorption.

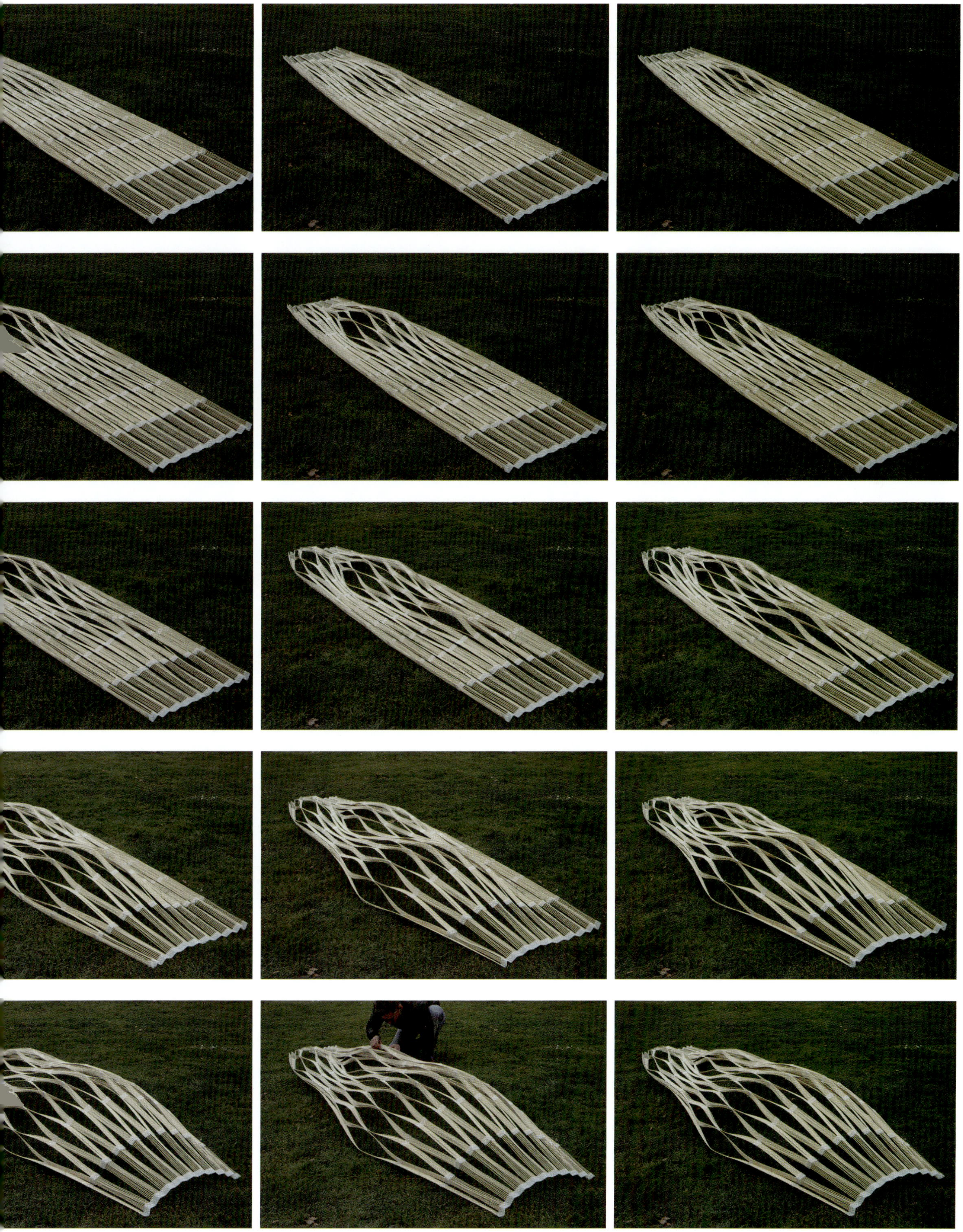

3.2.2 Physical models to develop sketch_03
Expansion by struts

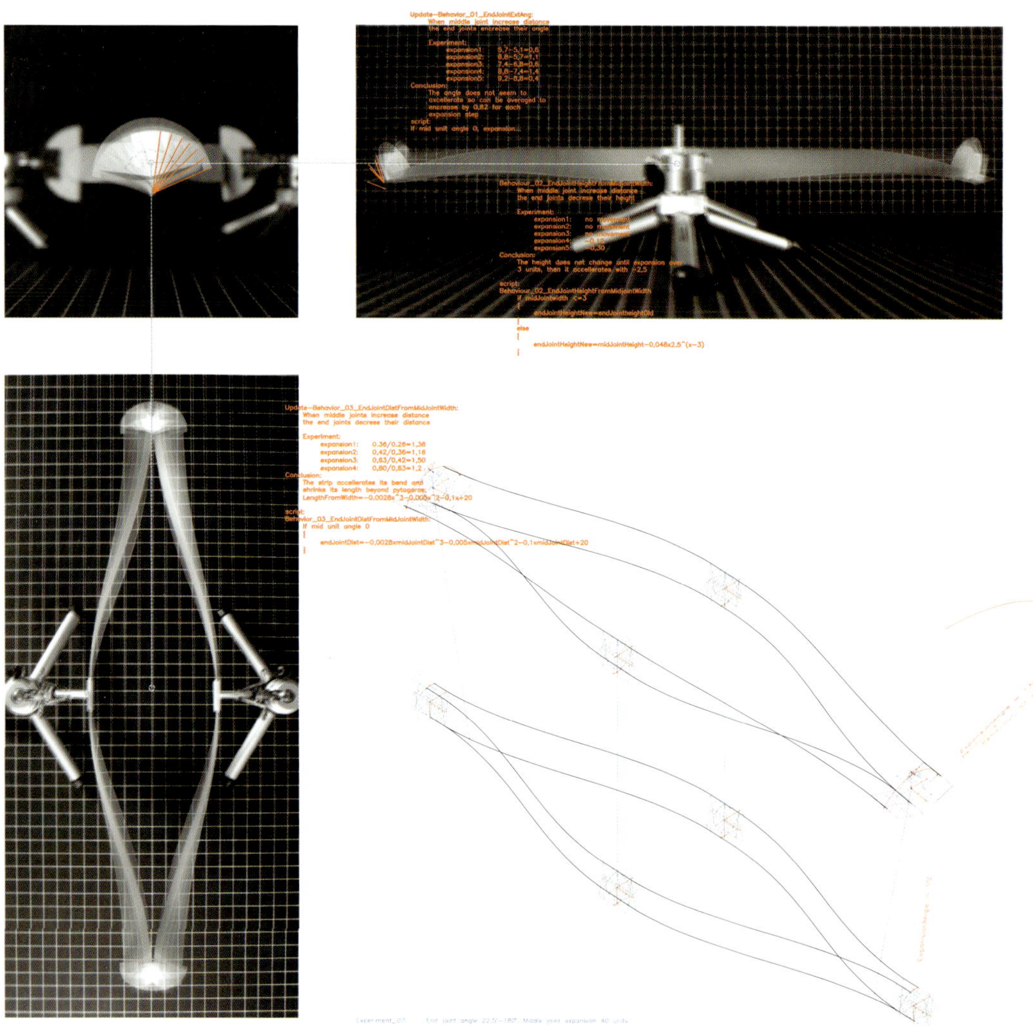

3.3.1 Experiment_07:
End joint angle 22,5°-180°, Middle joint expansion 40 units

This manipulation revealed itself to be the main factor for triggering the geometry to arch in the length direction. When expanding the gap between the strips, the connected end joint bends inwards towards the side of the connected edge. This means that by guiding the expanding gas towards modules for upwards or downwards arching, the geometry can be manipulated.

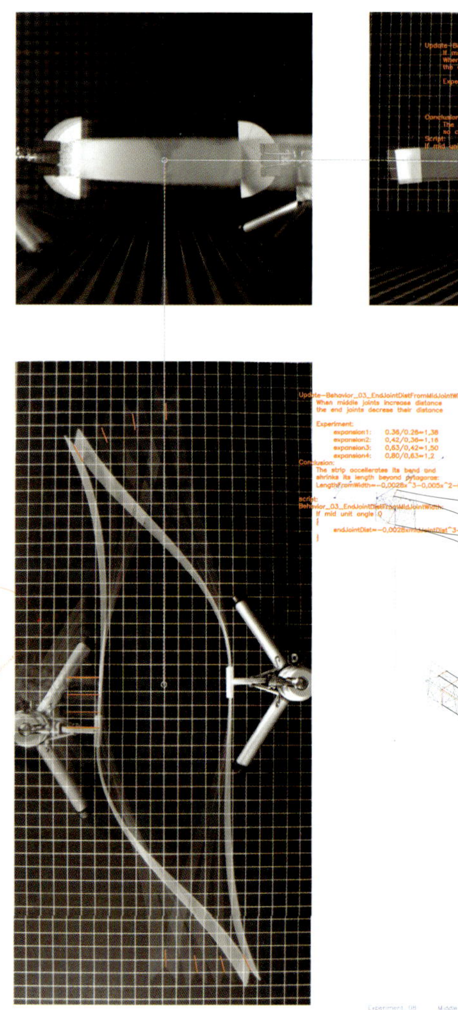

3.3.2 Experiment_08:
Middle joint shift 0-40 units, Middle joint expansion 40 units

The shift is not easily visual within a larger composition, but is in some degree present within every module, especially when multiple arches emerge and compete over the lower shared edge. From the physical model I would claim this effect as a positive contribution to form generation of the larger domes, making them optimise in relation ship to each other and the variable friction towards the ground. However, the maximum values would have to be monitored and controlled to stabilize the larger geometry.

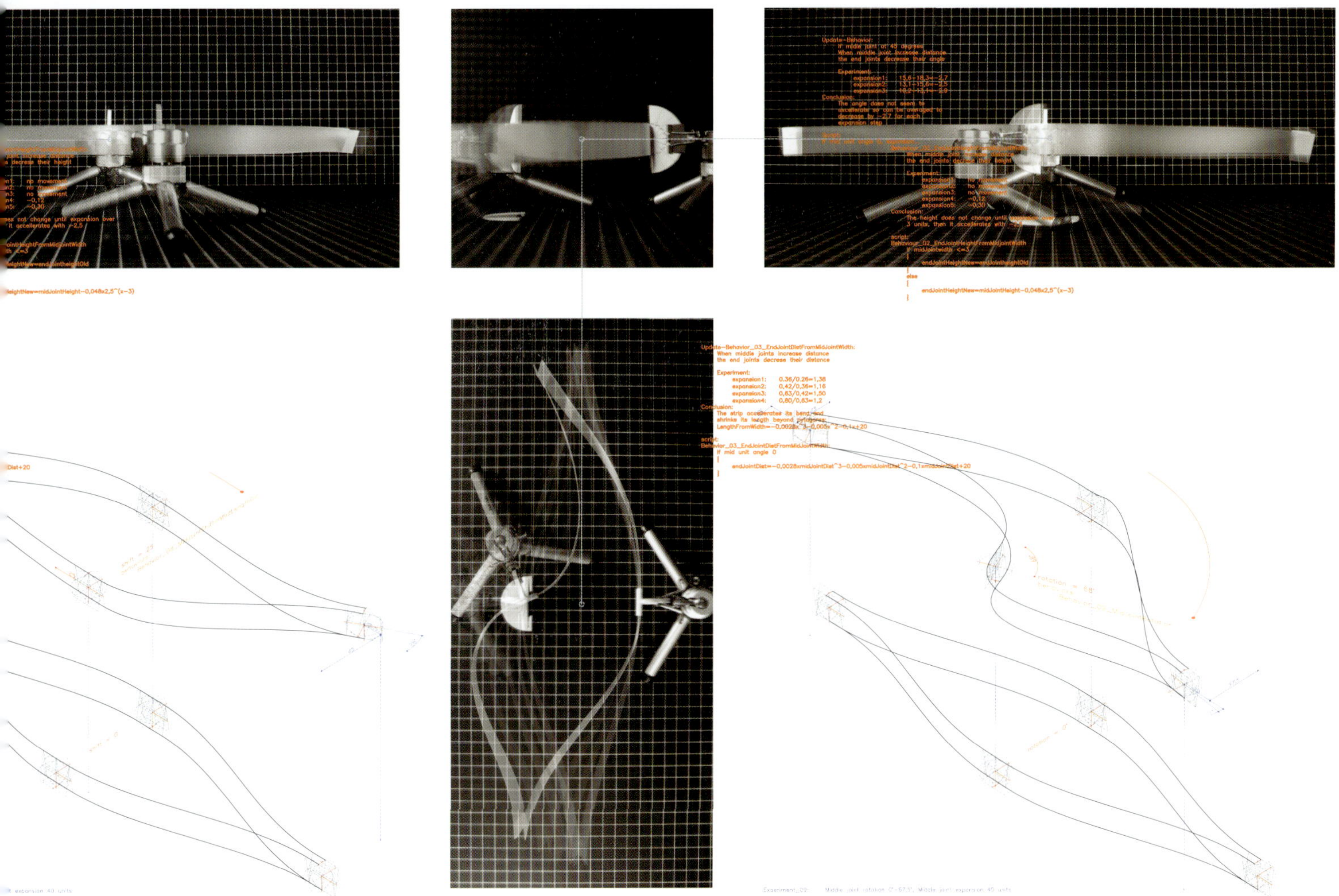

3.3.3 Experiment_09:
Middle joint rotation 0°-67,5°, Middle joint expansion 40 units

Rotating the middle joint shows a much larger impact on the geometry than the shift. It also triggers an asymmetrical form, which can be seen between the larger structure arches on the physical model. These points are exciting for form generation, but potential failure points and important to control by minimum and maximum values to maintain the larger structure.

3.3.1-3 Analysing the strip model
Load deformation experiments

To find the forms and scales achievable with this structure, I went through a series of load deformation experiments. These are the three most important ones. You can see that it is actually the expansion angle of the end joint which is making the structure arch. Then its the shift and rotation which manipulates the form

3.4 Strut expansion vs. pneumatic cushion actuators

Testing the theories

I expected some deviation between strut expansion and cushion expansion, the theory of the cushions supporting the strips without interfering with its formal language also needed to be tested.

In the current large scale model struts have been used to perform the expansion. The struts have a on or off performance, while the pneumatic cushions have a seamless transition towards maximum expansion. So there is expected some deviation in the behaviour when the actual pneumatic cushion actuators are implemented. The theory of how the cushions, cut by the natural form of the buckled strip, would support the strip formal language without too much influence also need to be tested. Another large scale model is needed. A prototype of the complete system.

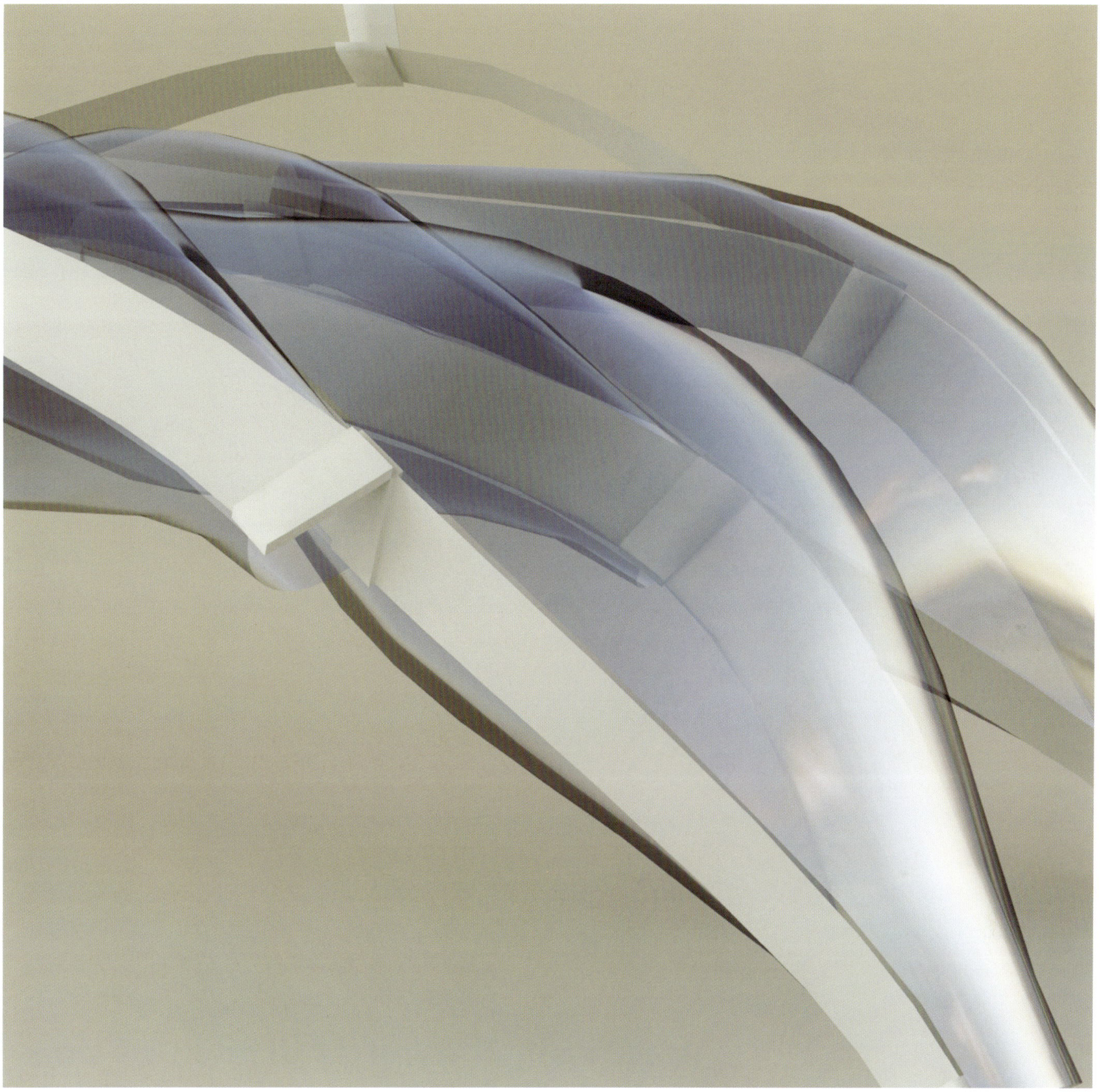

3.5 Physical pneumatic model animation
'Bird' view

So this is a prototype with all the components in place. Compared to the struts, the pneumatic cushions are now performing the expansion and acts as seamless actuators.

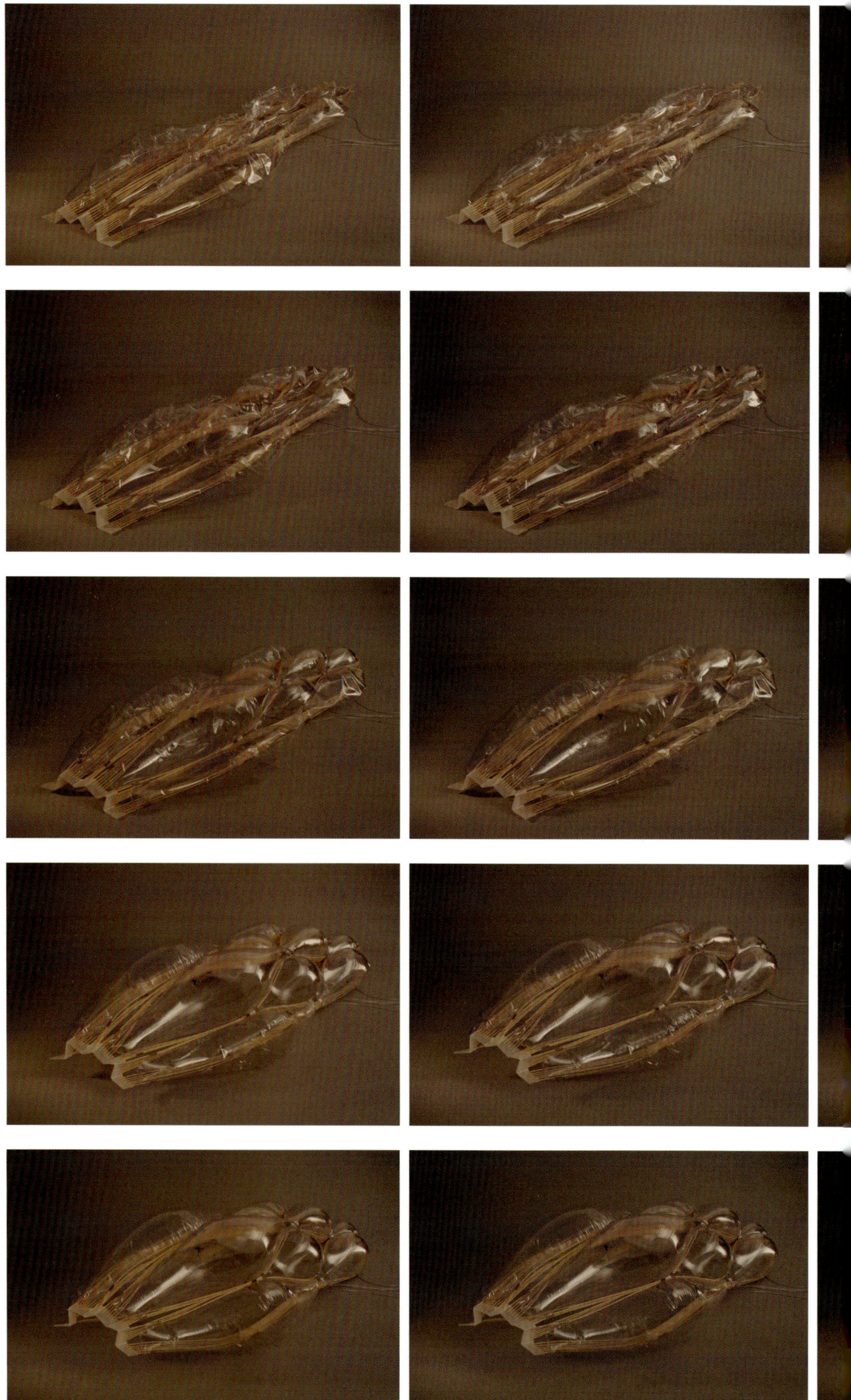

Photo Sue Barr

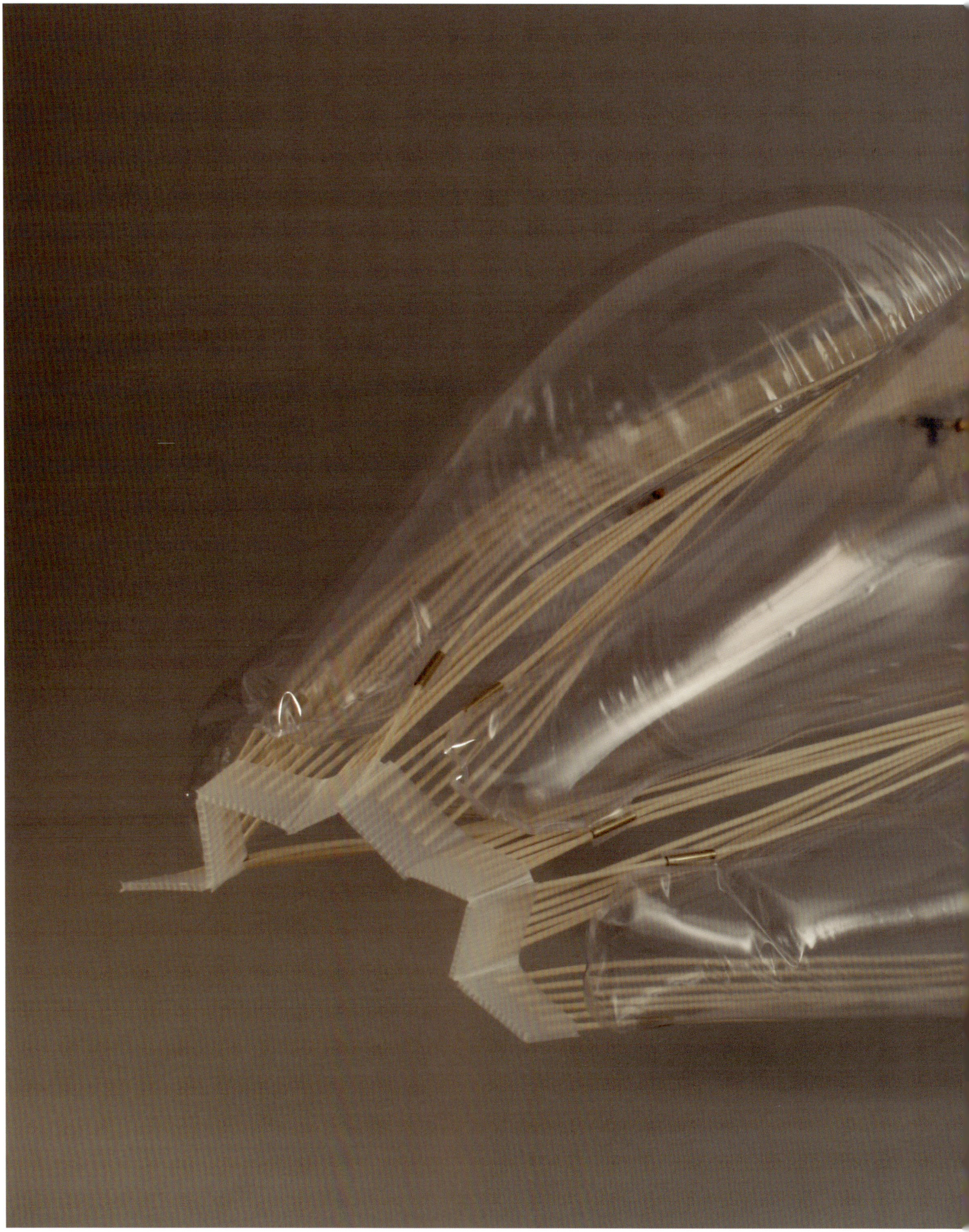

Cushion inflation makes the structure raise and span, while maintaining the strip behaviour. Photo Sue Barr

3.6 Analysing the pneumatic model
Impact of pneumatics to actuate and support the buckled strip behaviour

The cushions are cut from the form of the buckled strip, and seems to support the strip behaviour without too much influence. I would in that sense claim this experiment a success. This strengthens the claim of using pneumatics to support a structure, making it achieve larger scales.

In the animation you can see how the structure is sort of breathing from reacting to gas release.

Photo Sue Barr

4. Materialization, site context and predicted impact

4.1 In terms of materializing the project I looked a polyethylene, used in stratospheric balloons, for the cushions, and fibre reinforced acrylic for the algae tubes. Both materials can take the cold climate, while maintain translucency and flexibility.

4.2 I compared it to the material properties with the Eden project, and here is a load deflection graph on fibre reinforced tubes to find the force needed to actuate the structure.

4.3 In Khanty-Mansiysk I met the head of the chemistry department, who claimed to have produced several construction materials from pete-land, Including different types of plastics. This means that I could be able to find all the materials for the project locally.

4.4 The modules would be transported to site by snow mobiles, and possible to assemble for two people. The joint is designed as a light weight frame system, with two screws securing 8 tubes to minimize work labour and parts that could fail in the climate.

4.5 The ground connection is made by three steps.
Step 1: The lightweight structure arches and slides on top of the frozen ground as a reaction to gas release, until it settles towards optimal form.
Step 2: The increased temperature within the enclosure makes the frozen ground melt and vaporize. As it hits the surface it condenses and flows down to the drainage channel at the ground connection module. The raised structure, the stored drainage water, and transportation on the module path all add weight to the module. This weight and its role to soak the surrounding ground, makes it sink and ground itself, anchoring the structure to a stable edge condition.
Step 3: As the structure moves on to catch up with the moving melting edge, the lower part of the ground connection module is left behind. Leaving a developed water distribution system, fertile soil, and transportation arteries for future inhabitation.

4.6.1-7 I then contacted the senior agronomist at the Eden project, Professor Jim Dimmock, and together we developed a proposed agricultural time cycle. It's a succession of stronger crops, developing the ground. In year 6 we can grow wheat, by this time the ground should be able to grow almost anything. We also plant apple trees in preparation for next year's structure movement.
Year 7, the thawing edge indicated by the stippled line, has now moved beyond the enclosure, and as a result the structure is beginning to drop. The climate has also made its tow, some of the modules might need maintenance. So at wintertime the structure is moved to catch up with the thawing edge and begins a new time cycle. What's left behind is a field of apple trees. Their weakness is that the blossoming is killed by early frost, but if they are artificially planted, they would grow strong enough to take the winter while produce crops of fruit during the summer. So the vegetation, triggered by the structure, continues to grow after the structure moves on.

4.7 The city Khanty-Mansiysk is the most influential oil region in Russia, containing 230 of Russia's 300 billion tons total oil reserves. The region has the 20 last years tripled its workforce and are now over 1 000 000 people with primary knowledge within the oil industry. As the oil resources are declining, and are predicted to run out within the next 20 years, the proposal offers a future within a redefined industry, utilising their work knowledge, while taking part in reducing the carbon footprint and becoming more self sufficient.

4.8 The red arrows indicate the oil production in the county in 2008, while the green arrows coming in to the city shows the imported goods creating a huge carbon foot print.

4.9 In 2038 you can see the proposal taping into the thawing edge, leaving a trace every 7th year of vegetation and inhabitation. You still see the red arrows indicating the oil industry, but at the same time you see the green arrows from the agricultural crops. From the diagram you can see that the city is far more self sufficient now, and could even export some goods, making a huge reduction on their carbon footprint.

4.10 What I like about the project is the way it takes something perceived negative and tries to turn it into an opportunity. If agriculture in the rain forest leaves a trace of ash in fertile land, the proposal leaves a trace of vegetation and inhabitation in undeveloped land.
What is new is the way it challenges the use of double curvature and asymmetrical optimised forms within pneumatic structures. This is a symbiosis system where the pneumatic is used to actuate and support the main language making it challenge far larger scales.

4.11 Here you can see a comparison between the structural models, the first one is double curvature strut expansion, the second single curvature pneumatic expansion, while the final one which was made for the technical studies thesis challenges the double curvature with pneumatic expansion.
It spanned 4 meters and used 500 meters of acrylic tube to create the strip language, and was exhibited under the end of year show at the Architectural Association 2008.

4.1

4.7

4.6.1-2

4.10

4.11

4.11

4.1 Materialising the proposal
Cold climate, lightweight, translucency

In terms of materializing the project I looked a polyethylene, used in stratospheric balloons, for the cushions, and fibre reinforced acrylic for the algae tubes. Both materials can take the cold climate, while maintain translucency and flexibility.

I began by looking at the requirements of the bi-ofence algae farm and the pneumatic cushions used in the Eden project. However additional consideration has to be taken into the colder climate, low cost manufacturing to cover larger areas, while aiming towards modular light weight components for mobility in difficult terrain.

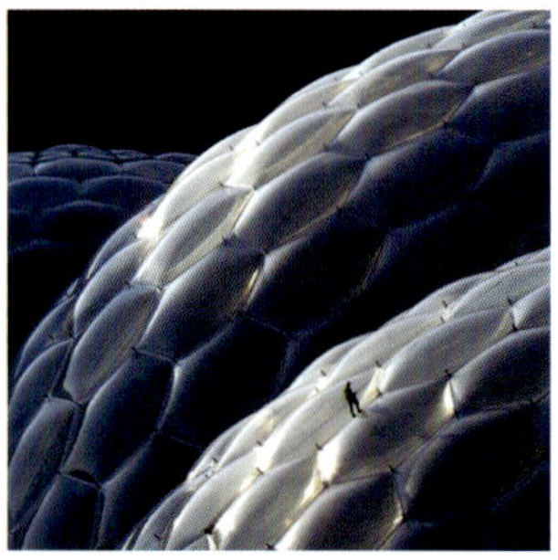

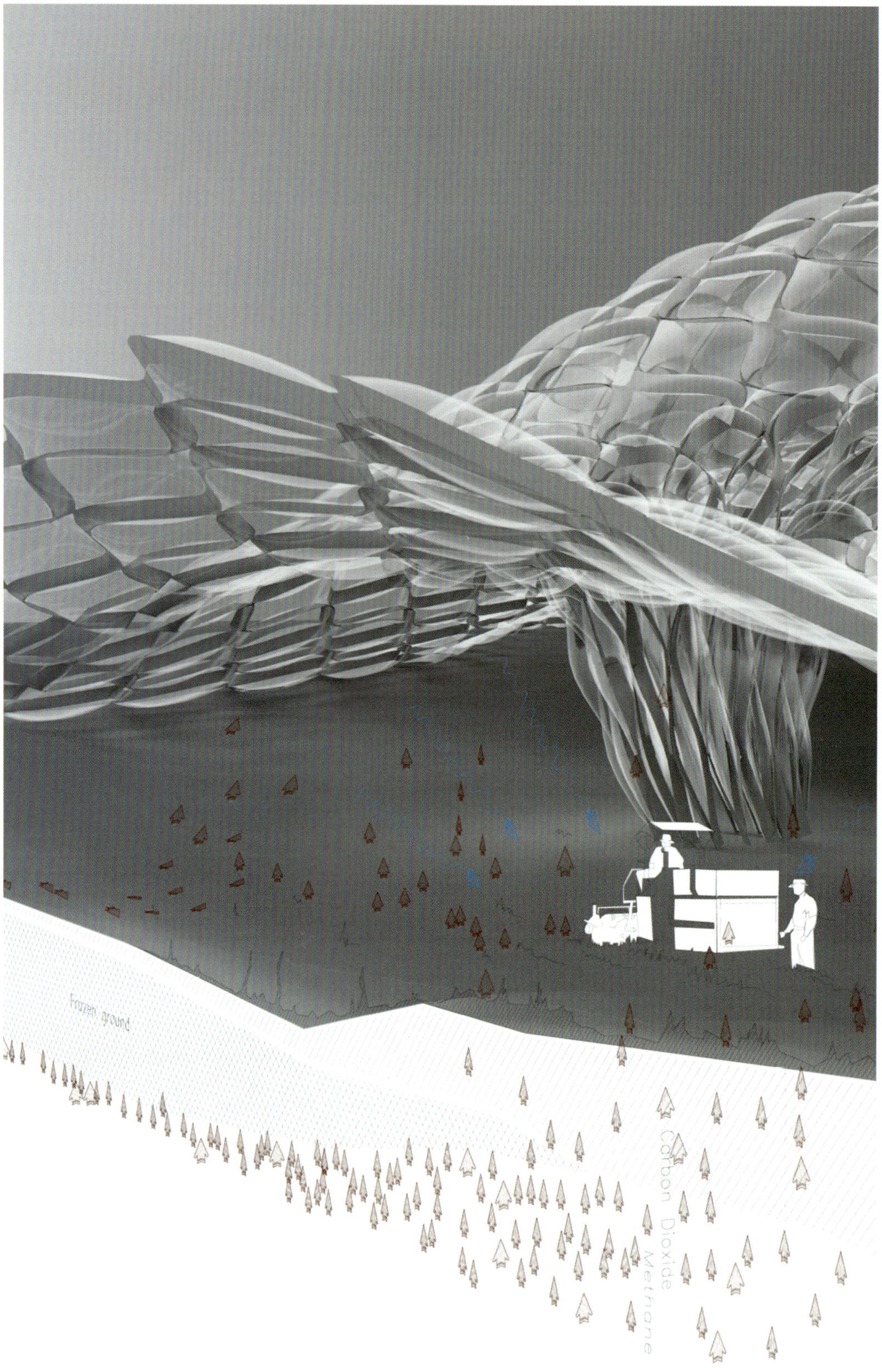

Left
Eden self supported with cushions in place r46
Eden construction supported by scaffolding r47

Right
Polyethylene used in stratospheric balloon r40
Silicone rubber r48
Trek oclv carbon fibre bicycle frame r49
Specimen glass in casted acrylic r50
Self-erecting Nasa space components r51
Three dimensional weaving machine r51

Cushion skin:
Polyethylene C2H4

Polyethylene is used in stratospheric balloons, making it suitable for the cold climate, the pressure, while maintaining high translucency.

Algae tubes 1:
Acrylic glass

Acrylic glass (polymethyl methacrylate)
(C5O2H8)n
Density 1.19 g/cm^3
Melting point 130-140 °C

Glass
SiO2
Density 2.52 g/cm^3
Melting point 1040 °C

Glass is heavier than acrylic, but can from increased stiffness produce longer spans with less material. However acrylic is chosen as it is more flexible to adopt to the strip buckling behaviour.

Dampening between tubes and joint:
Silicone [R2SiO]n

To avoid direct contact and sheer, softer members are placed between the joints and the tubes. Silicone might be suitable for this task, due to greater friction, and little reaction to temperature change

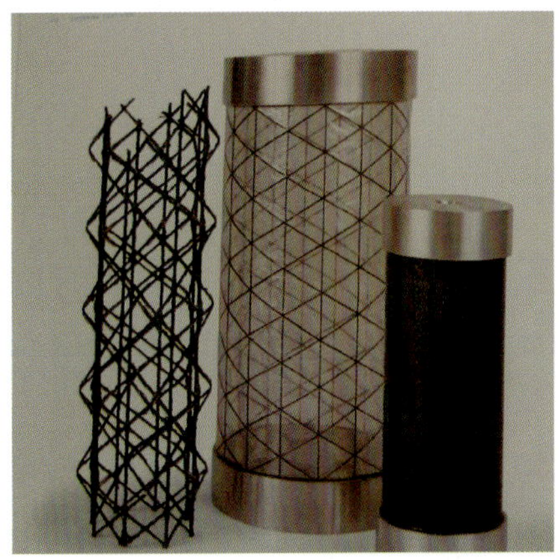

Algae tubes 2:
Acrylic tubes reinforced with carbon or glass fibre

Reinforcing the tubes would increase their stiffness, lower the plasticity, and risk of breaking.

Joint:
Reinforced thermoplastics

Carbon or glass fibre reinforced thermoplastics would be lightweight and strong for the rigid parts of the joint.

Algae tubes 3:
Production of acrylic tubes reinforced with carbon or glass fibre

Seamless circular non-crimp ultrasonic welded tube. by Pfaff Industriemaschinen AG 2001.

A similar method might be used to weave and weld intersections of reinforcement fibres into the production of the acrylic tubes.

Materials and weight

Hexagon tube

Material: Carbon Steel, weight 7.8 g/cm3
Quantity: 200mm outer, 190mm inner= 46kg/m
x 34540mm (11m x 3.14)= 1597kg

Substructure tube Domes

Material: Carbon Steel
Quantity: 200mm outer, 190mm inner= 46kg/m
x 34540mm (11m x 3.14)= 1597kg

Substructure tube Triangulation Offset

Material: Carbon Steel
Quantity: 100mm outer, 90mm inner= 11kg/m
x 60000mm (11m x 3.14 x2)= 693kg
Frame total = 3887kg

Joint

Frame

Material: Carbon Steel
Quantity: 10kg

Tube dampening

Material: Silicone
Quantity: 1kg

Joint bolts

Material: Carbon Steel
Quantity: 0.5kg
Joint total = 11.5 kg x 18 joints = 207kg

Cushion

Material: Efte 3.5kg/m2
Quantity: 34.5m2 x 2
Total = 241kg

Total:
4335kg /34.5m2 = 125.6kg
Module weight per m2 = 125.6kg/m2

Materials and weight

Tube

Material: Reinforced acrylic, weight 0.9g/cm3
Quantity: 80mm outer, 72mm inner = 1.1 kg/m
x 8 tubes x 16000mm x2 sides= 281kg

Joint

Frame

Material: Thermoplast weight 0.9g/cm3
Quantity: 1kg x 3 frames = 3 kg

Tube dampening

Material: Silicone
Quantity: 1 kg

Tube lock

Material: Thermoplast
Quantity: 0.5kg x 3 locks = 1.5 kg

Joint pivot cylinder

Material: Carbon Steel
Quantity: 2 kg
Joint total = 7.5 kg x 2 joints = 13 kg

Cushion

Material: Polyethylene 4.7 kg/m2 (5mm thickness)
Quantity: 30m2 x 2 = 282 kg

Assembled module dry

576 kg / 30m2 = 19.2 kg

Dry module weight per m2 = 19.2 kg/m2

Assembled module filled with algae

576 kg / 30m2 = 19.2 kg
4kg water per tube per meter x 16 meters x 16 tubes = 1024kg
1024kg /30m2 = 34 kg

Filled module weight per m2 = 53.3kg/m2

6.2 Material properties
Comparison to the Eden project and full scale deflection force

I compared it to the material properties with the Eden project, and here is a load deflection graph on fibre reinforced tubes to find the force needed to actuate the structure.

The closest deflection comparison I could find to the proposed tubing was BAC carbon fibre producer. This is one of the strongest and stiffest production techniques, making parts for cars and air planes.
This graph will have to serve as the maximum stiffness the project tubes can obtain, by reducing the carbon fibre reinforcement fibres, the flexibility and elastic deformation increases.

Material values are estimated from descriptions of the Eden Project by Grimshaw Architects web site (r51) and visual dimensions from images (r52), and Wikipedia values on material properties (r53)

"The biomes are an exercise in efficiency, both of space and of material. Structurally, each dome is a space frame reliant on two layers. The first, an icosahedral geodesic skin, is made up of hexagonal modules that range in diameter from 5m to 11m. Each comprises six straight, compressive, galvanised steel tubes that are light, relatively small and easily transportable. This makes it possible for each hexagon to be pre-assembled on the ground before it is craned into position and simply bolted to its neighbour by a standard cast steel node.

The primary layer is joined to a secondary one by diagonal Circular Hollow Section members at the node points. Structural stability is guaranteed by the shell action of the intersecting domes, that is, the meeting of inner and outer structural members to form pinned connections. These are anchored to reinforced concrete strip foundations at the perimeter." r51

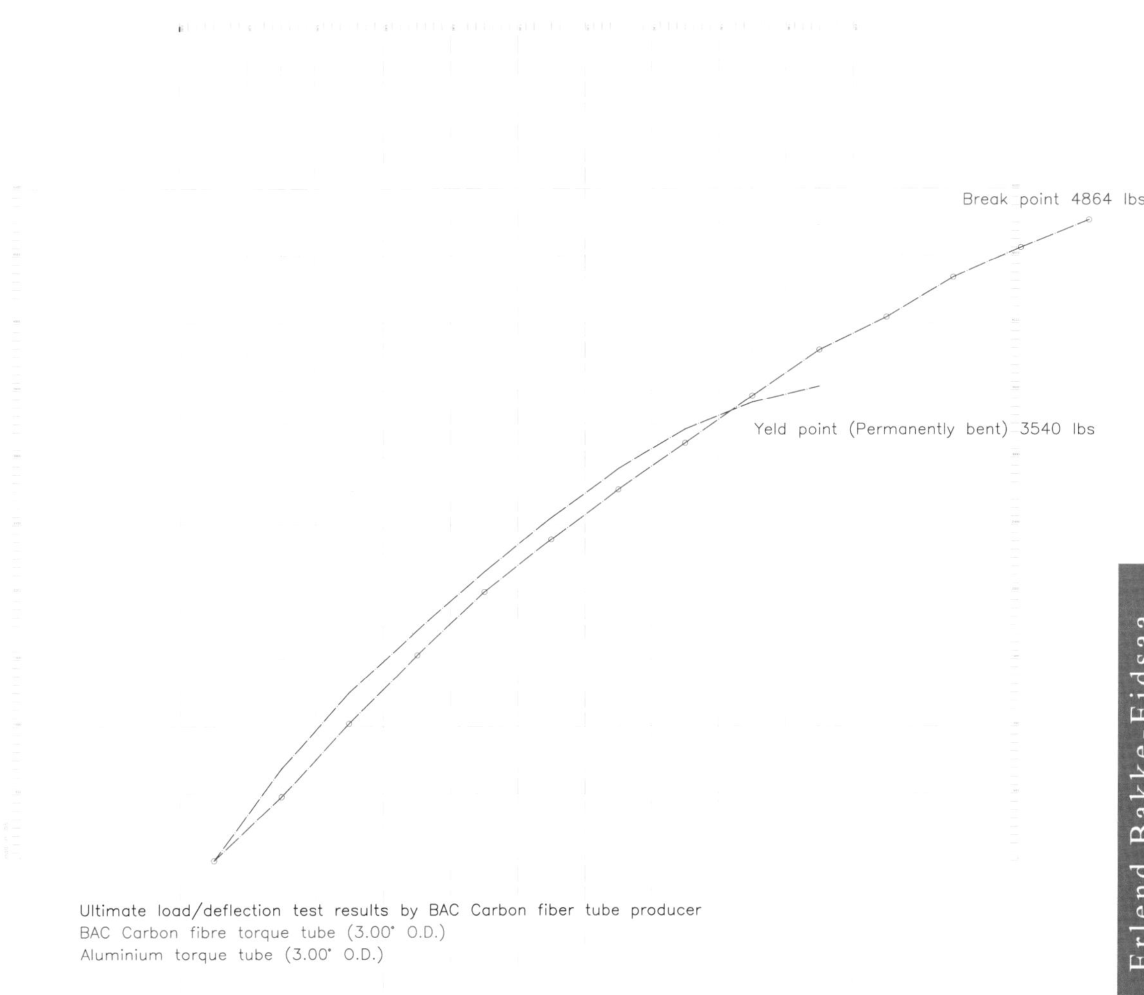

Left
Eden Project modules on a human scale r52
Pneumatic prototype, Photo Sue Barr

Right
Specimen glass in casted acrylic r50
Self-erecting Nasa space components r51
Three dimensional weaving machine r51
Machine measuring load deflection in tubes r55
Load deflection analysis on carbon fibre tubes r54

Erlend Bakke-Eidsaa

Professor Alexander Galochkin of Ugra University
Concrete replacement by Alexander Galochkin
Insulation replacement by Alexander Galochkin
Thermoplastics 'Saris Bones 2' r56
The Eden Project r46

4.3 Local resources

Chemist material development from pete-land

In Khanty-Mansiysk I met the head of the chemistry department, who claimed to have produced several construction materials from pete-land, including different types of plastics. This means that I could be able to find all the materials for the project locally.

Professor Alexander Galochkin is the dean of geology faculty at Yugra State University and head of the chemistry department. He referred to successful attempts to produce several construction materials from peat. Plastics including thermoplastics, a material with the same insulation properties as wood, and even a material which is close to having the same glue and compression properties as concrete.

His claims would have to be confirmed by further data, but it could mean I would be able to produce all materials for the module locally.

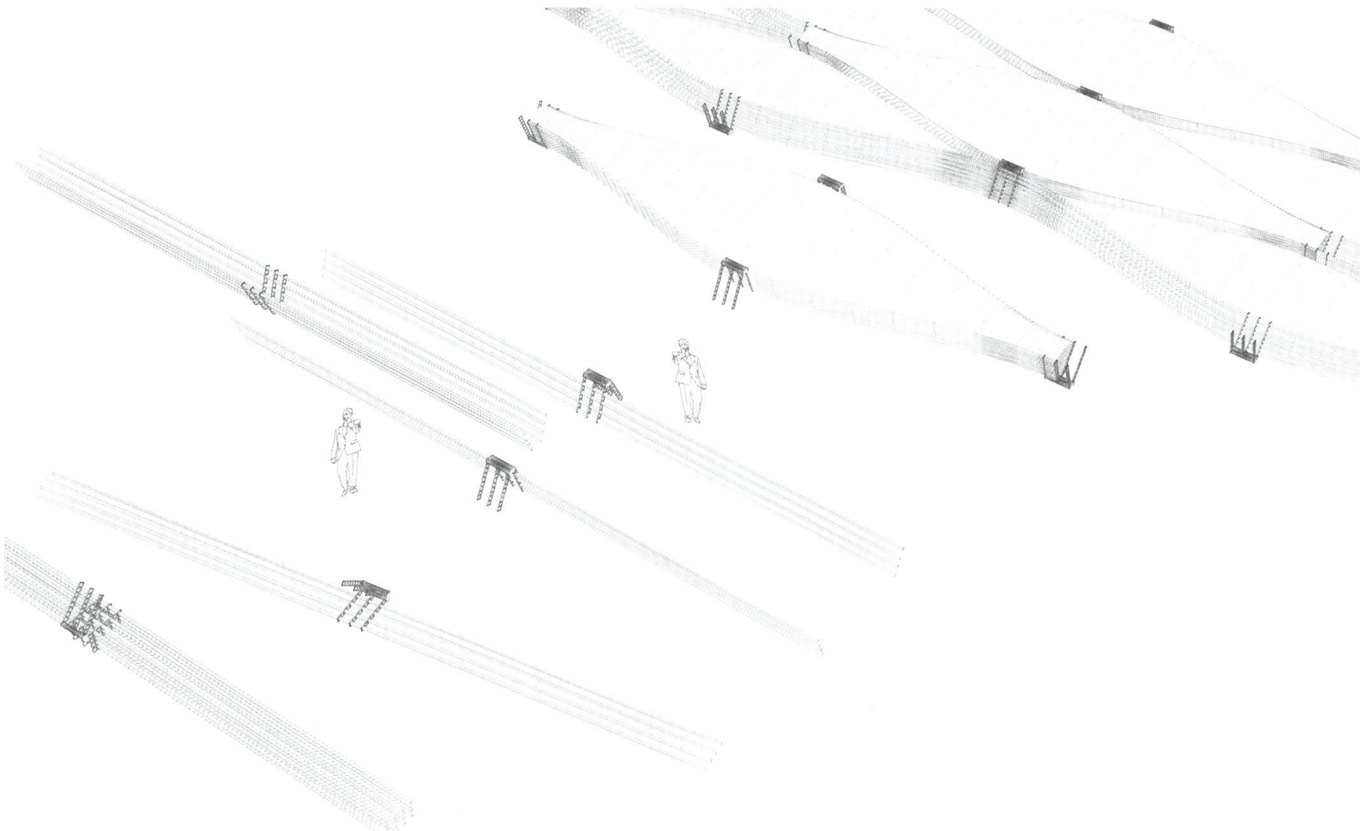

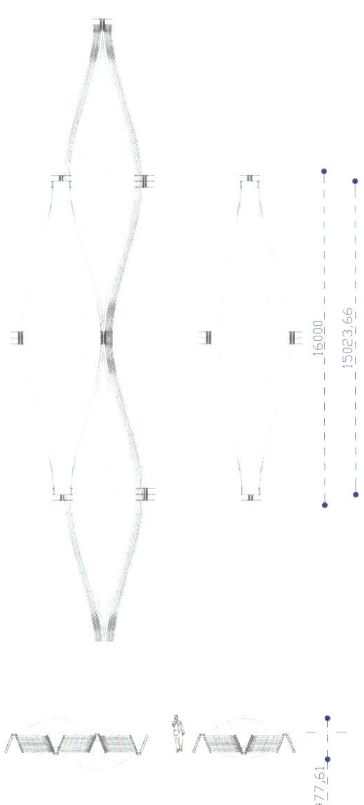

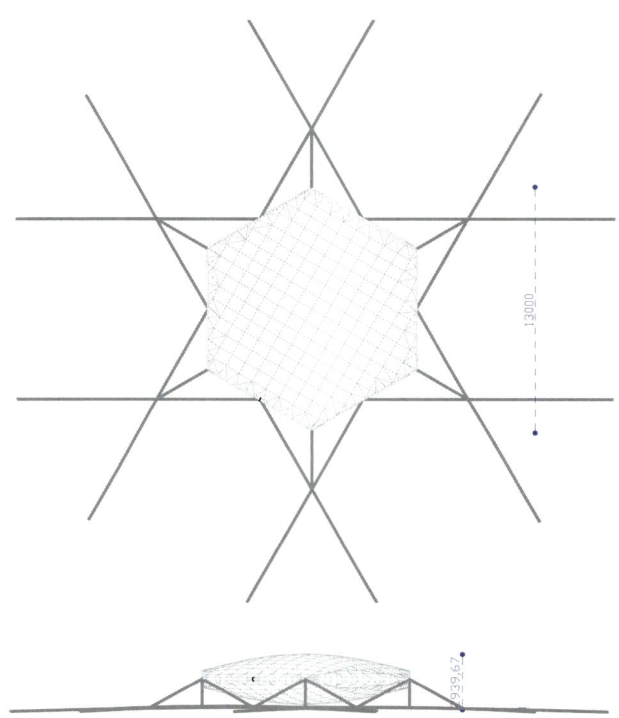

4.4 Transport and assembly

Snowmobiles and minimal parts

The modules would be transported to site by snow mobiles, and possible to assemble for two people. The joint is designed as a light weight frame system, with two screws securing 8 tubes to minimize work labour and parts that could fail in the climate.

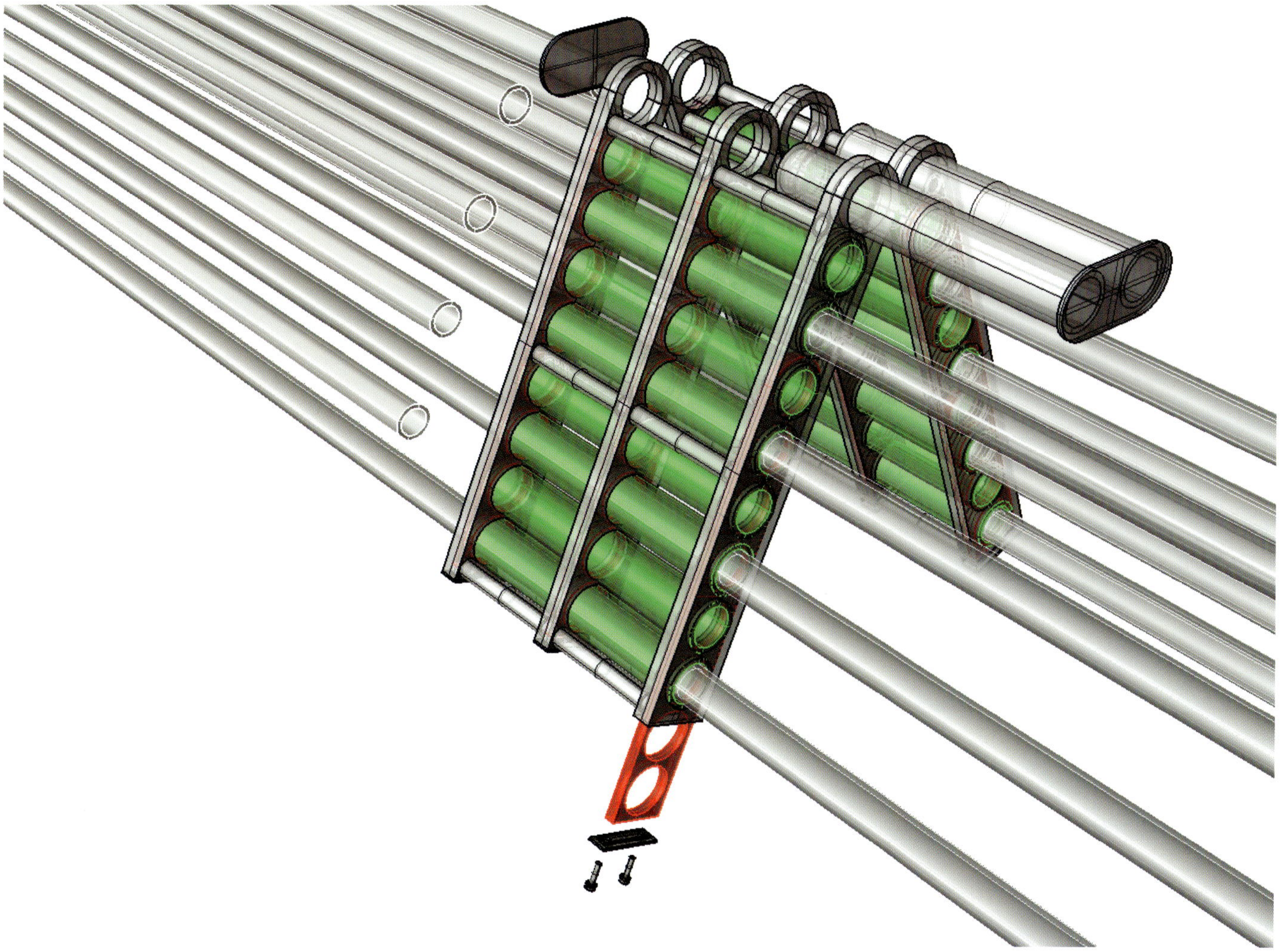

Transport
The modules measures 16 x 1 meters and is transported to site by river, train or trucks. Then to site by snowmobiles and possible to assemble by two people.

Module assembly
Light weight frame, with 2 screws to secure 8 tubes. 6 screws total on a joint to reduce labour and parts that can fail in the colder climate.

Left
The 'Taurus' snowmobile pulling heavy loads r57

Right
Silicone rubber r48
Specimen glass in casted acrylic r50
Self-erecting Nasa space components r51
Trek oclv carbon fibre bicycle frame r49
Polyethylene used in stratospheric balloon r40
Three dimensional weaving machine r51

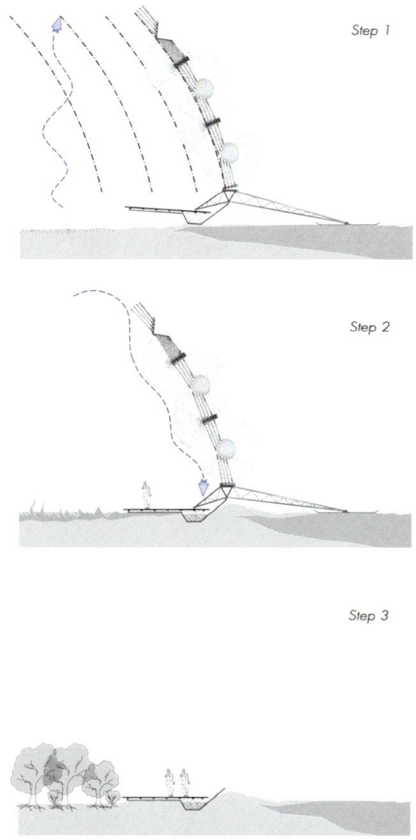

4.5 Ground connection from and drainage and water distribution

Utilising drainage to anchor the structure

The ground connection is made by three steps.

Step 1: The lightweight structure arches and slides on top of the frozen ground as a reaction to gas release, until it settles towards optimal form.

Step 2: The increased temperature within the enclosure makes the frozen ground melt and vaporize. As it hits the surface it condenses and flows down to the drainage channel at the ground connection module. The raised structure, the stored drainage water, and transportation on the module path all add weight to the module. This weight and its role to soak the surrounding ground, makes it sink and ground itself, anchoring the structure to a stable edge condition.

Step 3: As the structure moves on to catch up with the moving melting edge, the lower part of the ground connection module is left behind. Leaving a developed water distribution system, fertile soil, and transportation arteries for future inhabitation.

Weight on ground module: 50kg per module x 60 modules in the enclosure arch = 1500kg on each edge of the arch.

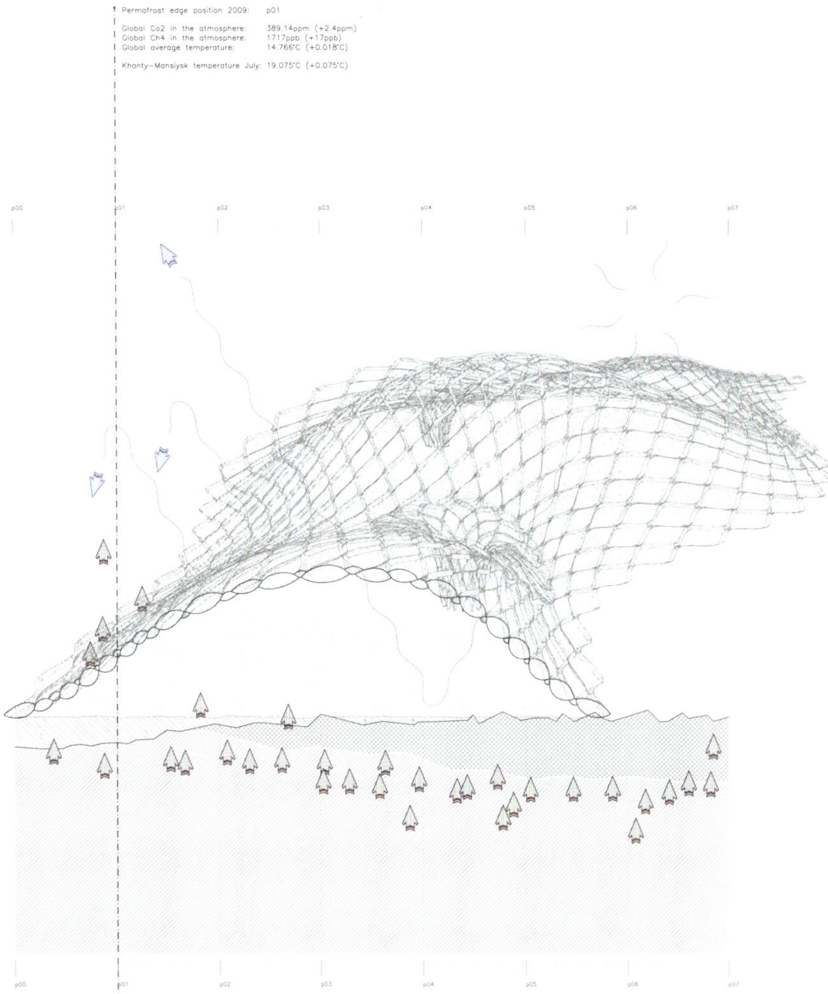

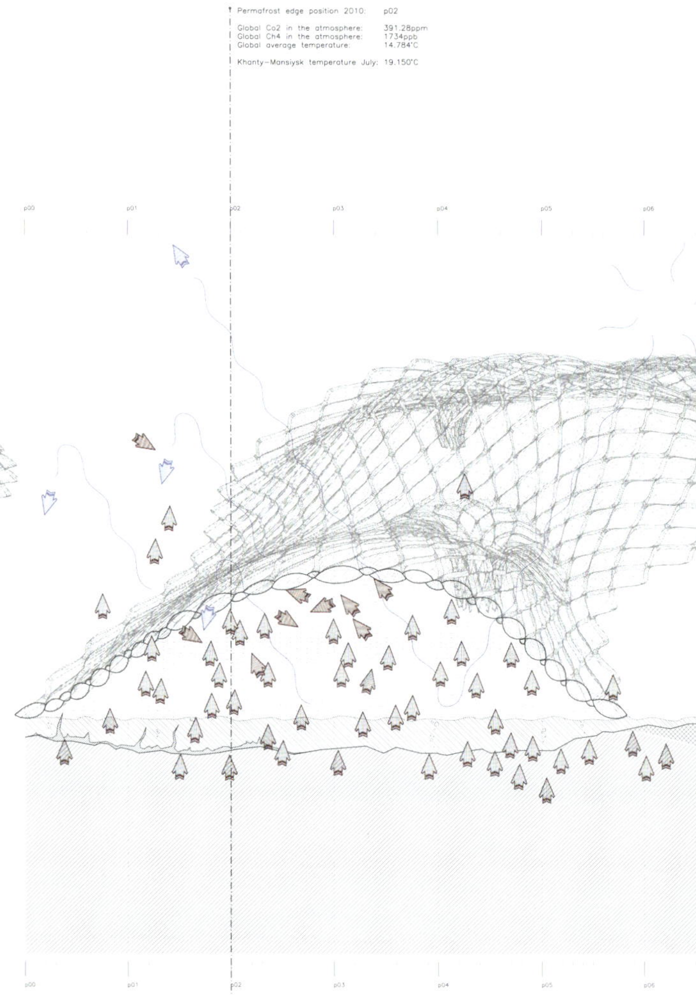

4.6.1 Predicted strategy year 1

Land
Frozen, hard and dry

Product

4.6.2 Predicted strategy year 2

Land
The melted ice has turned the ground to wetlands, while accelerating the methane release.

Product
Bio fuel and fertiliser

Frozen tundra r58
Melted tundra r58
Laboratory algae production r32
Biodiesel produced from algae r59
Barley crops r60
Pig, example of live stock feeding on barley r61
Leffe beer made from barley r62
Rye crops r63
Bread made from rye r64

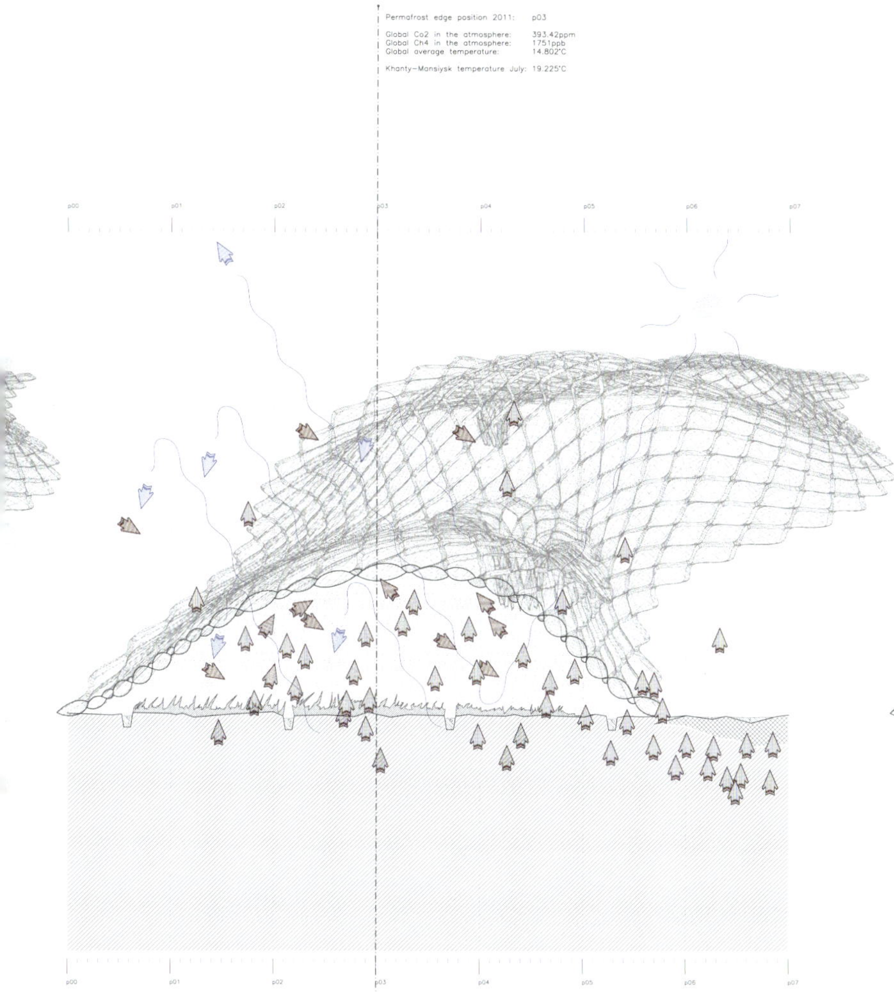

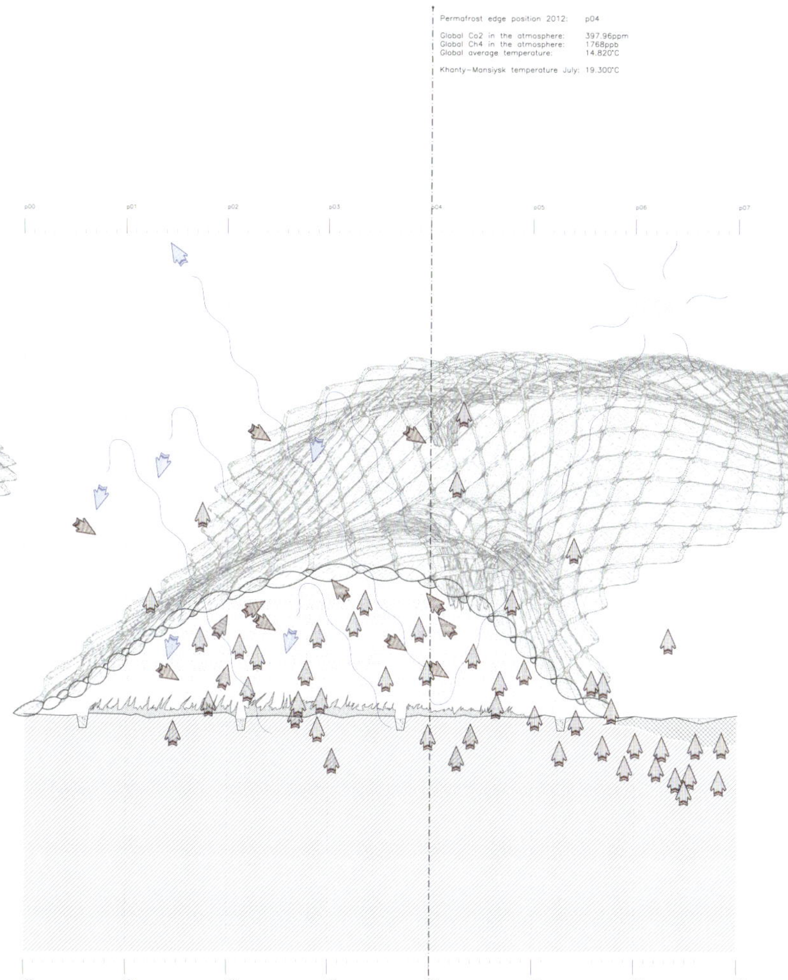

4.6.3 Predicted strategy year 3

Land
Drainage channels has been implemented, the ground has began to dry up and becomes increasingly fertile from the added fertiliser from last years production.

Product
Barley, one of the strongest cereals. Dr Dimmock is currently trying to make it grow in sea water. Mainly used for live stock food and beer production

4.6.4 Predicted strategy year 4

Land
Increasingly fertile, easier to plow after last years crop.

Product
Rye. Very strong cereal, grown in northern Europe due to its robustness.
Mainly used in bread production.

4.6.1-7 Proposed agricultural time cycle
Page 1

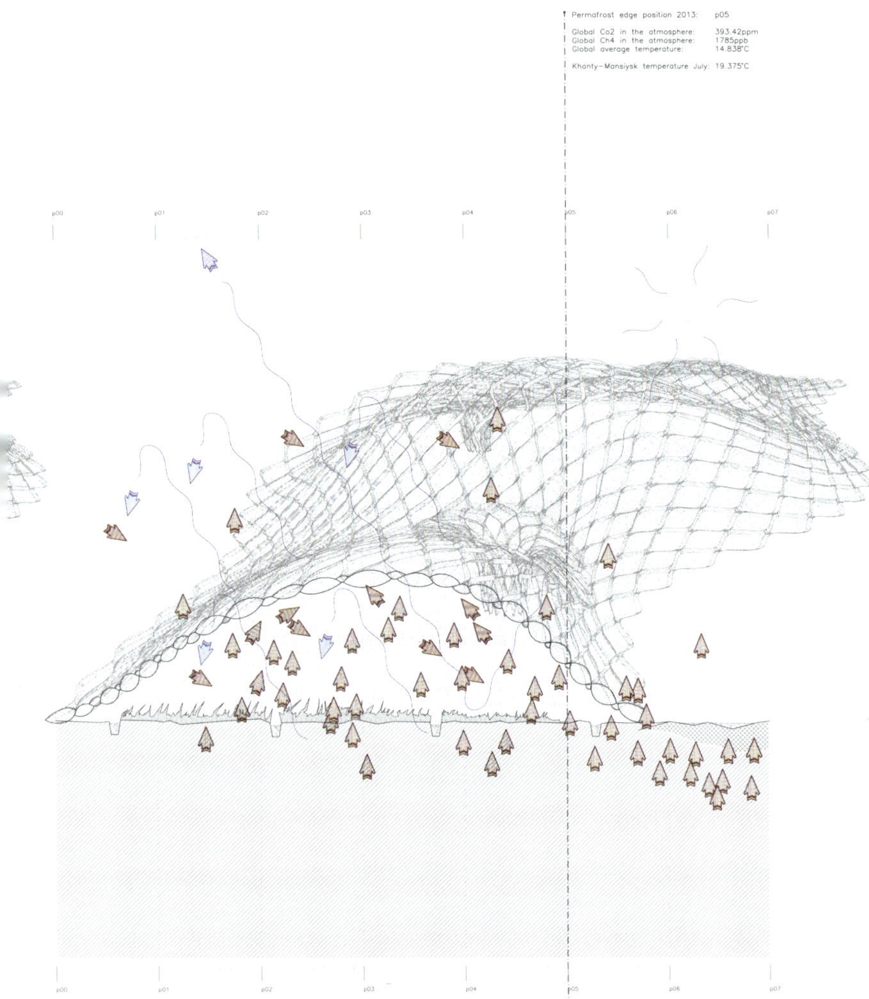

4.6.5 Predicted strategy year 5

Land
Like in British agriculture, two years of cereal crop needs to be broken up with a non cereal crop to maintain soil fertility.

Product
Fisher beam.

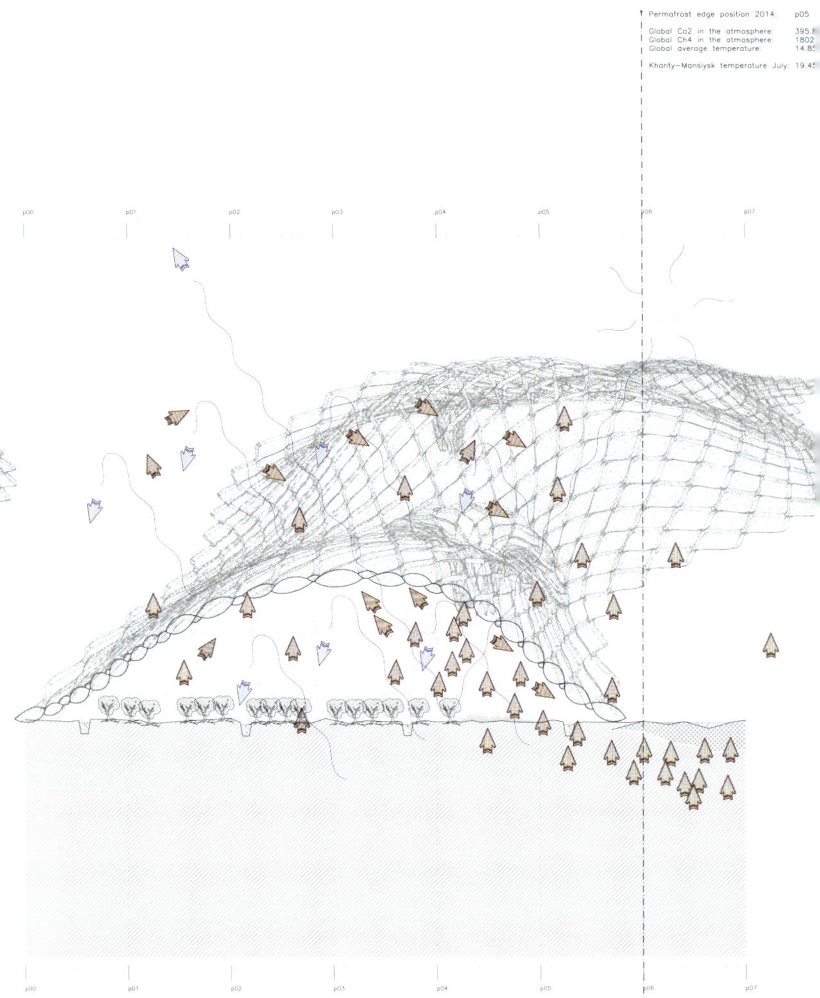

4.6.6 Predicted strategy year 6

Land
After the fisher beam crop, the soil should be close to optimal for agriculture in this area.

Product
Wheat, not the strongest cereal, but the soil should now handle it. Wheat is one of the most used cereals in baking and dairy products.

On the other hand is the gas beginning to diminish, which is slowing down co2 levels that boosts vegetation, as well as the algae farm bio fuel production, indicated by pressure loss in structure.

The climate has also made its impact and some of the modules might need repairs.

To prepare for moving the structure, apple trees are artificially planted to grow strong enough over the year to handle the exterior climate.

Wheat crop r65
Dairy products made from wheat r66
Apple trees surviving the cold climate r67
Apple trees producing fruit during the summer r68
Kid enjoying the fruits of the trees r69
Professor Jim Dimmock from the Eden Project r70

Through discussing the project and site conditions with the Senior Agronomist at the Eden project, Dr. Jim Dimmock,, we developed this agricultural time cycle strategy as a starting point. Agricultural strategies are normally achieved through trying several crops, to find the one that thrives the best. But in theory, given the ground conditions described by the casus geologist team, and a developed drainage pattern, did he not find any reason why this strategy should not work.
He added that drained pete land is one of the most fertile soils we have in the U.K.

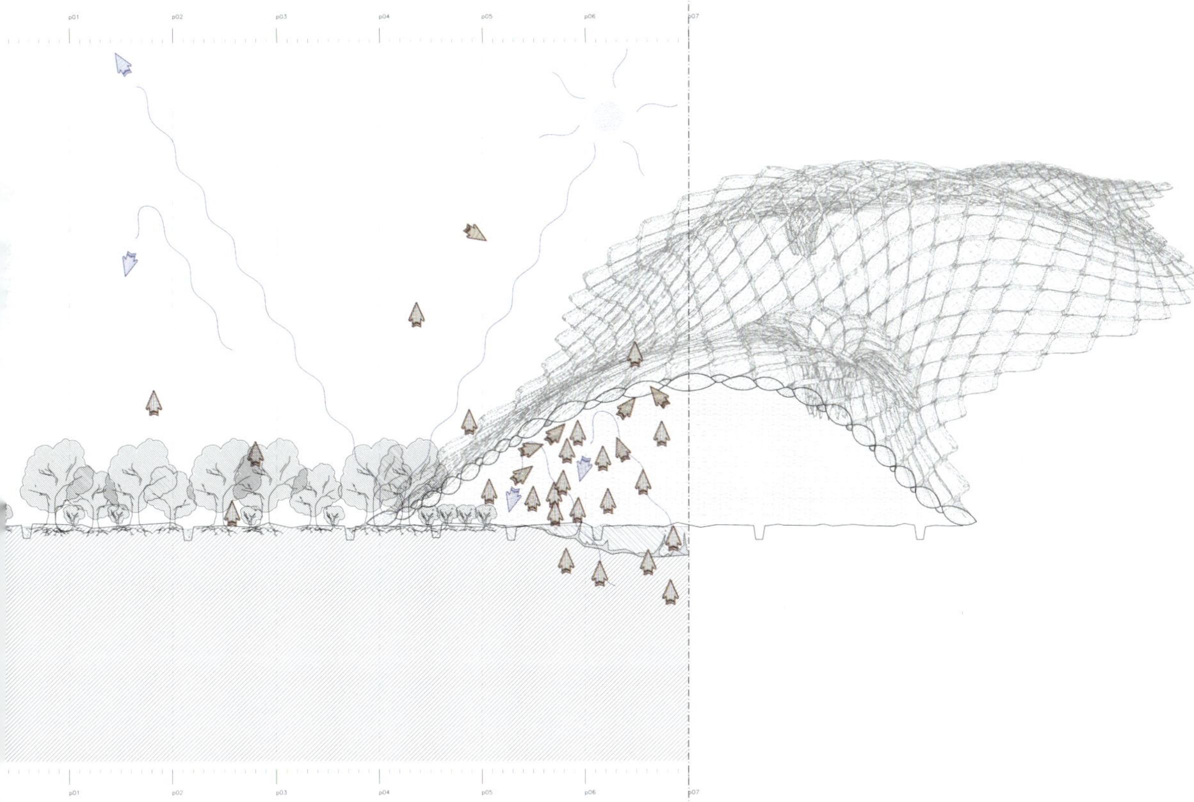

4.6.7 Predicted strategy year 7

Land
The structure is moved to catch up with the thawing edge, and begins a new ground development cycle.

The melted ice has turned the ground to wetlands, while accelerating the methane release.

Product
Bio fuel, fertiliser and apple trees

The apple trees that was left when the structure moved, continues to grow, producing fruit and absorbing gas from the atmosphere.

Apple trees are very strong trees, their weakness is that they blossom early on the year which restrains them in areas with early frost such as Siberia. However, if they are helped artificially in the planting process, they grow strong enough to take the winters, and produce fruit during the summer.

4.6.1-7 Proposed agricultural time cycle
Page 2

I then contacted the senior agronomist at the Eden project, Professor Jim Dimmock, and together we developed a proposed agricultural time cycle. It's a succession of stronger crops, developing the ground. In year 6 we can grow wheat, by this time the ground should be able to grow almost anything. We also plant apple trees in preparation for next year's structure movement.
Year 7, the thawing edge indicated by the stippled line, has now moved beyond the enclosure, and as a result the structure is beginning to drop. The climate has also made its tow, some of the modules might need maintenance. So at wintertime the structure is moved to catch up with the thawing edge and begins a new time cycle. What's left behind is a field of apple trees. Their weakness is that the blossoming is killed by early frost, but if they are artificially planted, they would grow strong enough to take the winter while produce crops of fruit during the summer. So the vegetation, triggered by the structure, continues to grow after the structure moves on.

4.7 Re-employing work force and city context

Labour force, infrastructure and way of life of the local industry

The city Khanty-mansiysk is the most influential oil region in Russia, containing 230 of Russia's 300 billion tons total oil reserves. The region has the 20 last years tripled its workforce and are now over 1 000 000 people with primary knowledge within the oil industry. As the oil resources are declining, and are predicted to run out within the next 20 years, the proposal offers a future within a redefined industry, utilising their work knowledge, while taking part in reducing the carbon footprint and becoming more self sufficient.

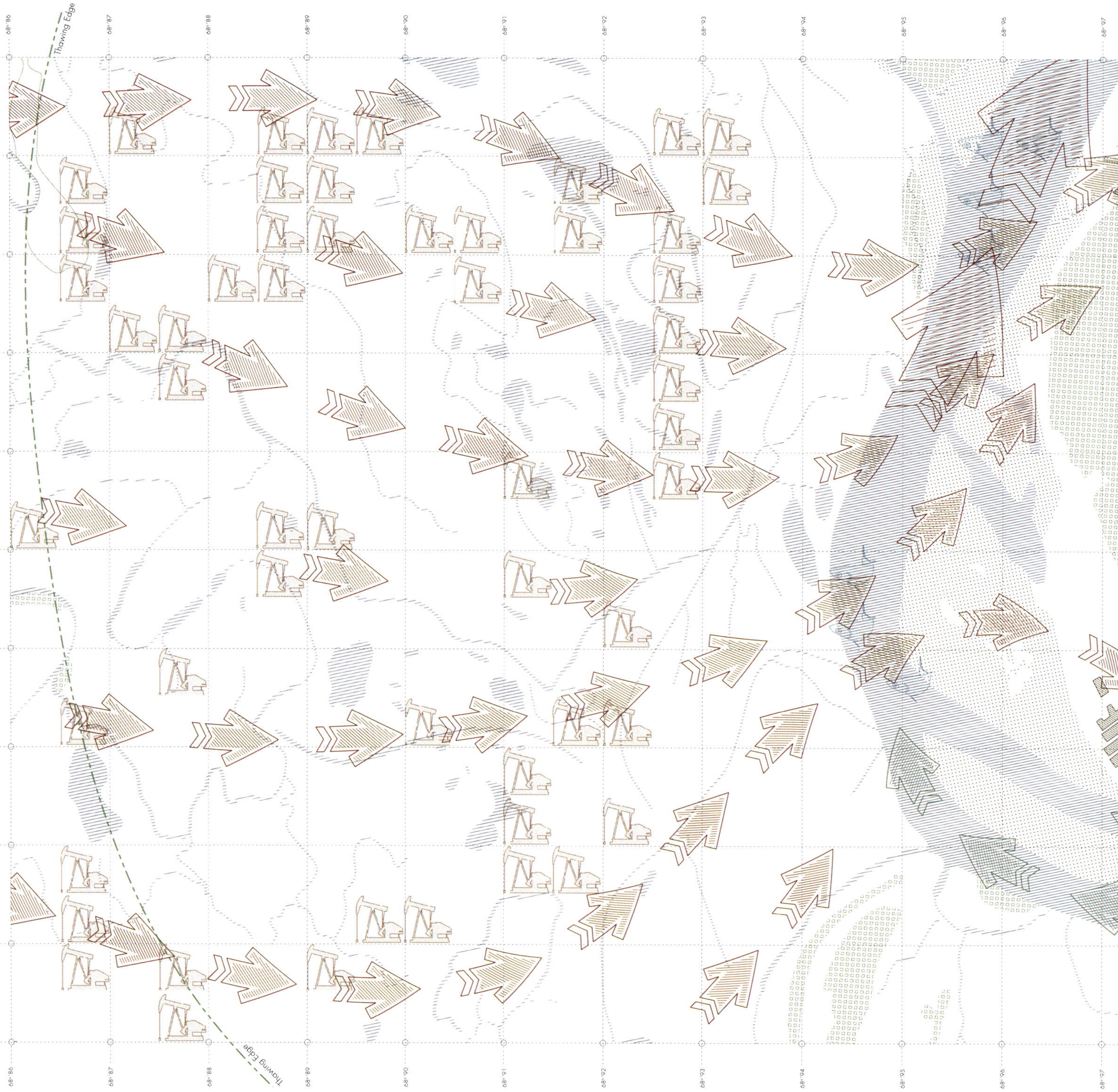

4.8 Current industry 2008

The red arrows indicate the oil production in the county in 2008, while the green arrows coming in to the city shows the imported goods creating a huge carbon foot print.

Large scale site coverage strategy.
Year 1 The project introduces 2-3 test enclosures.
Year 3 Attract further sponsors by referring to successful results of the test enclosures in terms of financial profit, workforce re-employment, and environmental impact. Implement further enclosures from new sponsor funds. (Site could sustain up to 100 new enclosures p.a)
Year 4 Another class of enclosures
Year 6 Attempt to move some of the test enclosures to see if the ground has developed enough to survive after this time span.
Year 7 If crops from the revealed soil continue to thrive, it will serve as proof of the proposed agricultural time span and that the system works. (Further clusters of enclosures will follow every year) If the soil improvements have not proved permanent, the remaining test units would be left longer to reach full conversion of the soil and refine the time span strategy for future enclosures to optimise productivity.

Oil reserve figures.
The oil figures below are often politically manipulated and considered national secrets. In Khanty-Mansiysk these numbers where confidently disputed, however the reservoirs are declining.

Summary of Reserve data as of 2007 (Russia)
Estimated reserves: 60 x 10^9 bbl. r71
Production rate: 9.5 x 10^6 bbl/d r72

17 years left with current production r73

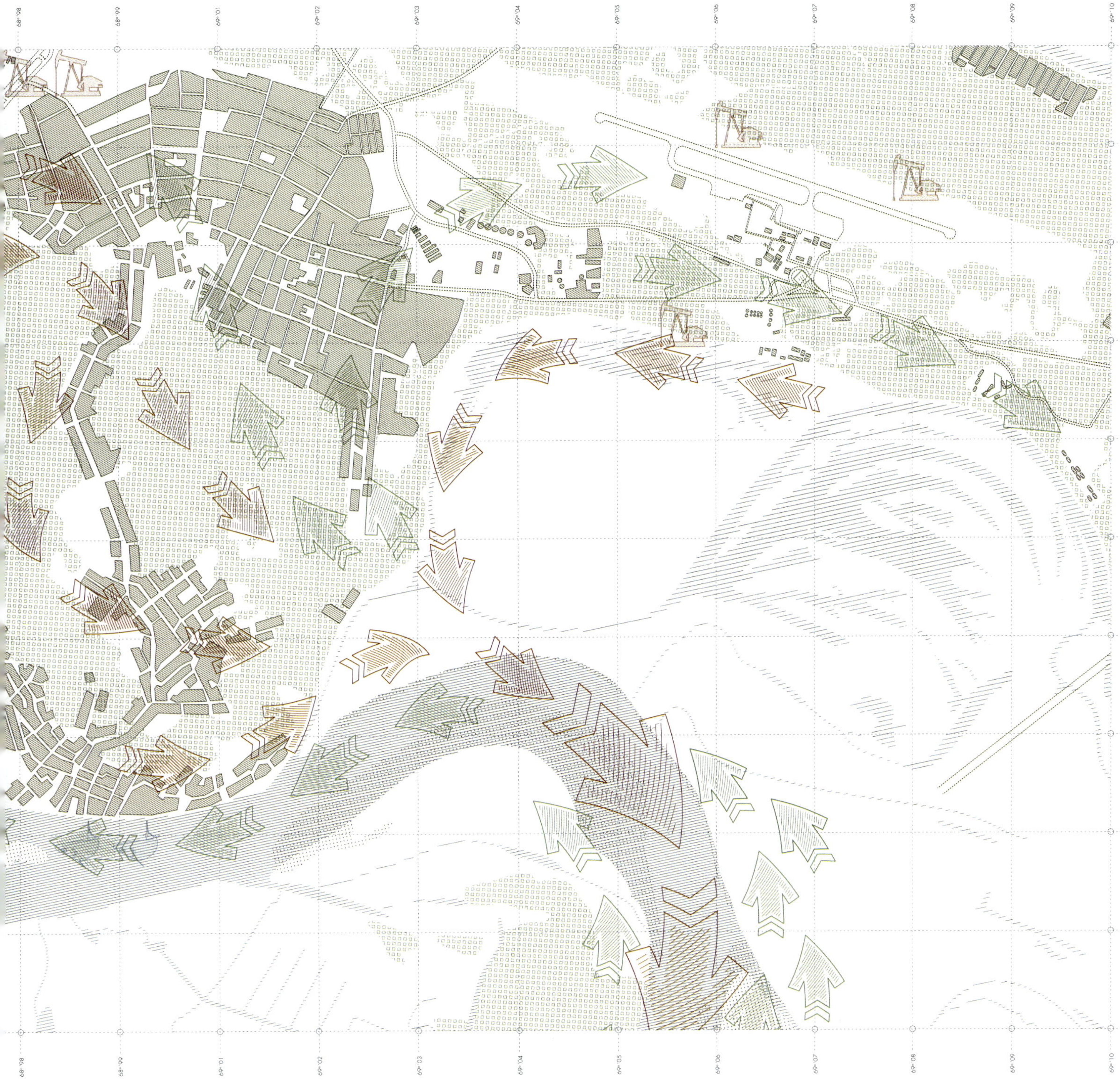

Current economy and carbon footprint.
The local economy is centred around a main industry - oil. The city is not self sufficient, and imports its food, consumer goods and industrial machinery from the south. Oil flows out via the river and a main road. While all other goods necessary for civilisation flow in the same way. This creates a very large carbon footprint - in addition to the methane gas being leaked out of the boggy ground, and the CO2 emitted by the local oil industry, there is also the energy used to transport food and goods across the vast waste of Siberia.

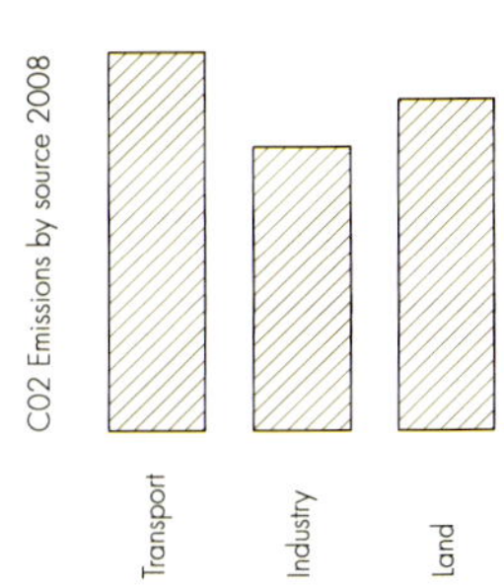

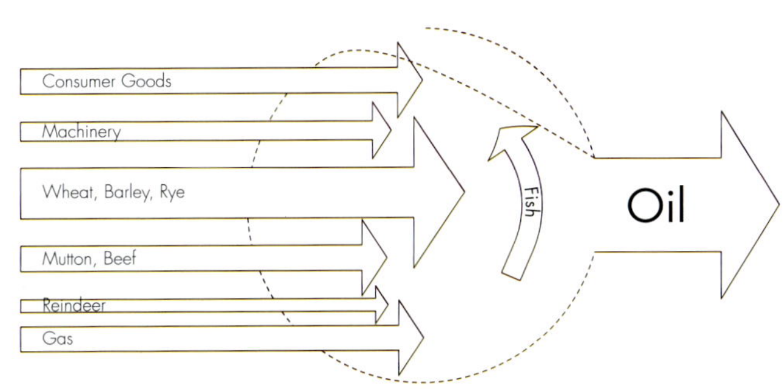

Thanks to Adam Johnston for helping with this page

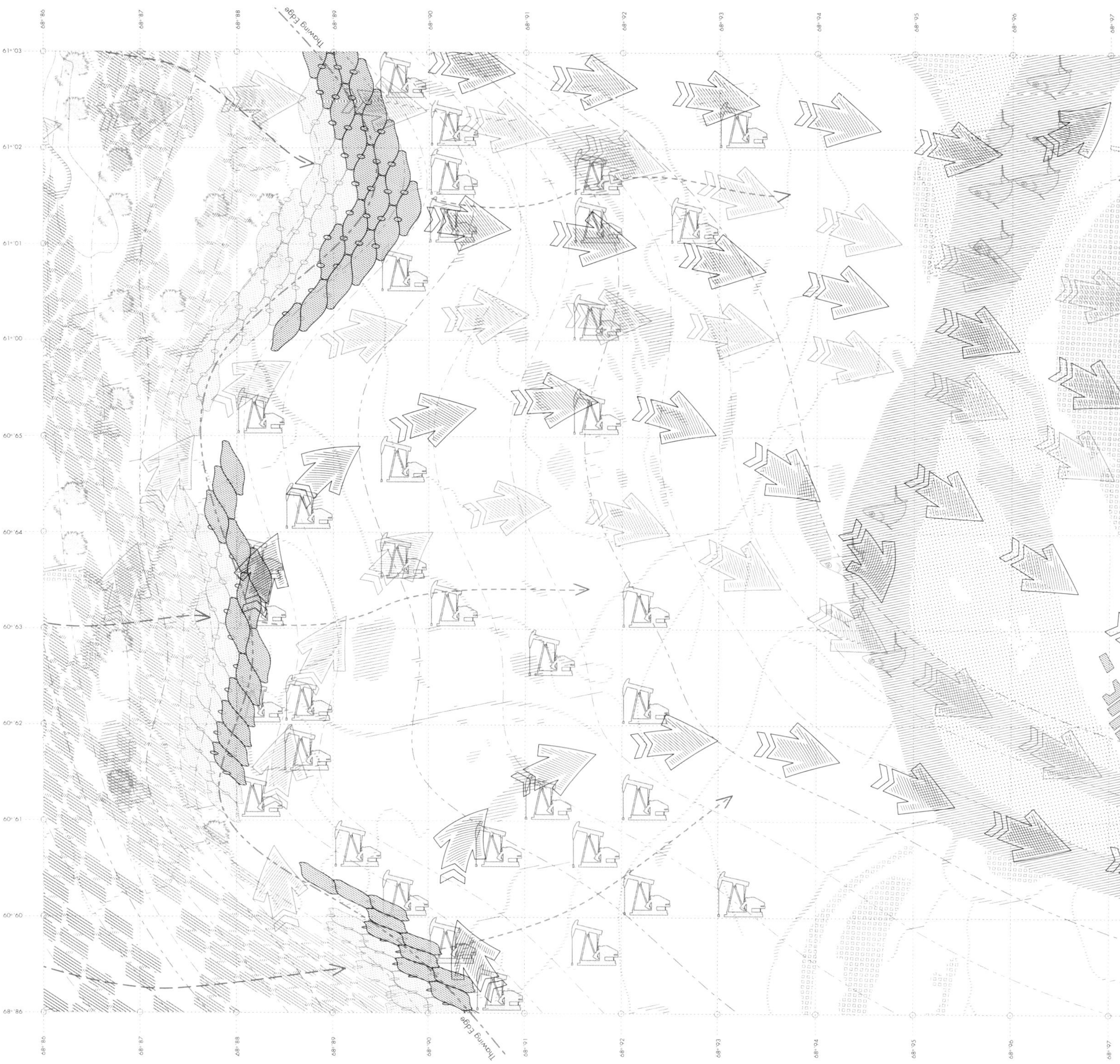

4.9 Proposed industry 2038

In 2038 you can see the proposal taping into the thawing edge, leaving a trace every 7th year of vegetation and inhabitation. You still see the red arrows indicating the oil industry, but at the same time you see the green arrows from the agricultural crops. From the diagram you can see that the city is far more self sufficient now, and could even export some goods, making a large reduction on their carbon footprint.

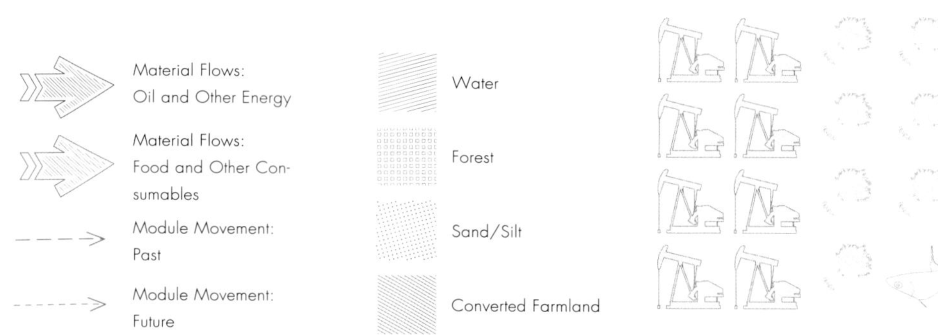

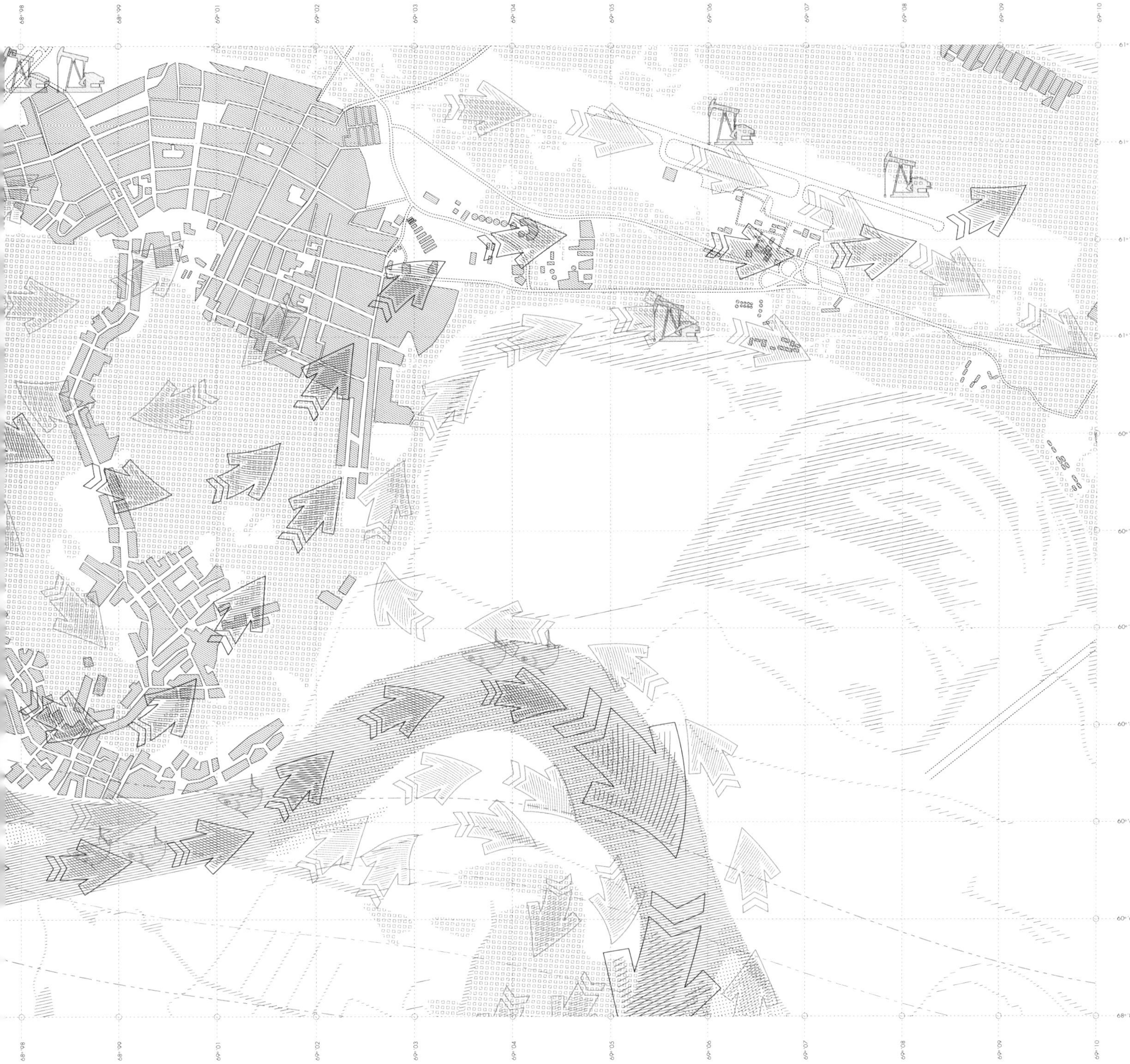

Long term economy strategy
The inventors of the algae greenhouse structures converts land into a form usable for agriculture, and over time enough land will be converted to make the town self sufficient in terms of food production. The addition of agriculture, as well as the alternative fuel and carbon credits provided by the algae greenhouse structures increases the size of local GDP and increases its diversity. Employment is now split between oil and agriculture

Long term carbon strategy
With the proposal soaking up methane and converting it first into CO2 and then into biofuel, the local emissions due to land will vastly decrease. The algae-based fuel production is far cleaner than the traditional oil industry, and with the newly created ability to grow food locally, CO2 due to transportation is also significantly reduced.

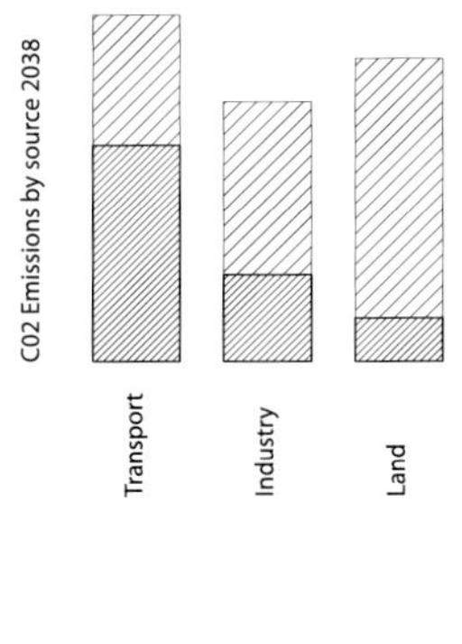

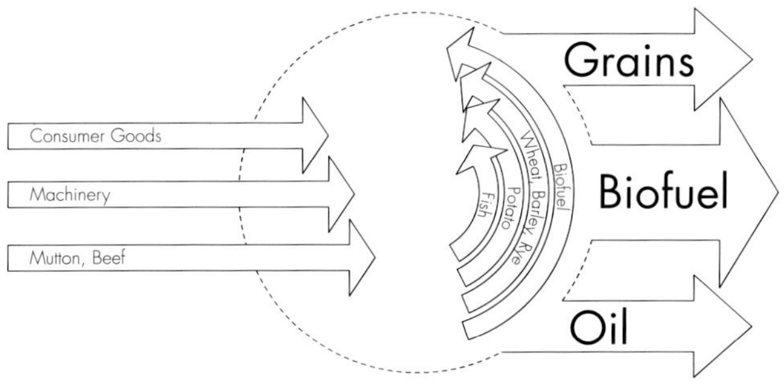

Erlend Bakke-Eidsaa

Thanks to Adam Johnston for helping with this page

4.10 Conclusive thoughts

What I like about the project is the way it takes something perceived negative and tries to turn it into an opportunity. If agriculture in the rain forest leaves a trace of ash in fertile land, the proposal leaves a trace of vegetation and inhabitation in undeveloped land.
What is new is the way it challenges the use of double curvature and asymmetrical optimised forms within pneumatic structures. This is a symbiosis system where the pneumatic is used to actuate and support the main language making it challenge far larger scales.

Proposal_03 - sited proliferation of system

4.11.1-3 Physical model evolution

Here you can see a comparison between the structural models, the first one is double curvature strut expansion, the second single curvature pneumatic expansion, while the final one which was made for the technical studies thesis challenges the double curvature with pneumatic expansion.
It spanned 4 meters and used 500 meters of acrylic tube to create the strip language, and was exhibited under the end of year show at the Architectural Association 2008.

4.11.1 Strip double curvature

Double curvature structure expanded by struts

Span 3500mm

Completed strip model	
tube diameter:	3 mm
tube length:	600 mm
joint depth:	10 tubes (35 mm)
joint width:	30 mm

90 modules of 10 tubes = 900 tubes

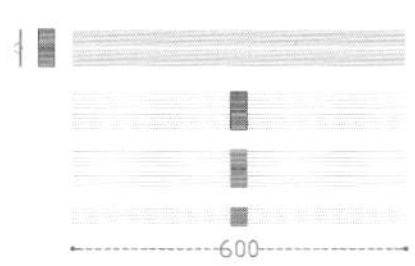

4.11.2 Pneumatic single curvature
Single dome curvature structure actuated and supported by pneumatic cushions

Span 2000mm

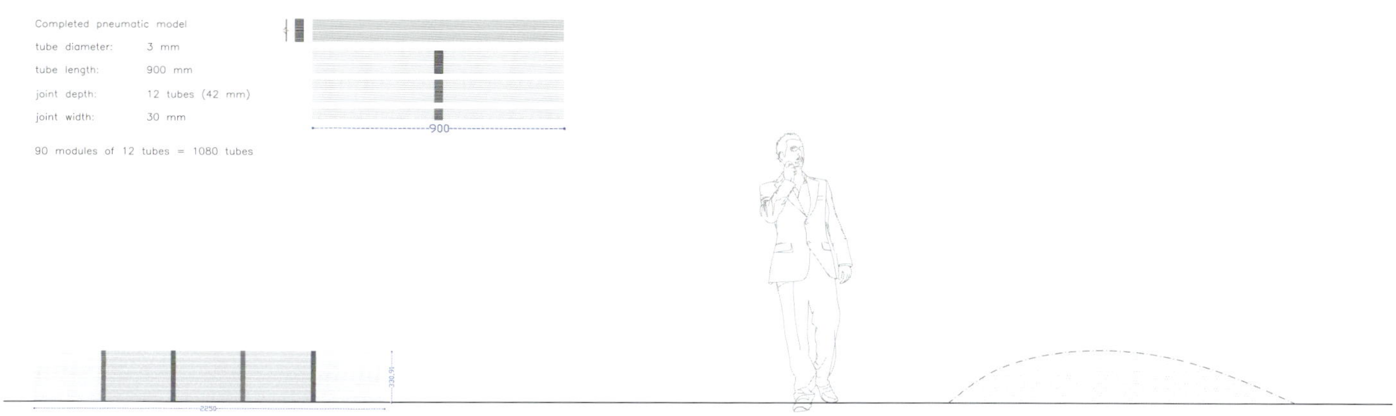

Photo Sue Barr

Erlend Bakke-Eidsaa

4.11.3 Pneumatic double curvature

Double curvature structure actuated and supported by pneumatic cushions

Span 4000mm

Proposed double curvature pneumatic model

compared to pneumatic model:
added 2 cushions in short direction
added 2.5 cushions in length direction

tube diameter: 5 mm

tube length: 1000 mm

joint depth: 10 tubes (50 mm)

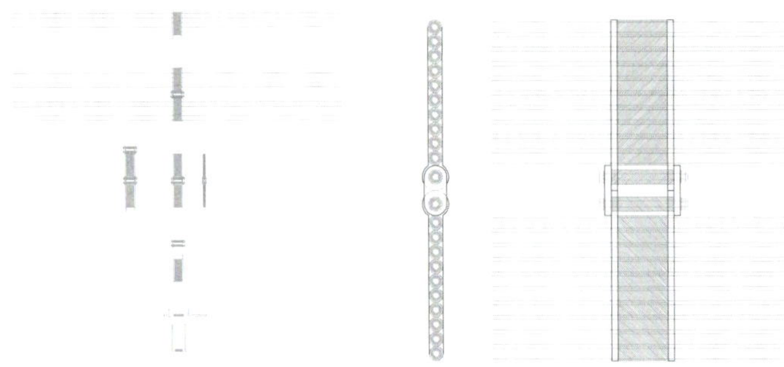

Material expenses and assembly time

42 modules
54 joints
29 cushions (39 slots ¾ filled)

Price:
acrylic tubes module 5mm outer diameter:
10 acrylic tubes per module x 42 modules = 420 tubes
420 tubes x 1000mm = 420 meter tubes x ...pounds per tube = ...pounds

acrylic tubes joints 7mm outer diameter 5mm inner diameter:
22 tubes per joint x 54 joints = 1188 tubes
1188 tubes x 24mm = 29 meter tubes x ...pounds per tube = ...Pounds

pvc skin
1 m2 per cushion x 29 cushions = 29 m2
29 m2 x 3 pounds pr m2 = 87 pounds

valves
2 valves, 2 nuts, 4 washers x 29 cushions = 58 valves, 58 nuts, 116 washers
ca 2 pounds per cushion = 58 pounds

air compressor = 40 pounds
acrylic 3mm sheet for joint frame = 10 pounds?
nuts and bolts:
(2 bolts, 4 washers, 2 nuts per joint)
ca 1 pound per joint x 54 joints = 54 pounds

Total price: 249 pounds + 5 mm tubes and 7 mm tubes

Assembly time:
42 modules:
Tubes and joint glued to one module: 10 min
42 modules x 10 min = 7 hours,
One day..

54 joints:
Draw template and laser cutting frames
One day
22 tubes per joint glued to frame and connected by bolt = 30 min
54 joints x 30 min = 27 hours = 2.25 days of 12 hours
3 days of 12 hours work

29 cushions: 39 slots reduced to (¾) 29 cushions to expose frame
cut welding and valves per cushion 1 hour
8 brass connectors cut sanded and glued: 1hour
29 cushions x 2 hours = 58 hours
4.8 days of 12 hours work

Total assembly time: 12 days total

Assembly support team: Mo Wongiwong, Deramore Hutchcroft, Matthew Hanrahan 'the security guard', Adam Johnston

Pneumatic model assembly technique

How I made the models

As I am interested in seeing how far this technique can be developed, I have decided to reveal my method for manufacturing the pneumatic models, but please make sure to reference the project and the Architectural Association if you find it useful.

For the cushions I used thin PVC fabric which you can buy at larger fabric stores. It is a bit toxic when you weld it so make sure you have a lot of fresh air around you if you do it for several hours. Then I used a hi-frequency welder, the type which is used in the catering industry to seal of plastic bags. You can buy this cheaply at eBay. To get the curved weld, I tried making a larger bent 'cast' which I redirected the current to, but this was less successful as I could not find thin enough metal for the welding tracks. My solution was to make different size 'stamps' which used the original metal tracks that comes with the welder. I layered wood, teflon film and then metal tracks, attached this to a handle which was insulated against the current. Then I redirected the current from the welder to this stamp. By adjusting the timer of the welder, I could get a succession of more or less perfect welds which overlapped into a longer curved weld. The curve was found through flattening the objects in rhino. General pneumatic rule is that all inflated objects go towards optimal circular shape. The air valves were made with brass fuel gauges from radio controlled cars, with two washers and a nut which mechanically sealed of the connection. To connect the cushions to the rods I made a secondary weld outside the first one which leaves a sleeve to fit the rods through. This had to be made at spread out points to allow movement. To inflate the structures silently I used fish tank air pumps.

The only issue I had was that as the metal tracks was bent to fit the stamp, at the bend the temperature increased and made a weaker weld. I tweaked this by curving the bend, which made it much better. It made the models work for several demonstrations, but as the model was transported around and kept inflated at the exhibition for two weeks, the weld at these points began to leak and I had to re-weld them.

So as I have been generous to pass on my hard work test and fail approach, I hope that you would do the same, and if you find a better solution to the stamp assembly, please email it back to me so this method can be developed even further and create some truly amazing pneumatic models in the future.

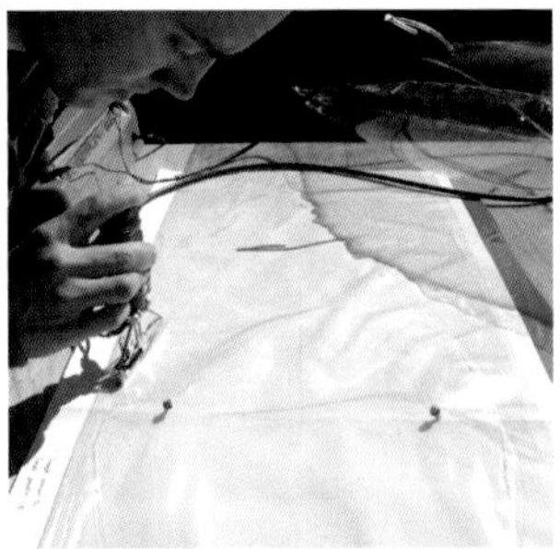

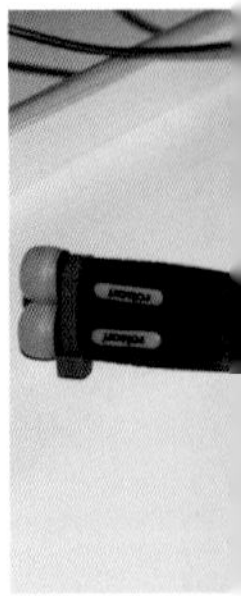

'Cast frame' from model train tracks, failed due to too much resistance in thick metal tracks.
Pneumatic form traced from strip model and flattened to 2d pattern in rhino.
Stamp to weld along flattened pattern.
Valve from radio controlled cars.
Hi frequency welder from the catering industry.
Silent air pump from fish tanks.
Assembly team helping to complete final model.

References

An inconvenient truth, Al Gore, Bloomsbury 2006

AD, Emergence: Morphogenic design strategies, Michael Hensel, Achim Menges, Michael Weinstock, Wiley-Academy 2004

Architectural Association Technical Studies Design Thesis, Proliferated Pneumatic Surfaces, Achim Menges, Architectural Association Technical Studies Archives

Architectural Association Emergent Technologies Design Thesis, Pneumatic Formfound Concrete Structures,Gabriel Sanchiz Garin, Architectural Association EmTech Archives 2007

Peter walker and partners, Peter Walker and Jane Brown Gillette, Thames & Hudson 2005

West Siberian Peatlands and Carbon Cycle, W. Bleuten and E. D. Lapshina with teams, Tomsk 2007

The Effect of climate change on the pristine peatland ecosystems and (sub)actual carbon balance of the permafrost boundary zone in Sub-Artic Western Siberia ("CASUS"), W. Bleuten, Pertti Marktikainen, Vyacheslav Zakharov, Elena D. Lapshina, Aleksei Naumov, Vladimir Golomolzin, INTAS project 03-51-6294 Final Report April 2007

Long Live Ugra!, author name, publisher 2000

http://www.geog.uu.nl/fg/casus/noframes.htm

http://www.thefraserdomain.typepad.com

http://www.grestech.com/index.php

http://www.variconaqua.com

http://www.energypowershift.com

http://www.algafarm.com/resources/IOE499reportalgaestrains.doc

http://www.sciencentral.com/articles/view.php3?type=article&article_id=218392957

r01 http://www.gtresearchnews.gatech.edu
r02 http://www.thecampingmachine.com
r03 http://www.claytoncramer.com
r04 http://www.kuci.org
r05 Nasa
r06 Photo by Norman Martin
r07 http://www.xplanet.dyndns.org
r08 United nations environment programme
r09 casus
r10 casus
r11 casus
r12 National Oceanic and Atmospheric Administration, USA today
r13 www.newscientist.com
r14 European environment agency:
r15 Sergei Kirpotin, of Tomsk State University
r16 apprentice alumni
r17 flickr galuth
r18 flickr dioedie
r19 www.greenwich-peninsula-chaplaincy.org.uk
r20 flickr warren_pilkington
r21 www.bravodacosta.com
r22 Achim Menges
r23 www.yeadondomes.com
r24 www.pools.org.uk
r25 www.msnbc.msn.com
r26 www.sciencentral.com
r27 www.algafarm.com/resources/IOE499reportalgaestrains.doc
r28 www.sciencentral.com/articles/view.php3?type=article&article_id=218392957
r29 Photo by Dr. Dennis Kunkel, http://nikoninstruments.mediaroom.com/file.php/135/IOD_Kunkel-10750-3.jpg
r30 thefraserdomain.typepad.com
r31 www.variconaqua.com
r32 http://www.projectrho.com/rocket/chlorella.jpg
r33 http://msnbcmedia2.msn.com/j/msnbc/Components/Photos/060517/060517_Module2_hmed_12p.h2.jpg
r34 www.energypowershift.com
r35 Photo by United nations environment programme
r36 Norman Martin
r37 Long live Ugra!
r38 http://www.fosterandpartners.com/Projects/1514/Default.aspx
r39 Biathlon World championship in Khanty-Mansiysk 2003
r40 http://www.cnes.fr/automne_modules_files/standard/public/p3395_9c9190e76382bba29e3e59c489195892P28882.jpg
r41 http://www.fabiofeminofantascience.org/RETROFUTURE/manhattandome2.jpg

r42 Google earth
r43 Photo Cornelis Verwaal
r44 Verkenningen Markerward deel 2
r45 travel.webshots.com/album/32016986NUVElQKdsZRemove frame
r46 Vector Foiltec http://www.orchidboard.com/community/gallery/data/539/Eden-02.jpg
r47 Photo Simon Burt http://lpmpjogja.diknas.go.id/kc/e/eden/eden-63.jpg
r48 http://www.global-b2b-network.com/direct/dbimage/50270983/Silicon_Rubber_Bands_2_In_1.jpg
r49 www.pbase.com
r50 Flickr estherase http://www.flickr.com/photos/estherase/33466667/
r51 Extreme Textiles, McQuaid, Matilda, Princeton Architectural Press, 2005
r52 Flickr tr33lo http://www.flickr.com/photos/tr33lo/242253453/
r53 Wikipedia, researching material properties on different materials
r54 http://www.bactechnologies.com/torque_tubes.htm
r55 www.CONBAM.de
r56 ww w.togoparts.com
r57 Flickr eir@si http://www.flickr.com/photos/eirasi/529532821/
r58 CASUS Base camp photographer
r59 Flickr jsbarrie, http://www.flickr.com/photos/jsbarrie/280557626/
r60 www.assuredukmalt.com/pictures.htm
r61 Flickr jere-me, http://www.flickr.com/photos/beeldenzeggenmeer/405092064/
r62 Flickr oldskooldave, http://www.flickr.com/search/?q=leffe&w=42656657%40N00
r63 Flickr Steve.wilson, http://www.flickr.com/search/?ss=2&ct=6&w=91242112@N00&q=rye&m=text
r64 http://www.breadsetcetera.com/graphics/rye_bread.jpg
r65 http://www.train4dev.net/images_jpg/image585.jpg
r66 http://farm1.static.flickr.com/215/470927658_1530e044a7_b.jpg
r67 Flickr Zenara, http://www.flickr.com/photos/zenera/archives/date-posted/2006/03/02/
r68 Flickr digitalsextant, http://www.flickr.com/photos/digitalsextant/archives/date-posted/2005/10/04/
r69 Flickr willgraham, http://www.flickr.com/photos/willgraham/archives/date-posted/2006/10/08/
r70 http://newsimg.bbc.co.uk/media/images/39274000/jpg/_39274793_jim_dimmock_cannabis203.jpg
r71 Oil & Gas Journal, January, 2007
r72 US Energy Information Authority,September,2007
r73 http://en.wikipedia.org/wiki/Oil_reserves

Acknowledgements

This work could not have been completed without the support of the following people: Architectural Association diploma unit 16 (2007/08), tutors Steve Hardy and Jonas Lundberg. Architectural Association technical staff, particularly Mike Weinstock and Wolf Mangelsdorf. The C.a.s.u.s. Siberian geologist and biologist team, Professor Elena Lapshina. Ugra University in khanty-Mansiysk Siberia Russia, and Oleg Deyev as Russian interpreter and guide. Thanks to Adam for help with final drawings, and Deramore, Mo, Matthew who helped me with the final Model.

The funds which enabled me to stay at the AA:
Eileen Gray fund for the 2006/2007 academic year.
R D Hammett Bursary for the academic year 2007/2008
Student forum travel bursary to Siberia
The Holloway Trust award 2007/2008
Marilyn Dyer and Sabrina Blackstad.

I would also like to deeply thank my love Silje for her support and patience over the last five years.

Erlend Bakke-Eidsaa